Scientific Internationalism in Cold War Central and Eastern Europe

New Perspectives on Central and Eastern European Studies

Published in association with the Herder Institute for Historical Research on East Central Europe, Marburg, Germany

Series Editors
Peter Haslinger, Herder Institute
Heidi Hein-Kircher, Martin-Opitz-Library and Ruhr-University Bochum

Decades after the political changes that accompanied the fall of the Soviet Union, Eastern Europe remains one of the most misunderstood parts of the world. With a special focus on the Baltic states, Poland, the Czech Republic, Slovakia, and Hungary, New Perspectives on Central and Eastern European Studies investigates the historical and social forces that have shaped the region, from ethnicity and religion to imperial legacies and national conflicts. Each volume in the series explores these and many other topics to contribute to a better understanding of Central and Eastern Europe today.

Volume 8
Scientific Internationalism in Cold War Central and Eastern Europe: The Age of Reciprocity
Edited by Jan Surman and Tomáš W. Pavlíček

Volume 7
Challenging Norms: Family Planning as a Reflection of Social Change in Twentieth-Century Eastern Europe
Edited by Heidi Hein-Kircher, Elisa-Maria Hiemer, and Denisa Nešťáková

Volume 6
Russian Historiography from 1880 to 1905: Pavel N. Miliukov and the Moscow School
Thomas M. Bohn

Volume 5
The Science of State Power in the Habsburg Monarchy, 1790–1880
Borbála Zsuzsanna Török

Volume 4
The Middle-Income Trap in Central and Eastern Europe: Causes, Consequences, and Strategies in Post-Communist Countries
Edited by Yaman Kouli and Uwe Müller

Volume 3
The World beyond the West: Perspectives from Eastern Europe
Edited by Mariusz Kałczewiak and Magdalena Kozłowska

Scientific Internationalism in Cold War Central and Eastern Europe

The Age of Reciprocity

Edited by
Jan Surman and Tomáš W. Pavlíček

berghahn
NEW YORK • OXFORD
www.berghahnbooks.com

First published in 2027 by
Berghahn Books
www.berghahnbooks.com

This book was written as a part of the joint research project of the National Science Centre, Poland (OPUS 2020/39/I/HS3/03589) and Czech Science Foundation (GA ČR 21-45624L), entitled "Czechoslovak-Polish Scholarly Entanglements in the Cold War: Between High Politics and Individual Strategies."

Library of Congress Cataloging-in-Publication Data

A C.I.P. cataloging record is available from the Library of Congress
Library of Congress Cataloging in Publication Control Number: 2026032630

British Library Cataloguing in Publication Data

A catalogue record for this book is available from the British Library

EU GPSR Authorized Representative

LOGOS EUROPE, 9 rue Nicolas Poussin, 17000, LA ROCHELLE, France
Email: Contact@logoseurope.eu

ISBN 978-1-80758-068-1 hardback
ISBN 978-1-80758-069-8 epub
ISBN 978-1-80758-070-4 web pdf

https://doi.org/10.3167/9781807580681

Contents

Illustrations

Preface

Historians of science and scholarship have long grappled with examining knowledge production beyond what sociologist Bruno Latour termed "centers of calculation"—institutional hubs, typically universities, that historically dictated the norms of academic science.[1] In recent decades, historiographical "turns" have broadened the scope of historical inquiry, redirecting focus from singular locations to the intricate networks through which knowledge circulated, often extending from local vicinities to global scales. However, the inclusion of new localities has not resulted in a critical re-evaluation of overarching narratives and the roles of previously omitted spaces, with postcolonial and decolonial studies being notable exceptions. Additive historical approaches, particularly those rooted in nationalistic frameworks, have failed to fundamentally alter field-dominating historical accounts. Cold War studies, characterized by the bipolarity of superpowers, serve as a prime example of such a field.[2]

The very concept of "science" is similarly constrained, predominantly analyzed through the lens of its Anglophone interpretation. While the broader semantic range of this term in Slavic languages (as well as in German and French) is often debated within local contexts, these discussions rarely challenge the prevailing international paradigm. This is, in part, because English has become the dominant language of contemporary global history. Consequently, providing alternatives to both Anglophone-centricity and "science"-centeredness is of paramount importance.

This volume stems from the project "Czechoslovak–Polish Scholarly Entanglements in the Cold War Between High Politics and Individual Strategies," which aimed to offer East-Central European perspectives on the circulation of knowledge during the Cold War. Sponsored by the Czech Science Foundation and the National Science Centre (Poland) and located at the Masaryk Institute and Archives of the Czech Academy of Science and the Institute for History of the Polish Academy of Sciences, the project united specialists from four countries (Austria, Czechia, Germany, and Poland), contributing research across a diverse range of disciplines, from the history of astronomy to the past of the scientific-technical revolution.[3] The aim was to represent the semantic breadth of how science was understood in socialist East-Central Europe and to explore the similarities and differences in the experiences of scholars at opposite ends of the spectrum between "soft" and "hard" sciences.

The book comprises six studies that examine selected disciplines and fields, encompassing both natural and social sciences, as well as various scales of knowledge circulation, from localized cross-border exchanges to global networks such as Pugwash. These studies demonstrate that the investigation of bilateral contacts is most fruitful when these contacts are analyzed not exclusively through transnational/global or national/local lenses, but rather at the intersection of these perspectives, within the "forgotten territory between the micro and the macro."[4]

The exploration of Czechoslovak–Polish academic entanglements under socialism reveals the complex interplay of formal structures, informal solidarities, and asymmetrical power dynamics. Moving beyond narratives of top-down imposition or heroic state resistance, this volume underscores the negotiated nature of knowledge circulation within the Eastern Bloc. By highlighting a diversity of actors, temporalities, and motivations, we challenge simplistic narratives and instead foreground the everyday practices, compromises, and opportunities that shaped socialist internationalism. Ultimately, the study of these entanglements opens new perspectives not only on the past of East-Central European science but also on the methodological challenges of writing its history in the present moment of resurging nationalism and global instability.

Summaries of all interviews, archival excerpts, and supporting documents collected during the preparation of this volume are available for consultation through the research repository of the Masaryk Institute and Archives of the Czech Academy of Sciences (MÚA AV ČR), under restricted access for academic use. For inquiries regarding data access or permissions, please contact the editors directly.

Jan Surman and Tomáš W. Pavlíček

NOTES

1. Jöns, "Centre of Calculation."
2. Nadkarni, "Bipolarism Redux."
3. For full results of the project, see the webpage "Czechosłowacko-polskie sieci akademickie podczas Zimnej Wojny." The project has been supported by Narodowe Centrum Nauki (no. 2020/39/I/HS3/03589) and Grantová agentura České republiky (no. 21-45624L).
4. De Landa, *A New Philosophy of Society*, 6; a similar approach, but focusing on circulation as a space in-between was proposed by Raj, "Beyond Postcolonialism . . . and Postpositivism."

BIBLIOGRAPHY

"Czechosłowacko-polskie sieci akademickie podczas Zimnej Wojny między wielką polityką i strategiami indywidualnymi" [Czechoslovak-Polish academic networks during the Cold War: Between high politics and individual strategies]. Retrieved on 2 February 2026 from https://entanglements.ihpan.edu.pl/.

De Landa, Manuel. 2009. *A New Philosophy of Society: Assemblage Theory and Social Complexity*. London: Continuum.

Jöns, Heike. 2011. "Centre of Calculation." In *The SAGE Handbook of Geographical Knowledge*, edited by John Agnew and David N. Livingstone, 158–70. London: SAGE Publications Ltd. https://doi.org/10.4135/9781446201091.n13.

Nadkarni, Vidya. 2010. "Bipolarism Redux? The Cold War and After." In *Oxford Research Encyclopedia of International Studies*, edited by Nalanda Roy. Oxford: Oxford University Press. https://doi.org/10.1093/acrefore/9780190846626.013.325.

Raj, Kapil. 2013. "Beyond Postcolonialism . . . and Postpositivism: Circulation and the Global History of Science." *Isis* 104, no. 2: 337–47. https://doi.org/10.1086/670951.

The Age of Reciprocity
An Introduction

Jan Surman, Lukas Becht, Maciej Górny, Kornelia Kończal, Doubravka Olšáková, Tomáš W. Pavlíček, Luboš Studený, Michaela Šmidrkalová

The 1968–69 period was a difficult time in Czechoslovak–Polish relations. In the early spring of 1968, the political crisis and antisemitic campaign affected many Polish scholars, forcing them to leave the country, while Czechoslovak scholars tried to utilize their only recently attained relative freedom to aid their Polish colleagues. In August of the same year, this period of relative freedom—known as the Prague Spring—came to an end when the armies of the five Warsaw Pact countries, including Poland, invaded Czechoslovakia. Although many Czechoslovak scholars would emigrate in the years that followed,[1] Poland was not on their itinerary. While the two scholarly communities continued to cooperate on a formal level, the archival records of the following few years read like a list of complaints from both sides. Nevertheless, through a combination of administrative measures and informal initiatives, academic contacts would return to their previous level.

Poland and Czechoslovakia—ruled by communist parties from 1948—had emerged as independent states with a common border after 1918. This border also became a subject of dispute, and the interwar period was marked by the Czechoslovak–Polish (Seven-Day) War over Těšín/Cieszyn Silesia in 1919 and Czechoslovak support for Ukrainian and Russian emigration, which Poland interpreted as support for Ukrainian separatism.[2] After the Munich Agreement in 1938, Poland annexed parts of Cieszyn Silesia, thereby further straining relations. These were restored by decree to reinforce the socialist alliance (which remained firm until Polish troops helped put an end to the Prague Spring). At the same time, the three nations involved—Czechs, Poles, and Slovaks—were in the process of nation-building, which involved mobilizing different, and sometimes conflicting, pasts.[3] Thus, some processes needed time even at the time of alliance—like the signing of the border treaty, which happened only 1958. Importantly, while the Czech and Slovak nation-builders invoked a revival—of an early modern (Czech) or medieval (Slovak) culture that had been suppressed under Habsburg rule and then revived in the nineteenth century—Polish nationalism cultivated a continuity with the Polish-Lithuanian Commonwealth, including its eighteenth-century cultural and scientific developments, thus claiming a place among the cultivated nations of the time. These differences created tensions and also shaped stereotypical perceptions. The cultures of Czechs and Slovaks were considered "small" in Poland, which, in turn, was viewed as an egotistical, self-centered culture from the Czech and Slovak sides (as Czechs and Slovaks were supposedly historically more open to cultural exchange with other Slavs).[4] Not unlike the situation between Poland and East Germany, whose relationship was often characterized by the term "forced friendship," while official relations were formally friendly, in practice they were tense and more often disrupted by individual acts than by politically structured approaches.[5]

The evolution of Czechoslovak–Polish relations in the midst of socialist rule in East-Central Europe is in many ways characteristic of relations between other socialist states. Oscillating between friendship and conflict, shaped—but not defined—by shifting political prerogatives, they also relied on a variety of informal encounters, some of which dated back to the interwar period. Throughout the four decades of the Eastern Bloc's existence, the politics of rapprochement dictated the rhetoric of interstate relations, while nationalisms still dominated the region—including the Russian-centric imperial politics of the Soviet Union. Czechoslovaks and Poles experienced contrasting dynamics: during the post-1968 "normalization" period in Czechoslovakia, when the Prague Spring reforms were rescinded and their proponents and supporters were persecuted,[6] the country became increasingly dependent on the Soviet Union, while Poland moved toward greater independence. However, differences existed even earlier; for example, Fulbright scholarships were available in Poland as early as 1959 but not in Czechoslovakia until the late 1970s.[7]

Because of their peripheral geographical position in the Eastern Bloc, both countries also had an important intermediary function. Their locations and traditions of intellectual contacts were favorable for international congresses, as they were equally accessible to West and East. Moreover, local conferences organized by the central institutes of their respective academy of sciences, which were often held in mountain regions, involved scholars from regional universities and research institutes (which was particularly welcomed by the Slovaks). Although the maneuvering between these different prerogatives, varying at the local, socialist, and global levels, was characteristic of the socialist period, the intensity of these opposing factors in the Czechoslovak–Polish example makes it an ideal case study of interstate "cooperation and competition."[8]

Almost four decades after the collapse of the socialist system, the question of how to write the history of socialist science continues to divide scholars. This has been exacerbated by the recent right-wing resurgence in East-Central Europe, which has often used nationalist discourses of othering, and also by the Russian invasion of Ukraine, which began in 2014 and has come to influence broader intellectual debates outside the two countries involved since the full-scale invasion in 2022. This has led to a wide range of debates on decolonization, from those that use nuanced theoretical tools to those that describe any kind of socialist entanglement as a form of imperialist oppression.[9] While the resurgence of black-and-white historiography mostly concerns relations with (Soviet) Russia, it has also left its mark on the subject of this book, as historians following this dualist approach were deliberately blurring the line between cooperation and coercion within the Eastern Bloc.

It is at this point that the concept of entanglement becomes very fruitful, as it has been developed to study colonial and postcolonial phenomena of knowledge exchange, thus including reflection on the effects of power asymmetry on often idealized relations.[10] These idealizations have included both the "West" and the "East" (that is, the Soviet Union), and it has often been the case that those who idealized one demonized the other. For many scholars, these hierarchies, especially with regard to the East, were overturned after the collapse of the Soviet Union, which also led to the cessation of ongoing collaborations and, in some cases, to the disappearance of memories of them.[11] This poses a methodological challenge to both conventional history, based on written sources, and oral history.[12]

It is important to note that, in our case, this reflection does not begin or end with the position of the Soviets or the "West" in socialism, nor with the position of Czechoslovakia and Poland in relation to each other (or with the different positions of Czechs, Poles, and Slovaks in this equation). For example, the Slovak Academy of Sciences (which existed in parallel with the Czechoslovak Academy of Sciences [ČSAV]) was able to organize numerous informal international workshops due to its position further away from the centers of power—especially after 1968, when Slovakia was less affected by restrictive normalization policies.[13] Slovak was also often the language in which translations of foreign literature (for example, by the French sociologist Pierre Bourdieu) could be published before they appeared in Czech. Thus, the case studies in this volume must focus on a variety of power differentials in situ and on a variety of actors with different legal statuses and identifications. And because of the shifting borders, these identifications may also differ from citizenship and may not fall into the categories into which the nationalizing states divided people.[14]

Due to the embeddedness of Czechoslovak–Polish history in the complex socialist power relations, the case studies presented in this volume move beyond the question of bilateral relations to reassess the circulation of knowledge across the Eastern Bloc and beyond. Socialist internationalism, émigré scientists, and science diplomacy situate these individual stories in broader global contexts. Our introduction follows the reverse path: from the global to the international to the socialist to the bilateral level, ending with the locality of the case studies. It was important to show how the actors in the various encounters mobilized different resources and thus co-created the different contexts in which they acted.[15] Although hierarchically governed, the socialist system allowed for a great deal of flexibility, which is a feature we wanted to call attention to. Therefore, the introduction does not offer a grand narrative, but rather focuses on fissures, interstices, and ruptures within master narratives—sites of tension and ambiguity that may open up new avenues for historical reflection.

From Global to Glocal

Recent decades have witnessed numerous shifts in the conceptualization of the history of knowledge networks, culminating—probably tentatively—in global and planetary histories. As the 2020s have been less a time of growing globalization and more one of disruptions to global flows (COVID-19, the war against Ukraine, Trump's tariff wars and the reconfiguration of transatlantic relations, etc.), historians are likely to respond further. However, global history is not only history on a new scale but also a perspective that has implications for thinking about the relationality of all history, including microhistory.[16]

While the history of science experienced a similar trend, an important intermediary in this respect was entangled history and the history of the circulation of knowledge.[17] It emerged as a response to a long series of developments, both international (transnational history, postcolonial history) and local Parisian globalized (*transfert culturel* and *histoire croisée*). It showed that a number of "Western" developments were the result of the circulation of knowledge originating in colonies, an origin that was rendered invisible in the process of its onward travel and codification. Methodologically, this meant, on the one hand, a decentering of the received narratives of the traditional history of science, and on the other, a necessity to consider local developments, regardless of their geographical location, as part of globally circulating knowledge. Importantly, however, the question of why certain knowledge does not circulate, or circulates only locally, also became crucial. This brought media and the mediality of transfer, as well as questions of power asymmetries, more into focus than in global history. It also accentuated the localist perspective on knowledge, as opposed to the assumed universality that had dominated the historiography of science (and science itself) since the last decades of the nineteenth century.[18]

Similarly, since the 2010s, the history of internationalism has abandoned the "sanitized version of internationalism, with its seemingly universal values and timeless institutions" in favor of viewing it as a "complex and contested process."[19] In the history of science, this went hand in hand with the abandonment of the idea that there could be such a thing as neutral, value-free internationalism.[20] This necessitated the addition of a historically nuanced analysis of the process, reflecting changing power relations between individual and collective actors, competing interests, and opposition to internationalism.[21] The interest in internationalism as practice also led to an investigation of what might be called the "dark side" of internationalism.[22] While this research began with the study of fascist transnationalism, with the rise of new authoritarianism and of the New Right around the world, there has been growing focus on illiberal internationalism, including the relationship between liberal and illiberal internationalism.[23] In recent years, much attention has been paid to the connection between the far right, illiberalism, and anti-gender and anti-feminist movements, revealing how they form productive networks that in many ways contradict the idea that such movements are nationalist and anti-globalist.[24] The focus on liberal internationalism also explains why one of its key alternatives, Catholic internationalism, has received little attention, even though it was hailed as a model international community as early as 1925.[25] Socialist internationalism, another alternative to liberal internationalism, will be discussed in more detail below.

In recent decades transnational historiography has also become critical of the promises of the Habermasian understanding of the transnational sphere as a counterweight to nationalist internationalism. The critique from decolonial studies that transnationalism (but also internationalism) has been constructed as a "Western" category has become increasingly salient. Looking at practices of transnational activism has been suggested as one way of "decolonizing" histories of internationalism.[26] This follows a general trend of writing "transnationalism from below," to focus on actors, their agencies, and the ways they shape the transnational sphere through their actions.[27]

The present volume explores the insights arising from this reflexivity. Specifically, it addresses socialist internationalism as a particular version of twentieth-century internationalism, and also the practice of "doing internationalism"[28] under non-democratic conditions, which it adopts as a point of departure. Writing "internationalism from below" and focusing on actors engaged in a variety of transnational activities led not only to a readjustment of scales, but also to a search for methodological innovations at the beginning of the project that resulted in this publication. Our aim, in particular, was to use ego documents and oral history to complement and counterbalance narratives rooted in institutional archives. This allowed us to more thoroughly explore the possibilities of maintaining agency within a socialist, Soviet-dominated, strictly hierarchical academic system, while at the same time not idealizing the actors' egocentric accounts, which are characteristic of post-socialist memory cultures.[29]

Socialist Internationalism

As Rachel Applebaum notes in her study of the socialist "empire of friends," "Soviet and Eastern European citizens were bound together not only by military force but also by transnational contacts and personal encounters."[30] Internationalism and transnationalism have long been a particular focus of historians of socialist science and culture, oscillating between critiques of its politically imposed limits and demonstrations of how it expanded the scope of traditional treaties.[31] Recently, the term "restricted internationalism" has been proposed to characterize the Soviet-imposed limits on the autonomy of the science policies of the socialist states, which corresponds to the Soviet idea of the "limited sovereignty" of non-Soviet Eastern Bloc countries (popularly known as the Brezhnev Doctrine).[32] This idea refers to the fact that when we compare experiences in Moscow, Kyiv, or Warsaw, we encounter differing degrees of freedom within academic circles. However, differences pertained also to different types of institutions, and even institutes' cadres were divided into different categories. This was also applicable to travel: the East German term *Reisekader* captures this difference very well, as it distinguishes between those who were allowed to travel outside the Eastern Bloc and those who were not.[33] Nevertheless, there were ways to get around the restrictions.[34] Scholars had their own specific interests and strategies, and they were also able to use internationalism as a resource for personal gain.[35]

The initial focus of the historiography of socialist internationalism was the study of inter-bloc contacts, with a dynamic that can be described as "from the Iron Curtain to the Nylon Curtain,"[36] that is, with increasing attention to the permeability of the "East–West" divide. Increasingly, this focus has shifted to work on "second world" engagement with the "third world" on the one hand,[37] and to questions of internal transfers and structures within the Soviet Union on the other.[38] While still Soviet-centric, the Eastern Bloc has increasingly been portrayed as a space of diverse interests, in which countries could pursue their own agendas, both in terms of contacts with socialist and non-socialist countries. This is in line with the ongoing discussion on the decolonization of Central and Eastern European history, as it concerns not only the question of multiperspectivity but, more importantly, the epistemic framework that shapes the debates on the Eastern Bloc.[39] In this regard, relational history offers an escape from the methodological nationalism that characterizes post-1989/1991 historiography.

Importantly, even research on organizations that were multilateral by design, such as the Council for Mutual Economic Assistance (COMECON),[40] has paid little attention to multilateralism within the alliance.[41] There is an ongoing debate about whether COMECON's emphasis on "socialist integration"—the organization's slogan—was merely a rhetorical tool, especially in symbolic rivalry with the European Economic Community,[42] or an actual policy. Nevertheless, compared to other socialist organizations, COMECON provided the greatest space for the expression of the agency of non-Soviet actors.[43] It was thus both an instrument of Soviet policy and a space in which the autonomy of individual states could be exercised. These multilateral cooperation spaces deserve particular attention in order to uncover the different dynamics that shaped them. They were not only intended to coordinate national research, but they also had the—often unfulfilled—ambition of jointly producing knowledge across the Eastern Bloc.

The effort for cooperation within the framework of COMECON intensified in the 1970s and was aimed at conducting joint research programs and establishing joint research institutes.[44] As the chapters by Luboš Studený and Lukas Becht or Maciej Górny in this volume demonstrate, the support of COMECON—not only financial but also organizational in providing a communicative platform—was crucial for bilateral cooperation, often replacing and building on it.

Another space of multilateral cooperation was created by the Academies of Sciences of the Socialist Countries. The first joint meeting of their representatives took place in Warsaw in 1962, building on earlier bilateral cooperation. A range of problem commissions was set up to coordinate joint research initiatives. In 1971, the Agreement on Multilateral Scientific Cooperation between the Academies of Sciences of the Socialist Countries was signed, providing a formal framework for expanded collaboration, including negotiations over participation in COMECON projects. Thus, most multilateral socialist cooperation had a direct connection to the tasks assigned by COMECON, contributing to the entanglement of these cooperative spaces.

Multilateralism also led to the creation of joint projects and institutes, the most famous of which were the Joint Institute for Nuclear Research in Dubna (founded in 1956 as an independent institute, from 1971 COMECON-affiliated) and INTERKOSMOS (founded in 1967).[45] In addition to the COMECON centers, cooperative centers were established throughout the Eastern Bloc, building on the expertise of existing institutes, especially those affiliated with the academies of sciences. The Stefan Banach International Mathematical Center for Socialist Countries in Warsaw (established in 1972) and the International Laboratory for Strong Magnetic Fields and Low Temperature Research in Wrocław (established in 1968) serve as illustrative examples.[46] As Tomáš Pavlíček shows in this volume examining the (ultimately unsuccessful) creation of the Center for Category Theory in Prague, these efforts were closely intertwined with everyday politics.

Importantly, while the socialist countries officially presented a united front while acting in the international arena, various allegiances and agendas of individual actors should not be understood as insignificant diversity that disappeared once the unity of the Bloc was required. At the 8th World Congress of Sociology in Toronto in 1974, for example, there were conflicts within the Polish delegation, some of whose members refused to support the pre-agreed socialist candidate, which caused the socialist countries to lose a place on the International Sociological Association board.[47] In order to understand the dynamics in socialist science, especially its international dimension, it is therefore necessary to also pay attention to the minor conflicts that are intertwined with the political prerogatives and are rarely recorded in official documents.

In the context of Czechoslovak–Polish science and technology relations between 1945 and 1989, the transnational sphere played a crucial role in facilitating exchange and cooperation beyond the constraints imposed by state-centered diplomacy. Neutral platforms where scientists from both countries could engage in dialogue, share knowledge, and contribute to global debates were provided by organizations such as the Pugwash Conferences on Science and World Affairs (est. 1957), the International Institute for Applied Systems Analysis (IIASA, Laxenburg near Vienna; est. 1972), and the European Coordination Centre for Research and Documentation in Social Sciences (CEUCORS, Vienna; est. 1963). These interactions allowed researchers to bypass political tensions and ideological divisions but at the same time foster a shared scientific identity within the Eastern Bloc, while also connecting with the international scientific community.

The general framework provided by the United Nations system for scientific and technological cooperation, which provided a global platform for multilateral cooperation, had a decisive influence on non-partisan internationalism. Through specialized agencies such as UNESCO, Czechoslovak and Polish scientists participated in international research programs, contributing to and benefiting from global knowledge networks. In both countries, non-state actors such as Pugwash and IIASA also influenced the development of scientific agendas and research priorities. These organizations encouraged the exchange of ideas and methodologies that were less constrained by national politics by focusing on issues of global importance such as disarmament, environmental challenges and technological modernization. In addition to facilitating the transfer of scientific expertise, this transnational engagement empowered individual scientists to act as diplomatic intermediaries, shaping the trajectory of scientific cooperation in ways that state institutions could not fully control. As a result, these networks contributed to the gradual erosion of rigid ideological boundaries, paving the way for more flexible and innovative approaches to science and technology in the region.

While internationalism focuses on the transcendence of various borders, it should not be forgotten that it was not possible for interstate relations to take place in a power-free space. Thus, an important concept in this volume is the "in-betweenness" of non-Soviet socialist actors.[48] This refers to the position in which, due to the political constitution of the Eastern Bloc, actors in countries such as Czechoslovakia or Poland were aware of their dependence on the Soviet centers and acted in accordance with this perceived dependence but nevertheless retained a significant degree of autonomy.[49] This in-between position distinguishes the East-Central European actors from their Soviet counterparts, even though power relations also had the potential to seriously complicate their everyday affairs, especially tensions with politicians within the respective country (which, however, were often instigated by other scholars). This in-betweenness can also be observed at all levels of international relations, from science diplomacy to interpersonal relations.[50] The evolution of this dynamic was influenced by changes in the power relations inside the Soviet hegemonic sphere.

Therefore, studying socialist internationalism with Czechoslovakia and Poland at its core reveals more than just the dynamics of bilateral relations in East-Central Europe. It involves reflection on centers and peripheries and their relations to each other. Given that the history of socialism has been politicized, bringing it back into a general reflection on power relations and their impact on knowledge production helps create a more fine-grained picture of twentieth-century East-Central Europe.

Internationalism after Socialism

Typically, post-socialist East-Central European literature portrayed the academic landscape of the Eastern Bloc as a competition between the domestic (national) and the foreign (Soviet), often focusing on national liberation and describing advocates of the Soviet system as having been "seduced" or as outright traitors.[51] It also used terms such as "*spętana Akademia*" (tethered academy)[52] to emphasize the externally imposed limits of freedom. The history of socialist rule in this narrative was described through the prism of Sovietization, defined as cultural and political imperialism, beginning with Stalinization in 1948 and continuing with varying intensity until the collapse of socialism. In Czechoslovakia, an important element of this narrative was the imposition of "normalization" following the invasion to suppress the Prague Spring in August 1968. In Poland, it was the imposition of martial law in 1981, although its consequences were less far-reaching. However, as Natalia Tsvetkova has convincingly argued, this narrative has largely overlooked the importance of the agency of local actors. She maintains that "the concept of cultural imperialism should be considered through the overt and covert resistance of local recipients," so as not to incapacitate the historical actors.[53]

In recent years, there has been a proliferation of literature challenging the Sovietization paradigm by emphasizing the agency of local actors. Rather than focusing on external influences, historians have emphasized the efforts of local socialist elites to maintain agency in the face of Soviet politicians' willingness to delegate power on the one hand, and the experience of the majority of the population who remained aloof from politics on the other.[54] In Czechoslovakia, for example, Zdeněk Nejedlý (1878–1962), Minister of Education (1948–1953) and the first head of the ČSAV (1952–1962), was able to implement a number of reforms that were incompatible with the Soviet system, even though, or perhaps because, he was a supporter of Stalinist policies.[55] University professors and students also negotiated the imposed policies and attempted to preserve the local characteristics of their universities.[56] However, it was not only elites that had agency: as Patryk Babiracki has argued, lower and middle-level officials were also able to use their agency in an effort to forge integration between the Soviet Union and Poland.[57]

With regard to intellectuals, recent scholarship has argued that the intelligentsia reproduced itself regardless of the political and social conditions of the system, suggesting that they were able to make good use of their cultural and social capital even when they faced repression as a class.[58] At the same time, the prism of changing contexts affected—but did not determine—long-standing intellectual contacts and trends.[59]

New narratives have also castigated the Sovietization paradigm for overlooking people's commitment to and belief in socialism. Using the case of the Czechoslovak Marxist philosopher Karel Kosík, Tomáš Hermann discussed how the historiography of scholars persecuted in 1968 tends to accentuate biographical breaks and overlook continuities in intellectual development, which in many cases included not only a belief in Marxism but also a belief in Stalinism, including explicit political engagement.[60] The biography of Kosík is not the only such case— for example, the Polish philosopher Leszek Kołakowski and sociologist Zygmunt Bauman fall into a similar category.[61] Other literature, most notably that of Agata Zysiak, has emphasized the socialist drive for modernization, which brought a degree of uniformity and conformity, but also mass education, which, the old intelligentsia considered destructive, but leftist scholars, workers, and many students actively supported.[62]

The shift from a narrative driven by political concepts to one focused on practices also occurred at the level of the relationship between socialist states. While socialist historiography followed a friendship-driven narrative, post-socialist historiography defined this friendship as "mandated."[63] A look at the groups for whom crossing the border was of paramount importance shows that, beyond the propagandistic tones, there was a plethora of informal contacts that connected socialist subjects. These included smugglers and dissidents, but also artists and scholars.[64] As Michaela Šmidrkalová's contribution to this volume argues, these informal contacts were easier to establish and maintain outside formal structures than within them. While being inside the Academy of Sciences had its obvious advantages, there were also drawbacks.

Socialist Academia

The formalization of international scientific relations in the socialist countries was a process associated with the prominent role of the academies of sciences. Throughout the Eastern Bloc, the academic system was divided into research, applied research and teaching. The role of universities was to specialize in teaching, with limited resources, both budgetary and temporal, allocated to research and travel. The departmental research institutes were very closely associated with applied research. They operated largely independently of the academies of sciences and the universities, and were directly subordinated to the individual ministries.

The academies of sciences played the preeminent role in research, which is why this volume predominantly focuses on their activities. The academies were responsible for research and were the institution that awarded the highest academic titles. They also often played an advisory role to political bodies and were required to coordinate a national plan for all research institutions in the country. In Czechoslovakia, the chairman of the Academy of Sciences was also a member of the Central Committee of the Communist Party (in Poland two first presidents of the Polish Academies of Sciences (further PAN) were not even members of the party). The lack of direct student contact and the high prestige of internationally recognized research meant that scientists within the academies of sciences were largely subject to a lesser degree of political control, and the academy's institutes often served as a refuge for those who had been excluded from universities. At the same time, the ČSAV and PAN replaced previously autonomous institutions in 1951 and 1952, respectively. Nevertheless, the division between teaching and research institutions was never clear-cut. Individual overlaps between the academies and universities persisted, as there were (male) professors who held key positions in both during the 1950s. The declared shortage of experts led to the double employment of PhD students and limited the access of younger people (especially women) to leading positions.

Socialist academia was also a highly dynastic system, characterized by limited upward mobility that largely depended on replacing those higher in the hierarchy, and by minimal exchange between different research units, laboratories, or departments.[65] This reinforced hierarchical dependency, but at the same time provided a degree of job stability (compared to the highly competitive, grant-dependent careers of twenty-first century academia).

However, this stability had limits, which were mostly related to political events. Most notably, every political rupture led to a wave of dismissals in academia, beginning with the communist takeovers and followed by the even more radical upheavals of 1968 in both countries. Research findings often faced censorship, especially in the humanities and social sciences.[66] Participation in conferences abroad was monitored by informants of the secret service (mostly fellow scholars).[67] Foreign literature was largely only available in central libraries, access to which often required special permission—especially in the Soviet Union, which was more restricted to international information than its satellite countries. Therefore, frequent visits to libraries in Warsaw or Budapest offered a way for Soviet scholars to keep abreast of new developments in their fields, especially those who were not permitted to travel outside the socialist countries. Polish libraries were frequently cited as being well stocked with non-socialist literature. Conversely, in disciplines such as mathematics, Western literature circulated from the East through Soviet reprints and translations.

Socialist academia was traditionally male-dominated and followed relatively typical patterns for the postwar global North: there were fewer women in the natural sciences than in the humanities, fewer women in the more prestigious disciplines, and a high proportion of women in fields considered feminine, such as literary studies or bibliography.[68] Socialism in both society and academia was characterized by equality in theory and the celebration of women in the public sphere, especially on festive occasions.

However, in practice gender discrimination was commonplace—and there was scant discussion of its shortcomings. A 1986 Polish study found that 56 percent of female academics felt disadvantaged compared to men when it came to opportunities to travel abroad on a scholarship or for work.[69] The reasons for this went unexplored, however, partly because the history of science has traditionally followed "pure" ideas rather than "people with bodies, situated in time, space, culture, and society, struggling for credibility and authority," and partly because research on gender issues has been slow to emerge in East-Central European historiography as a whole.[70] It is therefore necessary to find new ways of gathering information. Oral history is indispensable in this respect, as numerous projects have shown and as our interviews have confirmed.[71] They suggest that the visible disparity between men and women travelers was more the result of a general disparity at the level of research, rather than specific discrimination in the area of travel arrangements. However, other factors may have played a role, such as the traditional role of women in child care and thus the greater tendency for them to stay at home, even though female academics tend to have significantly fewer children than women working outside academia.[72]

Importantly, academia consisted of more than just scholars. It rested on the shoulders of a multitude of personnel, from secretaries to librarians to copyists to programmers, all of whom were predominantly female.[73] Although these workers were key to the functioning of the academy, they have often been outside the purview of historians of science. On many occasions they cooperated internationally, for example, in the collection of foreign literature, which was supposed to be complementary throughout the Eastern Bloc,[74] but these meetings produced fewer sources than scholarly meetings. (This is also true of cooperation between state research institutes and institutes for scientific information, which were tasked with collecting information from international sources, translating it, and disseminating it throughout the scientific community).

In international relations, although Sovietization might have been expected, it was often the case that traditionally close cooperation with countries such as West Germany or France prevailed over the politically prescribed and state-supported eastward orientation.[75] In 1959, the number of exchanges between Poland and Western countries was reportedly eight times higher than with the Soviet Union, and cooperation with the Soviet Union was more declarative than functional.[76] In Czechoslovakia, 1968 was the first year in which exchanges with capitalist countries exceeded those with socialist states—by a third. And while 1969 followed a similar trend, by the end of 1969 a series of measures had been taken to limit exchanges with non-socialist countries, and these exchanges fell drastically the following year, especially in terms of outward travel.[77]

While the framework of bilateral and multilateral agreements and the diversity of central and extramural institutes had an impact on exchanges, the aptitude of the institute heads also played a part in steering the direction and determining the intensity of contacts. The influential Czechoslovak microbiologist and politician Ivan Málek, for example, was able to argue in favor of developing cooperation with non-socialist countries as part of ideological competition, which allowed many researchers at his institute to complete fellowships at institutions in Western Europe and North America.[78] However, this autonomy had its limits. On the one hand, there were general international and inter-academic agreements that defined the cornerstones of cooperation, including the number of participants. On the other hand, every travel request had to be approved by a higher level of the academy, and the policy of approval depended on political prerogatives (and the financial means of the state).

Travel was not available to everyone, as it had been deemed considered a privilege, especially to countries outside the Eastern Bloc. This mainly stemmed from political reasons, as each request had to be approved by the directors and political authorities. Especially when traveling to the West, socialist scholars had to navigate tightly controlled passport regimes and bureaucratic and often politicized procedures.[79] Passports (which were state property) were kept at police stations and had to be requested for each trip (and immediately handed back after the return), which essentially excluded those who were politically suspect or considered likely to defect to the West. It was necessary to substantiate the importance of the travel, and with reference to scientific gains in the case of scientific trips (in the Soviet Union this applied to all travel, including private travel to Eastern Bloc countries).[80]

In addition, the cost of travel caused problems, as per diems were often so low that they barely covered a fraction of the necessary costs, and it was therefore essential to obtain invitations from abroad, with foreign institutions covering the scholar's expenses. Even travel within the Eastern Bloc could be financially burdensome, alleviated by institutional funds or often by informal arrangements (staying with friends instead of in hotels, etc.). Finding alternative funding was important because of the general shortage of hard currency (along with the concurrent abundance of non-convertible currency). This issue affected not only academics, but also the economy as a whole. Therefore, especially during the period of intensified non-socialist exchange, the receiving institutions covered the costs of the majority of outgoing fellowships.

In the case of the socialist academies, this led to reciprocal agreements. In order to minimize the transfer of money abroad, there had to be an equal number of outgoing and incoming scholars, although deviations were permitted for large discrepancies in country size, such as between Czechoslovakia and Poland. Moreover, the receiving institution always paid for the scholar's in-country expenses, while the sending institution paid for their travel. "Reciprocity" (PL: *wzajemność*; CZ: *reciprocita/vzájemnost*; SL: *vzájomnosť*), the term denoting this idea of equal exchange, became the key for understanding socialist academic exchange.[81] In contrast to the well-known socialist economic plans, which were designed to be exceeded, exceeding the plan of scholarly exchange was to be avoided, although it was possible with the agreement of all parties. On the other hand, underachieving, i.e., sending too few researchers abroad, could lead to reprimands, as noted by Jan Surman in this volume. This also led to a situation where institutes sent researchers abroad for no apparent reason, simply to meet quotas. While such travel may have been productive, many were more a kind of academic vacation.

Reciprocity agreements, whether considered productive or not, defined the formal framework of scholarly exchange with and within Eastern Bloc countries, and they occupied the administrators whose task it was to fulfill them, especially given the political restrictions and the limited number of people who were permitted to travel abroad. These agreements also shaped the strategies of those who wanted to utilize them for their own (professional or less scholarly) purposes. Although formal, reciprocal relations relied on informal contacts, they also structured the very conditions under which informal contacts could be made.

Czech–Polish Relations

The prehistory of modern Czech–Polish scholarly relations has often been written through the prism of Slavic contacts within the Habsburg Monarchy.[82] While these histories describe the unifying force of anti-Germanism as the glue that held Czechs and Poles together (excluding Slovaks, whose historical ties were bound up with the Hungarian part of the monarchy), border tensions cast a shadow over scholarly contacts after Czechoslovakia and Poland became independent. While institutional cooperation failed repeatedly in the 1920s and 1930s,[83] private contacts established before 1918 continued.[84] Some even became the basis for post–World War II contacts, helping to reestablish relations in the brief three years before the communist takeover. Some contacts were broken off in the 1950s and 1960s,[85] but others continued under changed and more formalized conditions.

After 1947, official academic relations were governed by bilateral agreements, first at the state level and then at the academy level.[86] An agreement between the two academies was reached shortly after the inauguration of the PAN in 1953 and formalized at the end of 1955 in an annex to the Intergovernmental Treaty on Cultural Cooperation.[87] The agreement was periodically renegotiated and updated. It included the main areas of research to be pursued and the number of scholars to be involved in academic exchanges. It defined topics and problems for cooperation at the level of the institutes, initially in a few selected fields, and expanded with each revision. Information about the work program and the participating scientists was to be sent to the other academy, although in practice both sides often failed to do this on time, and it was repeatedly cited as a problem. The areas of research corresponded to the research objectives of both states but increasingly also followed multilateral socialist agreements, especially within COMECON, although in the 1960s bilateral agreements remained the preferred method of the cooperating parties.[88] Later, with more intensive multilateral cooperation and an increasing focus on it from the 1970s onward, this even led to the siphoning off of quotas intended for bilateral exchange. While this volume cannot answer the question of to what extent "socialist integration" was more than just an aspiration, it was certainly taken into account in research planning, as several of the chapters in this book illustrate.

During the socialist period, Czechoslovak–Polish cooperation was characterized by a number of asymmetries. For example, although the reciprocal quota between the ČSAV and PAN continually increased, from 110 weeks in 1958, to 260 weeks in 1965,[89] 300 in 1971,[90] and 500 in 1981,[91] the relative size of the two academies meant that the importance of the collaboration was different in each. For ČSAV, PAN was among its most important collaborators throughout the period, whereas for PAN, ČSAV held a mid-level position among its partners. As many of the chapters in this volume show, there was also a symbolic asymmetry, as both Polish and Czechoslovak scholars tended to prefer Western or Soviet partners. This asymmetry was particularly marked after the Warsaw Pact invasion, when Polish scholars had many more opportunities to travel to non-socialist countries than their Czechoslovak comrades. This asymmetry was clear in many areas—for example, Polish books were widely translated and found in Czechoslovak libraries, but Czech and Slovak books were rarely translated into Polish. And like any asymmetry, it should be taken with a grain of salt, as it depends on the fields and disciplines involved in the cooperation. Nevertheless, the Polish side often played a more active role in the negotiations between the ČSAV and PAN—proposing more topics for cooperation, seeking to revise the agreed-upon division and hierarchy of scientific fields for cooperation, and suggesting initiatives such as joint awards and more.

Personal contacts, both within and outside formal structures, were crucial in overcoming institutional asymmetries. As the contributions to this volume demonstrate, collegiality and friendship played a key role in fostering collaboration where it was not initially anticipated or supported by institutions. Importantly, these relationships also transcended political ruptures and even generations.[92] Focusing on them helps us to consider alternative temporalities to those of political history, which may or may not correspond to scholarly dynamics.

Chapter Overview

This volume's seven chapters explore case studies of Czechoslovak–Polish scholarly interactions during the socialist period. To highlight the coexistence of formal, reciprocal relations and informal networks, we focused on disciplines closely tied to politics. This approach allows us to show that—even in fields subject to strict state oversight—there was room for individual initiative. Indeed, such efforts were often essential for effective cross-border cooperation. Reflecting ongoing discussion about scales of historical inquiry, the chapters move from the local level toward a broader, transnational perspective.

Michaela Šmidrkalová explores the entanglements of Czech, Slovak, and Polish historians of science and technology. She focuses on three lesser-known individuals—Juraj Voda, Stanisław Miczulski, and Stanisław Zahradnik—whose regional and national affiliations shaped their scholarly trajectories. Šmidrkalová emphasizes how cross-border scholarly cooperation was often driven by individuals in peripheral institutions who fostered meaningful intellectual exchanges. For these individuals, travel to neighboring countries was an opportunity to engage internationally. All three were pioneers in their fields but were sidelined once the history of science became institutionalized and centralized. Zahradnik, a member of the Polish minority in Czechoslovakia, also faced mistrust from the secret service, which hindered his career.

Jan Surman examines the unsuccessful collaboration between Czechoslovakia and Poland regarding the Scientific-Technological Revolution (STR) during the 1970s. Led by Radovan Richta's influential team, Czechoslovakia became a central site of STR theorization. Despite the enthusiasm of Polish scholars, particularly young researchers like Lech Zacher, structural mismatches between institutions, Richta's prioritization of Soviet partners, and the repercussions of the events of 1968 contributed to the growing distance between the two countries. Personal tensions, such as Zacher's public critique of Soviet–Czechoslovak STR publications, further complicated matters. Ultimately, the breakdown of this collaboration exemplifies how interpersonal trust, political allegiance, and Cold War hierarchies intertwined to restrict intellectual exchange within the socialist bloc.

Luboš Studený and Lukas Becht examine Czechoslovakia and Poland within the transnational network of futures studies. Senior scholars such as Ota Šulc and Andrzej Siciński played pivotal roles in fostering connections through bilateral exchanges, international conferences, and multilateral projects. The chapter discusses three main constellations of cooperation: ideological critique, engagement with global futures studies communities, and institutional collaboration, as exemplified by the Wrocław Center for Prognostic Research (OBP). Despite differences in political climates and national academic priorities, the Czech–Polish network enabled mutual support and information exchange. However, structural and political limitations, including censorship and self-censorship, institutional instability, and martial law in Poland in 1982, restricted the network's continuity and impact.

Maciej Górny focuses on the intersection of the economy and socialist central planning. His work highlights the shortcomings of centrally planned economies and the rise of Marxist reformism in both countries in response to the inefficiencies of Stalinist economic models. Polish economists, who benefited from a more open intellectual climate after the 1956 political thaw, initiated debates that influenced their Czechoslovak counterparts. Central to the narrative is Ota Šik, whose economic reforms in Czechoslovakia during the 1960s were inspired by, and received critical attention from, Polish economists such as Michał Kalecki and Oskar Lange. Despite the failure of the reform efforts and the suppression of the Prague Spring, the legacy of academic collaboration between the two countries persisted into the 1970s.

Tomáš W. Pavlíček's chapter explores the evolution of mathematical exchange from the end of World War II to the late socialist period. Using oral histories and archival sources, Pavlíček reveals how personal encounters, such as study visits, seminars, and local mountain conferences, served as vital nodes for scholarly communication. While Polish mathematicians generally had greater international visibility and institutional support (for example, through the Banach Center in Warsaw), Czech scholars such as Eduard Čech and his younger colleague Miroslav Katětov played pivotal roles in transnational cooperation, notably through the Prague Topological Symposia. Overall, Czech–Polish entanglements in mathematics were marked by informal exchanges, in which personal agency often compensated for structural asymmetries in socialist academia. However, informality's role changed across time—from a basis for formal contacts to a supplement to them.

Kornelia Kończal explores practices of cooperation between Polish and Czechoslovak sociologists. Between 1945 and 1989, the development of sociology in both countries followed similar trajectories, shaped by shared intellectual traditions and later disrupted by Stalinist repression. Although sociology was revived earlier in Poland (1956) than in Czechoslovakia (1964), bilateral cooperation remained limited and structurally asymmetric: Czech and Slovak scholars engaged more actively with Polish sociology than the other way around. The chapter examines these encounters through a praxeological lens, focusing on the everyday practices of researchers within their institutional and ideological contexts. It distinguishes between "apparent activities" and "inspiring interactions," showing that meaningful exchange often relied on personal initiative. Case studies of Jiří Kolaja and Jan Szczepański illustrate how individual agency could transcend structural constraints.

Doubravka Olšáková's chapter examines the Pugwash Conferences' role as a unique transnational channel of science diplomacy and dialogue between Czechoslovakia and Poland. She details how scientists from both countries circumvented rigid state control to influence disarmament and security discussions, particularly through initiatives like the Rapacki Plan for a nuclear-free Central Europe. Czechoslovakia initially led Pugwash efforts in the 1960s, but after the 1968 Warsaw Pact invasion, Poland assumed leadership, hosting key conferences in the 1970s and 1980s, even under martial law. Notable figures, such as František Janouch and Andrei Sakharov, challenged Pugwash to address human rights issues, pushing the organization beyond traditional disarmament dialogue. A turning point occurred at the 1982 Warsaw Pugwash Conference, where scientists met informally with Solidarity opposition figures. The chapter emphasizes the dual roles of scientists as experts and activists, as well as the challenging balance they maintained between fostering East–West dialogue and addressing repression within the Eastern Bloc.

Finally, Simone Turchetti discusses the implications of the ideas expressed in this volume for the history of science, particularly the history of science diplomacy. He argues that including the perspective of socialist countries allows for a more detailed picture of international scientific relations, which have thus far been analyzed through Western European and Anglo-American perspectives. Thus, the volume not only helps us understand the sciences in the Eastern bloc but also the emergence of the global scientific system.

Author Biographies

Lukas Becht is a political scientist and historian focusing on the history of future studies, political time, and knowledge production in Central and Eastern Europe. He studied political science, sociology, and philosophy, obtained a master's degree at the University of Freiburg and interned at the

Polish Sejm. In a binational PhD program between LMU Munich and the University of Vienna, he researches scientific futurology in socialist Poland. He is an Associated researcher at RECET, University of Vienna, where he also coordinated the project Economic Collectivism Old and New.

Maciej Górny is a historian specializing in the intellectual history and historiography of Central and Eastern Europe. He holds a PhD (2006) and habilitation (2015) from the Institute of History, Polish Academy of Sciences (IH PAN), where he now serves as a full professor and vice-director. His research explores national stereotypes, Polish–German relations, and Marxist historiography. Górny has led and contributed to several major research projects and is the author of influential works on World War I, historical geography, and the history of historical thought.

Kornelia Kończal is a historian specializing in contemporary European history, with a particular focus on Central and Eastern Europe, France, and Germany. She holds a PhD in History from the European University Institute in Florence, Italy (2017) and serves as Assistant Professor of Public History at Bielefeld University, Germany. She is currently completing her first monograph, *Politics of Plunder: How Expropriating German Expellees Helped and Harmed the Postwar Reconstruction of East Central Europe*, and editing a volume on *Doing Heritage without Heirs? Practices of Cultural Succession in Post-German Spaces*..

Doubravka Olšáková is a historian and political scientist, specializing in the history of science and technology and the Cold War. She works currently at the Institute of Political Sciences of the Faculty of Social Sciences at Charles University. She is the founding member of the Standing Commission on Science, Technology and Diplomacy, Division of History of Science and Technology of the International Union of History and Philosophy of Science and Technology, deputy editor-in-chief of *Centaurus*, the journal of the European Society for the History of Science, and a founding member of the Oxford University Press Science, Technology and International Relations series.

Tomáš W. Pavlíček specializes in the social and cultural history of Central Europe, oral history, and the history of science. He earned his PhD through a Czech–German program (LMU Munich–Charles University). As a historian at the Institute of History, Polish Academy of Sciences he leads international research on Czech–Polish scholarly networks and has published extensively on science, religion, and memory in nineteenth- and twentieth-century Europe.

Michaela Šmidrkalová is a historian at the Masaryk Institute and Archives of the Czech Academy of Sciences, specializing in twentieth-century Central European history and the history of science. She earned her PhD in Modern History from Charles University in 2015. Her research focuses on scientific institutions, international scholarly exchanges, and the cultural history of technology under socialism.

Luboš Studený is a historian specializing in the late socialist and postsocialist period of Czechoslovakia, with a focus on the history of future studies, postcommunist transformation, cybernetics and historiography theory. He holds a master's degree from Charles University, where he is currently completing his PhD in modern economic and social history. He has contributed to research on socialist enterprise transformation and Cold War scientific networks.

Jan Surman is a historian specializing in the intellectual and scientific history of Central and Eastern Europe. He earned his PhD in History at the University of Vienna and has held fellowships in Vienna, Moscow, Princeton, and Marburg before joining the Masaryk Institute and Archives of the Czech Academy of Sciences. His research explores the intersections of science, language, and politics. He has published widely on multilingualism in academia, the history of positivism, and knowledge circulation, and serves on editorial boards of several international journals in the history of science.

List of Abbreviations

ČSAV	Československá akademie věd (Czechoslovak Academy of Sciences)
MÚA AV ČR	Masarykův ústav a Archiv Akademie věd České republiky (Masaryk Institute and Archives of the Czech Academy of Sciences)
PAN	Polska Akademia Nauk (Polish Academy of Sciences)

Notes

1. Štrbáňová and Kostlán, *Sto českých vědců v exilu*.
2. Böhler, *Civil War in Central Europe, 1918–1921*, 115–21. On discussions in academia, see Ďurčanský, "Od niezgody do bratniej serdeczności i z powrotem," 144.
3. Górny, *The Nation Should Come First*; Hudek, Kopeček, and Mervart, *Czechoslovakism*.
4. Szczepańska, "Wybrane stereotypy Czecha i Polaka z perspektywy historycznej."

5. Mehlhorn, "Zwangsverordnete Freundschaft?"; Kerski, *Przyjaźń Nakazana?* For a narrative that follows the same topos for describing Czechoslovak–Polish relations, see Przeperski, *Nieznośny ciężar braterstwa*.
6. McDermott and Stibbe, *Czechoslovakia and Eastern Europe*.
7. See Samuilova, "Scientific and Educational Programs"; Avramchuk, *Budując Republikę Ducha*. The Czechoslovak Fulbright Program was established after the USSR's program, as the latter was founded 1973/74.
8. For the cooperation and competition approach, see Nickelsen and Krämer, "Introduction." See also Soutschek, *Wissenschaftskooperation*.
9. For a more nuanced approach, see, for example, Kravčenko and Stech, *The Unpredictable Past?*; Byford, Doak, and Hutchings, "Decolonizing the Transnational, Transnationalizing the Decolonial."
10. Conrad and Randeria, *Jenseits des Eurozentrismus*; Surman and Kaps, "Postcolonial or Post-Colonial?"
11. Surman and Petushkova, "Between Westernization and Traditionalism," for example, 462.
12. See the discussion on materials from this project in Surman and Pavlíček, "The Formal and the Informal."
13. Hudek, "An Uncommon Course."
14. See the chapter by Michaela Šmidrkalová in this volume.
15. On resources, see Ash, "Wissenschaft und Politik als Ressourcen füreinander."
16. Gänger and Osterhammel, "Introduction," 9.
17. Raj, "Beyond Postcolonialism"; Raj, *Relocating Modern Science*.
18. Shapin, "Placing the View from Nowhere."
19. Reinisch and Brydan, "Introduction," 2.
20. Wolfe, *Freedom's Laboratory*; on the Russian counterpart, see the discussion in Gordin, "Is Science Political?"
21. Holvikivi, Holzberg, and Ojeda, *Transnational Anti-Gender Politics*; Goetz and Mayer, *Global Perspectives on Anti-Feminism*; Laqua and Papadogiannis, "Youth and Internationalism in the Twentieth Century"; on a similar change in transnationalism, see Laqua, "Internationalism."
22. Patel and Reichardt, "The Dark Side of Transnationalism"; Bauerkämper and Rossolinski, *Fascism without Borders*; Brydan, *Franco's Internationalists*; Grutza et al., "Rethinking the Dark Side of Transnationalism"; Hedinger, "Fascist Internationalism."
23. Stoeckl and Ayoub, "Transnational Illiberal Networks"; Hetherington and Sluga, "Liberal and Illiberal Internationalisms."
24. Winkel, Roth, and Scheele, "Analytical Framing"; Zürn et al., "Towards a Typology of Contestation."
25. Green and Viaene, *Religious Internationals in the Modern World*; Engel, Kennedy, and Reynolds, "Editorial"; Brown, "Catholic Internationalism" quoted in Reinisch, "Introduction," 199–200.
26. Lewis, "Decolonising the History of Internationalism."
27. Tedeschi, Vorobeva, and Jauhiainen, "Transnationalism."
28. Reinisch and Brydan, "Introduction," 7.
29. Surman and Pavlíček, "The Formal and the Informal."
30. Applebaum, *Empire of Friends*, 16.

31. Doel, "Internationalism in Science"; Babiracki and Jersild, "Editors' Introduction"; Defrance and Kwaschik, *La guerre froide*; Autio and Miklóssy, *Reassessing Cold War Europe*; David-Fox, "The Implications of Transnationalism"; on a more transnational perspective, see Solovey and Dayé, *Cold War Social Science*.
32. Olšáková, "Expert Cultures and Infrastructural Globalism"; Olšáková, "Science, (Anti-)Communism and Diplomacy." On the Brezhnev Doctrine, see Ouimet, *Rise and Fall of the Brezhnev Doctrine*.
33. Wolle, "Das System der Reisekader."
34. Interview of Jan Surman with Dieter Hoffmann on 16 December 2022, stored at MÚA AV ČR.
35. Tõnismann, "Practices and Discourses of Internationalisation"; Surman "Internationalism."
36. Péteri, "Nylon Curtain."
37. For example, Mazurek, "Polish Economists in Nehru's India"; Stanek, *Architecture in Global Socialism*; Iacob, "Paradoxes of Socialist Solidarity"; Buzássyová, *The Eastern Bloc and Sub-Saharan Africa*.
38. For example, Rindzevičiūtė, *The Power of Systems*; Kilias, *Goście ze wschodu*.
39. See, for instance, Kravčenko and Stech, *The Unpredictable Past?*; Byford et al., "Decolonizing the Transnational."
40. Established in 1949, COMECON was primarily an organization for the coordination of economic activities between the member countries of the Eastern Bloc, although it increasingly included the issue of scientific and technological cooperation.
41. Crump and Godard, "Reassessing Communist International Organisations," 86.
42. Dautova, "Integratsiia."
43. Crump and Godard, "Reassessing Communist International Organisations."
44. This was highlighted in the Comprehensive Program for the Further Deepening and Perfection of Cooperation and for the Development of Socialist Economic Integration among the COMECON Countries from 1971.
45. On Dubna, see Da Silva Neto and Kojevnikov, "Socialist Internationalism and Science Diplomacy"; on INTERKOSMOS see Ben-Itzhak, "Network Analysis of International Cooperation." More on this cooperation in Wilczynski, *Technology in Comecon*, 260–95.
46. As they were politically supported, receiving generous funding, and facilitated bureaucracy, these centers became key hubs of socialist cooperation whose activities often survived beyond 1989/1991, sometimes with minimal changes. On the Banach Institute, see Duda, *Historia matematyki*, 345–46; on the Wrocław laboratory, see Tomaszewski, "Geneza Instytutu Niskich Temperatur."
47. Report on Czechoslovak participation in 7th World Sociological Congress in Toronto (Canada), 19–24 August 1974 (p. 5). MÚA AV ČR, ÚFS ČSAV, sociolog. sekce 63–4.
48. Lebow, Mazurek, and Wawrzyniak, "Making Modern Social Science."
49. See Babiracki and Jersild, "Editors' Introduction," 4–5; Applebaum, *Empire of Friends*, 13–15.
50. Olšáková, "A Matter of Courtesy"; Surman and Pavlíček, "The Formal and the Informal."
51. Lochmatow, *Public Knowledge in Cold War Poland*, 3.

52. Pleskot and Rutkowski, *Spętana Akademia.*
53. Tsvetkova, *The Cold War in Universities,* 18.
54. Connelly, *Captive University*; McDermott and Stibbe, *Czechoslovakia and Eastern,* 5.
55. Křesťan, *Zdeněk Nejedlý.* As in many similar cases, the venerable communist was supported by carefully selected young communists.
56. Tsvetkova, *The Cold War in Universities,* 18.
57. Babiracki, *Soviet Soft Power in Poland.*
58. Connelly, *Captive University*; Tromly, *Making the Soviet Intelligentsia*; Zysiak, *Limiting Privilege.*
59. Kosicki, *Catholics on the Barricades*; Kula, *Mimo wszystko bliżej Paryża niż Moskwy*; Shore, *Caviar and Ashes.*
60. Hermann, "Karel Kosík and His 'Radical Democrats.'"
61. Mentzel, *Kołakowski*; Wagner, *Bauman.*
62. Zysiak, "For Whom University?"; Zysiak, *Limiting Privilege.*
63. Mehlhorn, "Zwangsverordnete Freundschaft?"
64. Borodziej, Kochanowski, and von Puttka, "Einleitung"; Bazin, Glatigny, and Piotrowski, *Art beyond Borders*; Wasiak, "Schleichwege in der Galerie"; Kemp-Welch, *Networking the Bloc.*
65. Linková, "The Czech Research Landscape," 85. On the distinction between dynamic and dynastic laboratories on which this typology is based, see Linková and Červinková, "Vlastní Laboratoř."
66. Oates-Indruchová, *Censorship in Czech and Hungarian Academic Publishing.*
67. Kulecka and Rutkowski, *Oczami agenta.*
68. Dadej, *Habilitacja—bariera czy kariera?*; Linková, "The Czech Research Landscape."
69. Iwaszczyszyn, "Kariery naukowe kobiet w PRL."
70. Shapin, *Never Pure*; Fábián, Johnson, and Lazda, *Routledge Handbook of Gender.*
71. See here especially the interview of Luboš Studeny and Michaela Šmidrkalova with Ivan Hlaváček and Ludmila Hlaváčková on 29 June 2022, stored at MÚA AV ČR.
72. Siemienska, "Academic Careers in Poland," 73.
73. Derra, Kola, and Piasek, *Niewidzia(l)ne.*
74. *Mezhdunarodnoe soveshchanie.*
75. Szumski, *Polityka a historia,* 330; Dybiec, "Warunki rozwoju nauk humanistycznych," 256–82.
76. Szumski, "Wpływ elity władzy ZSRR," 137–43. On the tensions between declarative and practices in CSR, see also Šmidrkalová, "Mezinárodní kontext vývoje ČSAV 1963–1967," 171–72.
77. Šmidrkalová, "Mezinárodní kontext vývoje ČSAV 1968–1970," 321; Hálek, *Ve znamení "bdělosti a ostražitosti,"* 16; MÚA AV ČR, Zahraniční odbor Úřadu prezidia ČSAV; Proposals and Recommendations for Stays in Certain Socialist States, 1952–61.
78. Franc, *Ivan Málek a vědní politika 1952–1989,* 145–64.
79. Pleskot, "Kontakty naukowe Polski"; Pleskot, "Jak wyjechać na Zachód?"
80. Hálek, *Ve znamení "bdělosti a ostražitosti,"* 10; Pleskot, "Jak wyjechać na Zachód?" We thank Slava Gerovitch for information concerning the Soviet Union.
81. To an extent this is also true for US-Socialist exchange, see Sher, *From Pugwash to Putin,* 14.

82. For example, Stasiewicz-Jasiukowa and Janko, *Z dziejów polsko-czeskich*.
83. Ďurčanský, "Od niezgody do bratniej serdeczności i z powrotem," 147.
84. For example, Ďurčanský, "Władysław Mieczysław Kozłowski"; Laskosz and Chodějovský, *Korespondence Waleryho Goetela s Radimem Kettnerem*.
85. Górny, "Writing History"; Surman and Petushkova, "Between Westernization and Traditionalism."
86. See the Agreement on Friendship and Mutual Assistance in 1947. "Smlouva č. 154/1947 Sb., Smlouva o přátelství a vzájemné pomoci mezi Československou republikou a republikou Polskou," *Zákony pro lidi*. Retrieved 7 February 2025 from https://www.zakonyprolidi.cz/cs/1947-154. The protocol of the Agreement on Scientific Cooperation between ČSAV and PAN, signed in Prague, 24 November 1955.
87. Balcerak and Melichar, *Dokumenty i materiały do historii stosunków polsko-czechosłowackich*, 200–2, 272–75.
88. Brádlerová and Kmochová, "Nástin Zahraničních Styků," 108
89. Malecki, "Spolupráce Československé akademie," 353.
90. MÚA AV ČR, Spolupráce mezi ČSAV a PAN; here the numbers are explicitly 350 for ČSAV and 150 for SAV.
91. MÚA AV ČR, Coll. P ČSAV, box 186, Návrh plánů vědeckých styků ČSAV včetně SAV se zahraničím na rok 1981, 1980-14-10, p. á.
92. Pavlíček, "'The Gentlemen Did Not Like Each Other.'"

Bibliography

Applebaum, Rachel. 2019. *Empire of Friends: Soviet Power and Socialist Internationalism in Cold War Czechoslovakia*. Ithaca, NY: Cornell University Press.

Ash, Mitchell G. 2002. "Wissenschaft und Politik als Ressourcen für einander" [Science and politics as resources for each other]. In *Wissenschaften und Wissenschaftspolitik: Bestandaufnahmen zu Formationen, Brüchen und Kontinuitäten im Deutschland des 20. Jahrhunderts*, edited by Rüdiger vom Bruch and Brigitte Kaderas, 32–51. Stuttgart: Franz Steiner-Verlag.

Autio, Sari, and Katalin Miklóssy, eds. 2011. *Reassessing Cold War Europe*. Routledge Studies in the History of Russia and Eastern Europe 14. Abingdon: Routledge.

Avramchuk, Oleksandr. 2024. *Budując Republikę Ducha: historia Programu Fulbrighta w Polsce w latach 1945–2020* [Building a republic of spirit: The history of the Fulbright Program in Poland, 1945–2020]. Warsaw: PWN.

Babiracki, Patryk. 2015. *Soviet Soft Power in Poland: Culture and the Making of Stalin's New Empire, 1943–1957*. The New Cold War History. The University of North Carolina Press.

Babiracki, Patryk, and Austin Jersild. 2016. "Editors' Introduction." In *Socialist Internationalism in the Cold War*, edited by Patryk Babiracki and Austin Jersild, 1–16. Cham: Springer International Publishing. https://doi.org/10.1007/978-3-319-32570-5_1.

Balcerak, Wiesław, and Václav Melichar. 1985. *Dokumenty i materiały do historii stosunków polsko-czechosłowackich. T. 1, 1944–1960. Cz. 2, 1949–1960* [Documents and Materials on the History of Polish-Czechoslovak Relations. Vol. 1, 1944–1960.

Part 2, 1949–1960]. Wrocław: Zakład Narodowy im. Ossolińskich. Wydawnictwo Polskiej Akademii Nauk.

Bauerkämper, Arnd, and Grzegorz Rossolinski, eds. 2017. *Fascism without Borders: Transnational Connections and Cooperation between Movements and Regimes in Europe from 1918 to 1945*. New York: Berghahn Books.

Bazin, Jérôme, Pascal Dubourg Glatigny, and Piotr Piotrowski, eds. 2016. *Art beyond Borders: Artistic Exchange in Communist Europe (1945–1989)*. Budapest: Central European University Press.

Ben-Itzhak, Svetla. 2024. "Network Analysis of International Cooperation in Space 1958–2023: Evidence of Space Blocs." *Journal of Peace Research* 62, no. 3: 517–34. https://doi.org/10.1177/00223433231225162.

Böhler, Jochen. 2018. *Civil War in Central Europe, 1918–1921: The Reconstruction of Poland*. Oxford: Oxford University Press.

Borodziej, Włodzimierz, Jerzy Kochanowski, and Joachim von Puttkamer. 2010. "Einleitung: 'Schleichwege': Fragestellungen Und Probleme" [Introduction: "Back Channels": Questions and Issues]. In *Schleichwege: Inoffizielle Begegnungen Sozialistischer Staatsbürger zwischen 1956 Und 1989*, edited by Włodzimierz Borodziej, Jerzy Kochanowski, and Joachim von Puttkamer, 9-21. Böhlau.

Brádlerová, Daniela, and Nataša Kmochová. 2001. "Nástin zahraničních styků ČSAV 1962–1970" [An Overview of the Czechoslovak Academy of Sciences' international relations, 1962–1970]. In *Česká Věda a Pražské Jaro, 1963–1970: Sborník z Konference, Vyd. 1*, edited by Blanka Zilynská and Petr Svobodný. Práce z dějin vědy, 107–118. Studies in the History of Sciences and Humanities, sv. 2. Karolinum.

Brown, Stephen J. 1925. "Catholic Internationalism." *Studies: An Irish Quarterly Review* 14, no. 55: 476–79.

Brydan, David. 2019. *Franco's Internationalists: Social Experts and Spain's Search for Legitimacy*. Oxford Studies in Modern European History. Oxford: Oxford University Press.

Buzássyová, Barbora. 2024. *The Eastern Bloc and Sub-Saharan Africa: Czechoslovakia, UNESCO and Development Aid from the 1960s and Beyond*. Routledge Histories of Central and Eastern Europe. Abingdon: Routledge. https://doi.org/10.4324/9781003439813.

Byford, Andy, Connor Doak, and Stephen Hutchings. 2024. "Decolonizing the Transnational, Transnationalizing the Decolonial: Russian Studies at the Crossroads." *Forum for Modern Language Studies* 60, no. 3: 339–57. https://doi.org/10.1093/fmls/cqae038.

Connelly, John. 2000. *Captive University: The Sovietization of East German, Czech and Polish Higher Education, 1945–1956*. Chapel Hill, NC: University of North Carolina Press.

Conrad, Sebastian, and Shalini Randeria, eds. 2002. *Jenseits des Eurozentrismus: postkoloniale Perspektiven in den Geschichts- und Kulturwissenschaften* [Beyond Eurocentrism: Postcolonial perspectives in history and cultural studies]. Frankfurt am Main: Campus-Verlag.

Crump, Laurien, and Simon Godard. 2018. "Reassessing Communist International Organisations: A Comparative Analysis of COMECON and the Warsaw Pact in Relation to Their Cold War Competitors." *Contemporary European History* 27, no. 1: 85–109. https://doi.org/10.1017/S0960777317000455.

Dadej, Iwona. 2022. *Habilitacja—bariera czy kariera? porządek płci w polskiej kulturze akademickiej pierwszej połowy XX wieku* [Habilitation—Barrier or Career? Gender Order in Polish Academic Culture in the First Half of the 20th Century]. Warsaw: Instytut Historii PAN, Wydawnictwo Neriton.

Da Silva Neto, Climério Paulo, and Alexei Kojevnikov. 2024. "Socialist Internationalism and Science Diplomacy Across the Iron Curtain: Geneva, Dubna, IUPAP." In *Globalizing Physics*, edited by Roberto Lalli and Jaume Navarro, 175–191. Oxford: Oxford University Press. https://doi.org/10.1093/oso/9780198878681.003.0010.

Dautova, E. A. 2022. "Integratsiia: Slovo i poniatie v diskussiiakh partiinykh rukovoditelei evropeiskikh stran-chlenov Soveta Ekonomicheskoi Vzaimopomoshchi" [Intgration: Word and concept in discussions of party officials of European member-states of the COMECON]. *Nauchnyi Dialog (Rossiia, Ekaterinburg)* 11, no. 7: 7.

David-Fox, Michael. 2011. "The Implications of Transnationalism." *Kritika: Explorations in Russian and Eurasian History* 12, no. 4: 885–904.

Defrance, Corine, and Anne Kwaschik. 2016. *La guerre froide et l'internationalisation des sciences acteurs, réseaux et institutions* [The Cold War and the Globalization of Science: Actors, Networks, and Institutions.]. Paris: CNRS éditions. http://books.openedition.org/editionscnrs/28809.

Derra, Aleksandra, Anna Maria Kola, and Wojciech Piasek, eds. 2020. *Niewidzia(l)ne: kobiety i historia Uniwersytetu Mikołaja Kopernika w Toruniu* [The Invisible: Women and the history of Nicolaus Copernicus University in Toruń]. Scientia, Universitas, Memoria 2. Toruń: Wydawnictwo Naukowe Uniwersytetu Mikołaja Kopernika.

Doel, Ronald E. 2020. "Internationalism in Science after 1940." In *The Cambridge History of Science*, edited by Hugh Richard Slotten, Ronald L. Numbers, and David N. Livingstone, 60–74. Cambridge: Cambridge University Press. https://doi.org/10.1017/9781139044301.006.

Duda, Roman. 2019. *Historia matematyki w Polsce na tle dziejów nauki i kultury* [History of mathematics in Poland in Context of the history of science and culture]. Warsaw: Instytut Historii Nauki im. L. i A. Birkenmajerów Polskiej Akademii Nauk.

Ďurčanský, Marek. 2015. "Od niezgody do bratniej serdeczności i z powrotem: Wydział Filozoficzny Uniwersytetu Karola w Pradze a współpraca z nauką polską (1918–1938)" [From discord to fraternal cordiality and back again: The Faculty of Philosophy at Charles University in Prague and its collaboration with Polish scholarship (1918–1938)]. *Historia Slavorum Occidentis* 8, no. 1: 141–67.

———. 2005. "Władysław Mieczysław Kozłowski a česká filosofie" [Władysław Mieczysław Kozłowski and the Czech philosophy]. In *Slovanství a Věda v 19. a 20. Století*, edited by Marek Ďurčanský, 55–76. Prague: Archiv Akademie věd ČR.

Dybiec, Julian. 2015. "Warunki rozwoju nauk humanistycznych i społecznych" [The conditions of development of human and social sciences]. In *Historia nauki polskiej. T. 10, Cz. 2: 1944–1989. Instytucje*, edited by Joanna Schiller-Walicka and Leszek Zasztowt, 177–298. Warsaw: Instytut Historii Nauki im. Ludwika i Aleksandra Birkenmajerów PAN, Oficyna Wydawnicza Aspra-JR.

Engel, Elisabeth, James Kennedy, and Justin Reynolds. 2018. "Editorial: The Theory and Practice of Ecumenism; Christian Global Governance and the Search for World Order, 1900–80." *Journal of Global History* 13, no. 2: 157–64. https://doi.org/10.1017/S1740022818000025.

Fábián, Katalin, Janet Elise Johnson, and Mara I. Lazda, eds. 2022. *The Routledge Handbook of Gender in Central-Eastern Europe and Eurasia*. Routledge Handbooks of Gender and Sexuality. New York: Routledge. https://doi.org/10.4324/9781138347762.

Franc, Martin. 2018. "Doing One's Laundry in a Hotel Room: The Everydayness of Josef Charvát and Ivan Málek of Congresses and Scientific Conventions in Western Countries in the 1960s." In *Science Overcoming Borders*, edited by Věra Dvořáčková and Martin Franc, 207–228. Prague: Masarykův ústav a Archiv AV ČR, v.v.i.

———. 2010. *Ivan Málek a vědní politika 1952–1989, aneb, Jediný opravdový komunista?* [Ivan Málek and the science policy 1952–1989, or: the only true communist?]. Prague Masarykův ústav: Archiv AV ČR.

Gänger, Stefanie, and Jürgen Osterhammel. 2024. "Introduction: Rethinking History, Globally." In *Rethinking Global History*, edited by Jürgen Osterhammel and Stefanie Gänger, 1–20. Cambridge: Cambridge University Press. https://doi.org/10.1017/9781009444002.001.

Goetz, Judith, and Stefanie Mayer, eds. 2024. *Global Perspectives on Anti-Feminism: Far-Right and Religious Attacks on Equality and Diversity*. Edinburgh: Edinburgh University Press.

Gordin, Michael D. 2019. "Is Science Political?" *Boston Review*, August 21. Retrieved 3 April 2026 from https://www.bostonreview.net/articles/michael-d-gordin-science-political/.

Górny, Maciej. 2013. *The Nation Should Come First: Marxism and Historiography in East Central Europe*, trans. Antoni Górny. Warsaw Studies in Contemporary History, volume 1. Lausanne: Peter Lang.

———. 2020. "Writing History in Twentieth-Century Eastern Europe." In *The Routledge History Handbook of Central and Eastern Europe in the Twentieth Century*, vol. 3, edited by Włodzimierz Borodziej, Ferenc Laczó, and Joachim von Puttkamer, 194–255. Routledge.

Green, Abigail, and Vincent Viaene. 2012. *Religious Internationals in the Modern World: Globalization and Faith Communities since 1750*. The Palgrave Macmillan Transnational History Series. Basingstoke, UK: Palgrave Macmillan.

Grutza, Anna, Janka Kovács, Vojtěch Pojar, and Anastassiya Schacht. 2024. "Rethinking the Dark Side of Transnationalism from East Central and Eastern Europe." *European Review of History: Revue Européenne d'histoire* 31, no. 1: 1–16. https://doi.org/10.1080/13507486.2023.2288103.

Hálek, Jan, ed. 2011. *Ve znamení "bdělosti a ostražitosti": zahraniční styky a emigrace pracovníků ČSAV v dobových dokumentech (1953–1971)* [Under the motto of "vigilance and caution": Foreign relations and the emigration of Czechoslovak Academy of Sciences researchers in contemporary documents (1953–1971)]. Prague: Masarykův ústav a Archiv AV ČR.

Hedinger, Daniel. 2025. "Fascist Internationalism: From a Vanished Institution to a Failed Concept?" *Journal of Modern European History* 23, no. 2: 147–66. https://doi.org/10.1177/16118944251331427.

Hermann, Tomáš. 2021. "Karel Kosík and His 'Radical Democrats': The Janus Face of Dialectics of the Concrete." In *Karel Kosík and the Dialectics of the Concrete*, edited by Ivan Landa, Joseph Grim Feinberg, and Jan Mervart, 39–54. Leiden: Brill. https://doi.org/10.1163/9789004503243.

Hetherington, Philippa, and Glenda Sluga. 2020. "Liberal and Illiberal Internationalisms." *Journal of World History* 31, no. 1: 1–9. https://doi.org/10.1353/jwh.2020.0000.

Holvikivi, Aiko, Billy Holzberg, and Tomás Ojeda, eds. 2024. *Transnational Anti-Gender Politics: Feminist Solidarity in Times of Global Attacks*. Thinking Gender in Transnational Times. Basingstoke, UK: Palgrave Macmillan.

Hudek, Adam. 2022. "An Uncommon Course: Normalisation in Slovakia." In *Czechoslovakia and Eastern Europe in the Era of Normalisation, 1969–1989*, edited by Kevin McDermott and Matthew Stibbe, 97–119. Cham: Springer International Publishing. https://doi.org/10.1007/978-3-030-98271-3_5.

Hudek, Adam, Michal Kopeček, and Jan Mervart. 2022. *Czechoslovakism*. Routledge Histories of Central and Eastern Europe. New York: Routledge.

Iacob, Bogdan C. 2021. "Paradoxes of Socialist Solidarity: Romanian and Czechoslovak Medical Teams in North Korea and Vietnam (1951–1962)." *History: Monde(s)* 20, no. 2: 117–40. https://doi.org/10.3917/mond1.212.0117.

Iwaszczyszyn, Krystyna. 1986. "Kariery naukowe kobiet w PRL (komunikat z badań)" [Women's Academic Careers in the People's Republic of Poland (Research Report)]. In *Kobiety polskie: praca zbiorowa*, edited by Elżbieta Konecka, 272–289. Warsaw: Książka i Wiedza.

Kemp-Welch, Klara. 2018. *Networking the Bloc: Experimental Art in Eastern Europe 1965–1981*. Cambridge, MA: The MIT Press.

Kerski, Basil, ed. 2009. *Przyjaźń Nakazana? Stosunki między NRD i Polską w latach 1949–1990* [Forced Friendship? Relations between FGR and GDR 1949–1990]. Wrocław: Oficyna Wydawnicza Atut—Wrocławskie Wydawnictwo Oświatowe.

Kilias, Jarosław. 2017. *Goście ze wschodu: socjologia polska lat sześćdziesiątych XX wieku a nauka światowa* [Visitors from the East: Polish sociology of the 1960s and international social sciences]. Warsaw: Nomos.

Kosicki, Piotr H. 2018. *Catholics on the Barricades: Poland, France, and "Revolution," 1891–1956*. The Yale-Hoover Series on Authoritarian Regimes. New Haven, CT: Yale University Press.

Kravčenko, Volodymyr Vasyl'ovyč, and Marko Robert Stech, eds. 2024. *The Unpredictable Past? Reshaping Russian, Ukrainian and East European Studies*. Toronto: Canadian Institute of Ukrainian Studies Press.

Křesťan, Jiří. 2013. *Zdeněk Nejedlý: politik a vědec v osamění* [Zdeněk Nejedlý: A lonely politician and scholar]. Prague: Paseka.

Kula, Marcin. 2010. *Mimo wszystko bliżej Paryża niż Moskwy: książka o Francji, PRL i o nas, historykach* [Closer to Paris than to Moscow, after all: A book about France, the People's Republic of Poland, and us, the historians]. Warsaw: Wydawnictwa Uniwersytetu Warszawskiego.

Kulecka, Alicja, and Tadeusz Paweł Rutkowski. 2012. *Oczami agenta: środowiska naukowe i archiwalne w doniesieniach TW "Zyg"—Zygmunta Kolankowskiego* [Through the eyes of an agent: The academic and archival communities in the reports of informant "Zyg"—Zygmunt Kolankowski]. Warsaw: Wydawnictwo Neriton.

Laqua, Daniel. 2021. "Internationalism." In *EGO | Europäische Geschichte Online*. Leibniz Institute of European History (IEG). Retrieved 3 April 2026 from https://www.ieg-ego.eu/en/threads/transnational-movements-and-organisations/internationalism/daniel-laqua-internationalism.

Laqua, Daniel, and Nikolaos Papadogiannis. 2023. "Youth and Internationalism in the Twentieth Century: An Introduction." *Social History* 48, no. 1: 1–16. https://doi.org/10.1080/03071022.2023.2146895.

Laskosz, Joanna, and Jan Chodějovský, eds. 2019. *Korespondence Waleryho Goetela s Radimem Kettnerem: Korespondencja Walerego Goetla z Radimem Kettnerem; Correspondence between Walery Goetel and Radim Kettner*. Prague, Warsaw: Masarykův ústav a Archiv AV ČR, v.v.i, Polska Akademia Umiejętności a Archiwum Nauki PAN i PAU.

Lebow, Katherine, Małgorzata Mazurek, and Joanna Wawrzyniak. 2019. "Making Modern Social Science: The Global Imagination in East Central and Southeastern Europe after Versailles." *Contemporary European History* 28, no. 2: 137–42. https://doi.org/10.1017/S0960777318000474.

Lewis, Su Lin. 2024. "Decolonising the History of Internationalism: Transnational Activism across the South." *Transactions of the Royal Historical Society* 2 (December): 345–69. https://doi.org/10.1017/S0080440123000233.

Linková, Marcela. 2017. "The Czech Research Landscape: Shifts in Research Organization after 1989." In *Gender and Neoliberalism in Czech Academia*, edited by Marcela Linková and Marta Vohlídalová, 69–92. Sociologické nakladatelství (SLON); Sociologický ústav AV ČR, v.v.i.

Linková, Marcela, and Alice Červinková. 2013. "'Vlastní laboratoř': Akademické trajektorie a gender v současných biovědách" ["A lab of one's own": Academic trajectories and gender in contemporary biosciences]. *Gender a Výzkum / Gender and Research* 14, no. 1: 15–26.

Lochmatow, Alexej. 2024. *Public Knowledge in Cold War Poland: Scholarly Battles and the Clash of Virtues, 1945–1956*. Poland: Transnational Histories. New York: Routledge. https://doi.org/10.4324/9781003428251.

Malecki, Ignacy. 1967. "Spolupráce Československé akademie věd a Polské akademie věd" [Cooperation between the Czechoslovak Academy of Sciences and the Polish Academy of Sciences]. *Věstník Československé akademie věd* 3, no. 76: 352–55.

Mazurek, Małgorzata. 2018. "Polish Economists in Nehru's India: Making Science for the Third World in an Era of De-Stalinization and Decolonization." *Slavic Review* 77, no. 3: 588–610.

McDermott, Kevin, and Matthew Stibbe. 2022. *Czechoslovakia and Eastern Europe in the Era of Normalisation, 1969–1989*. Basingstoke, UK: Palgrave Macmillan.

Mehlhorn, Ludwig. 2001. "Zwangsverordnete Freundschaft? Die Entwicklung der Beziehungen zwischen der DDR und Polen 1949–1990" [A Forced Friendship? The evolution of relations between the GDR and Poland, 1949–1990]. In *Die deutsch-polnischen Beziehungen 1949–2000*, edited by Wolf-Dieter Eberwein and Basil Kerski, 61–73. Wiesbaden: VS Verlag für Sozialwissenschaften. https://doi.org/10.1007/978-3-322-99380-9_4.

Mentzel, Zbigniew. 2020. *Kołakowski: czytanie świata biografia* [Kołakowski: Reading the world. A biography]. Cacow: Wydawnictwo Znak.

Mezhdunarodnoe soveshchanie po voprosu o sotrudnichestve mezhdu bibliotekami v oblasti komplektovaniia fondov zarubezhnoi literatury [An international consultation on the cooperation between libraries concerning the fonds of foreign literatures]. 1965. Prague: Gosudarstvennaia biblioteka ChSSR-Universitetskaia biblioteka.

Nickelsen, Kärin, and Fabian Krämer. 2016. "Introduction: Cooperation and Competition in the Sciences." *NTM Zeitschrift Für Geschichte Der Wissenschaften, Technik Und Medizin* 24, no. 2: 119–23. https://doi.org/10.1007/s00048-016-0145-4.

Oates-Indruchová, Libora. 2020. *Censorship in Czech and Hungarian Academic Publishing, 1969–89: Snakes and Ladders*. London: Bloomsbury Academic.

Olšáková, Doubravka. 2020. "A Matter of Courtesy: The Role of Soviet Diplomacy and Soviet 'System Safeguards' in Maintaining Soviet Influence on Czechoslovak Science before and after 1968." *Berichte Zur Wissenschaftsgeschichte* 43, no. 4: 542–59. https://doi.org/10.1002/bewi.202000023.

———. 2018. "Expert Cultures and Infrastructural Globalism: Socialist Experts under Restricted Internationalism." In *Experts and Expertise in Science and Technology in Europe since the 1960s: Organized Civil Society, Democracy and Political Decision-Making*, edited by Christine Bouneau and David Burigana, 53–68. Lausanne: Peter Lang.

———. 2020. "Science, (Anti-)Communism and Diplomacy: The Pugwash Conferences on Science and World Affairs in the Early Cold War." In *Science, (Anti-)Communism and Diplomacy: The Pugwash Conferences on Science and World Affairs in the Early Cold War*, edited by Alison Kraft and Carola Sachse, 259–285. History of Modern Science, vol. 3. Leiden: Brill.

Ouimet, Matthew J. 2003. *The Rise and Fall of the Brezhnev Doctrine in Soviet Foreign Policy*. New Cold War History. Chapel Hill: University of North Carolina Press.

Patel, Kiran Klaus, and Sven Reichardt. 2016. "The Dark Side of Transnationalism Social Engineering and Nazism, 1930s–40s." *Journal of Contemporary History* 51, no. 1: 3–21. https://doi.org/10.1177/0022009415607956.

Pavlíček, Tomáš W. 2024. "'The Gentlemen Did Not Like Each Other': Experiencing Relationships in the Mathematical Community That Are Not Disclosed in Textbooks." *Wrocławski Rocznik Historii Mówionej* 14: 142–67. https://doi.org/10.26774/wrhm.369.

Péteri, György. 2004. "Nylon Curtain: Transnational and Transsystemic Tendencies in the Cultural Life of State-Socialist Russia and East-Central Europe." *Slavonica* 10, no. 2: 113–23. https://doi.org/10.1179/sla.2004.10.2.113.

Pleskot, Patryk. 2010. "Jak wyjechać na Zachód? Procedury wyjazdów polskich uczonych do państw kapitalistycznych z ramienia uczelni wyższych i PAN w latach 1955–1989" [How to travel to the West? Procedures of travelling to capitalist countries for Polish scholars from universities and the PAN 1955–1989]. In *Naukowcy władzy, władza naukowcom. Studia*, edited by Piotr Franaszek, 292–325. Warsaw: Instytut Pamięci Narodowej.

———. 2005. "Kontakty naukowe Polski z krajami kapitalistycznymi w okresie stalinizmu na przykładzie Francji" [Poland's scientific contacts with capitalist countries during the Stalinist era, on the example of France]. *Nauka* no. 2: 93–117.

Pleskot, Patryk, and Tadeusz Paweł Rutkowski, eds. 2009. *Spętana Akademia: Polska Akademia Nauk w Dokumentach Władz PRL. 2. Dziennikarze, Twórcy, Naukowcy* [The Shackled Academy: The Polish Academy of Sciences in Government Documents, vol. 2: Journalists, creators, scientists]. Warsaw: Instytut Pamięci Narodowej. Komisja Ścigania Zbrodni przeciwko Narodowi Polskiemu.

Przeperski, Michał. 2016. *Nieznośny ciężar braterstwa: konflikty polsko-czeskie w XX wieku* [The unbearable lighness of brotherhood: Polish-Czech conflicts in the 20th century]. Wydawnictwo Literackie.

Raj, Kapil. 2013. “Beyond Postcolonialism . . . and Postpositivism: Circulation and the Global History of Science.” *Isis* 104, no. 2: 337–47. https://doi.org/10.1086/670951.

———. 2010. *Relocating Modern Science: Circulation and the Construction of Knowledge in South Asia and Europe, 1650—1900*. Palgrave Macmillan.

Reinisch, Jessica. 2016. “Introduction: Agents of Internationalism.” *Contemporary European History* 25, no. 2: 195–205. https://doi.org/10.1017/S0960777316000035.

Reinisch, Jessica, and David Brydan. 2021. “Introduction.” In *Internationalists in European History: Rethinking the Twentieth Century,* edited by Jessica Reinisch and David Brydan, 1–13. London: Bloomsbury Publishing. https://doi.org/10.5040/9781350118546.

Rindzevičiūtė, Eglė. 2016. *The Power of Systems: How Policy Sciences Opened Up the Cold War World*. Ithaca, NY: Cornell University Press.

Samuilova, Simona. 2014. “Scientific and Educational Programs between the US and the Eastern Bloc during the Cold War.” *Bulgarian Historical Review* nos. 1–2: 162–82.

Shapin, Steven. 2010. *Never Pure: Historical Studies of Science as If It Was Produced by People with Bodies, Situated in Time, Space, Culture, and Society, and Struggling for Credibility and Authority*. Baltimore, MD: Johns Hopkins University Press.

———. 1998. “Placing the View from Nowhere: Historical and Sociological Problems in the Location of Science.” *Transactions of the Institute of British Geographers* 23, no. 1: 5–12. https://doi.org/10.1111/j.0020-2754.1998.00005.x.

Sher, Gerson S. 2019. *From Pugwash to Putin: A Critical History of US-Soviet Scientific Cooperation*. Bloomington, Ind.: Indiana University Press.

Shore, Marci. 2008. *Caviar and Ashes: A Warsaw Generation's Life and Death in Marxism, 1918–1968*. New Haven, CT: Yale University Press. https://doi.org/10.12987/9780300128628.

Siemienska, Renata. 1992. “Academic Careers in Poland: Does Gender Make a Difference?” *Higher Education in Europe* 17, no. 2: 60–84. https://doi.org/10.1080/0379772920170212.

Šmidrkalová, Michaela. 2024. “Mezinárodní kontext vývoje ČSAV 1963–1967” [The international contexts of the development of the Czechoslovak Academy of Sciences 1963–1967]. In *Dějiny Československé akademie věd II: 1963–1970,* edited by Martin Franc and Věra Dvořáčková, 170–189. Prague: Academia; Masarykův ústav a Archiv AV ČR, v. v. i.

———. 2024. “Mezinárodní kontext vývoje ČSAV 1968–1970.” In *Dějiny Československé akademie věd II: 1963–1970* [The international contexts of the development of the Czechoslovak Academy of Sciences 1968–1970], edited by Martin Franc and Věra Dvořáčková, 312–323. Prague: Academia; Masarykův ústav a Archiv AV ČR, v. v. i.

Solovey, Mark, and Christian Dayé, eds. 2021. *Cold War Social Science: Transnational Entanglements*. Basingstoke, UK: Palgrave Macmillan.

Soutschek, Liza. 2024. *Wissenschaftskooperation und -konkurrenz im Kalten Krieg* [Science cooperation and competition during the Cold War. Wissenschaftskulturen Reihe 3, Pallas Athene, vol. 59. Stuttgart: Franz Steiner Verlag.

Stanek, Łukasz. 2020. *Architecture in Global Socialism: Eastern Europe, West Africa and the Middle East in the Cold War*. Princeton, NJ: Princeton University Press.

Stasiewicz-Jasiukowa, Irena, and Jan Janko, eds. 1990. *Z dziejów polsko-czeskich i polsko-słowackich kontaktów naukowych* [From the history of Polish-Czech and Polish-Slovak academic relations]. Warsaw: Komitet historii nauki i techniki.

Stoeckl, Kristina, and Phillip Ayoub. 2024. "Transnational Illiberal Networks." In *The Oxford Handbook of Illiberalism,* edited by Marlene Laruelle, 827–848. Oxford: Oxford University Press. https://doi.org/10.1093/oxfordhb/9780197639108.013.29.

Štrbáňová, Soňa, and Antonín Kostlán, eds. 2011. *Sto českých vědců v exilu: encyklopedie významných vědců z řad pracovníků Československé akademie věd v emigraci* [One hundred Czech scientists in exile: An encyclopedia of prominent scientists from the Czechoslovak Academy of Sciences in the emigration]. Prague: Academia.

Surman, Jan. Forthcoming. "Internationalism, Normalisation and the Socialist Social Sciences: Radovan Richta and the Scientific-Technological Revolution after the Prague Spring."

Surman, Jan, and Klemens Kaps. 2012. "Postcolonial or Post-Colonial? Post(-)Colonial Perspectives on Habsburg Galicia." *Historyka Studia Metodologiczne* 42 (English-German Version): 7–35.

Surman, Jan, and Tomáš W. Pavlíček. 2024. "The Formal and the Informal in the History of Socialist Scholarly Interconnectedness in East Central Europe." *NTM Zeitschrift Für Geschichte Der Wissenschaften, Technik Und Medizin* 32, no. 2: 137–66. https://doi.org/10.1007/s00048-024-00387-0.

Surman, Jan, and Daria Petushkova. 2022. "Between Westernization and Traditionalism: Central and Eastern European Academia during the Transformation in the 1990s." *Studia Historiae Scientiarum* 21 (August): 435–83. https://doi.org/10.4467/2543702XSHS.22.014.15980.

Szczepańska, Elżbieta. 2015. "Wybrane stereotypy Czecha i Polaka z perspektywy historycznej" [Chosen stereotypes of Czechs and Poles in historical perspective]. *Bohemistyka,* no. 4: 366–77.

Szumski, Jan. 2016. *Polityka a historia: ZSRR wobec nauki historycznej w Polsce w latach 1945–1964* [Politics and history: The Soviet Union's approach to the study of history in Poland, 1945–1964]. Warsaw: Oficyna Wydawnicza ASPRA-JR.

———. 2015. "Wpływ elity władzy ZSRR na rozwój nauki w Polsce 1945–1964" [The influence of the Soviet ruling elite on the development of science in Poland, 1945–1964]. In *Historia nauki polskiej,* vol. 10, part 2: *1944–1989. Instytucje,* edited by Joanna Schiller-Walicka and Leszek Zasztowt, 123–175. Warsaw: Instytut Historii Nauki im. Ludwika i Aleksandra Birkenmajerów PAN. Oficyna Wydawnicza Aspra-JR.

Tedeschi, Miriam, Ekaterina Vorobeva, and Jussi S. Jauhiainen. 2022. "Transnationalism: Current Debates and New Perspectives." *GeoJournal* 87, no. 2: 603–19. https://doi.org/10.1007/s10708-020-10271-8.

Tomaszewski, Paweł E. 2018. "Geneza Instytutu Niskich Temperatur i Badań Strukturalnych PAN we Wrocławiu" [The genesis of the Institute of Low Temperature and Structure Research of the Polish Academy of Sciences in Wrocław, Poland]. *Studia Historiae Scientiarum* 17 (December): 175–203.

Tõnismann, Teele. 2022. "Practices and Discourses of Internationalisation: Sociology in the Baltic States before and after the Collapse of the Soviet Union." *Revue d'histoire des sciences humaines* no. 40 (June): 181–206. https://doi.org/10.4000/rhsh.7072.

Tromly, Benjamin. 2014. *Making the Soviet Intelligentsia: Universities and Intellectual Life under Stalin and Khrushchev.* New Studies in European History. Cambridge: Cambridge University Press.

Tsvetkova, N. A. 2022. *The Cold War in Universities: U.S. and Soviet Cultural Diplomacy, 1945–1990: New Perspectives on the Cold War,* vol. 9. Leiden: Brill.

Wagner, Izabela. 2020. *Bauman: A Biography*. Cambridge: Polity Press.

Wasiak, Patryk. 2010. "Schleichwege in der Galerie: Ausstellungen verbotener Künstler aus Ungarn, der Tschechoslowakei und der DDR in polnischen Autorengalerien" [Back channels in the gallery: Exhibitions of banned artists from Hungary, Czechoslovakia, and the GDR in Polish artist-run galleries]. In *Schleichwege: Inoffizielle Begegnungen sozialistischer Staatsbürger zwischen 1956 und 1989*, edited by Włodzimierz Borodziej, Jerzy Kochanowski, and Joachim von Puttkamer, 293–312. Cologne: Böhlau.

Wilczynski, Jozef. 1974. *Technology in Comecon*. Basingstoke UK: Palgrave Macmillan. https://doi.org/10.1007/978-1-349-01794-2.

Winkel, Heidemarie, Julia Roth, and Alexandra Scheele. 2022. "Analytical Framing: Three Paradigmatic Arenas of Global Contestations of Gender Rights." In *Global Contestations of Gender Rights*, edited by Heidemarie Winkel, Julia Roth, and Alexandra Scheele, 23-44. Bielefeld: Transcript Verlag. https://www.jstor.org/stable/jj.11425489.4.

Wolfe, Audra J. 2018. *Freedom's Laboratory: The Cold War Struggle for the Soul of Science*. Baltimore, MD: Johns Hopkins University Press.

Wolle, Stefan. 1999. "Das System der Reisekader als Instrument der DDR-Wissenschaftspolitik" [The system of travel delegations as an instrument of GDR science policy]. In *Materialien der Enquete-Kommission*, edited by Enquete-Kommission "Überwindung der Folgen der SED-Diktatur im Prozeß der deutschen Einheit," vol. 4, no. 2, 1597–1688. Baden-Baden: Nomos Verlag.

Zürn, Michael, Nieves Fernández Rodríguez, Lena Röllicke, Maximilian Weckemann, Alexander Schmotz, and Stefan Gosepath. 2024. "Towards a Typology of Contestation: Four Types of Contestants." *SCRIPTS Working Paper No. 44*. Berlin: Cluster of Excellence 2055 "Contestations of the Liberal Script (SCRIPTS)." Retrieved 3 April 2026 from https://www.scripts-berlin.eu/publications/working-paper-series/Working-Paper-44-2024/index.html.

Zysiak, Agata. 2020. "For Whom University? Rising Educational Desires via the Daily Press in Post-War Poland." *History of Education* 49, no. 6: 819–38. https://doi.org/10.1080/0046760X.2020.1764637.

———. 2023. *Limiting Privilege: Upward Mobility within Higher Education in Socialist Poland. Central European Studies*. West Lafayette, IN: Purdue University Press.

CHAPTER 1

National Internationalism in Socialist Science

Entanglements of Czech, Slovak, and Polish Historians of Science and Technology

Michaela Šmidrkalová

In 1981, a book entitled *Problems of Teaching the History of Science* was published in Prague.[1] It was the result of many years of cooperation between historians of science from the Institute of Czechoslovak and World History of the ČSAV (Československá akademie věd; Czechoslovak Academy of Sciences) and the Institute of the History of Science, Education and Technology of the PAN (Polska Akademia Nauk; Polish Academy of Sciences). Three years later and more than 300 kilometers further east, but still in Czechoslovakia, another Czechoslovak–Polish encounter took place. The Czech–Polish historian Stanisław Zahradnik (1932–2023), who for many years was director of the company archive of Třinec Iron and Steel Works (Třinecké železárny)[2] in Czech Silesia, became a "criminal suspect." One of the reasons was his participation in an exhibition on Polish education in Cieszyn Silesia—a region on the border between Czechoslovakia and Poland, the Czechoslovak part of which was home to a large Polish minority. The example of Zahradnik shows that Czechoslovak–Polish scientific contacts took place not only at the level of the highest scientific institutions but also in more remote or peripheral areas, and they were not always viewed favorably.

Other personalities were involved in such contacts, some of whom have almost been forgotten in the history of science and technology. One was Juraj Voda (1908–?), an employee of the SAV (Slovenská akadémia vied; Slovak Academy of Sciences) in Bratislava, whose contacts with Poland were particularly important in the second half of the 1950s. Another is Stanisław Miczulski (1923–2010), a historian of science specializing in the history of metallurgy, working at the Institute of the History of Science and Technology of the PAN in Cracow, who established contacts with Czech and Slovak scholars particularly in the 1960s. Miczulski, who never achieved the prominence of his superior, Polish metallurgist and historian Mieczysław Radwan, undertook a work-related trip to Czechoslovakia in 1965, where he visited many individuals, institutions, and cities. At first glance, his visit is described only in a short report printed in the Polish quarterly *Kwartalnik Historii Nauki i Techniki* (Quarterly of the History of Science and Technology),[3] but archival materials preserved in the AN PAN i PAU (Archiwum Nauki Polskiej Akademii Nauk i Polskiej Akademii Umiejętności w Krakowie; Archive of Science of the Polish Academy of Sciences and Polish Academy of Arts and Sciences in Cracow) allow us to reconstruct his journey and networking in much greater detail.

This contribution thus brings together the stories of three historians of science and technology who represent "regional" or "peripheral" science, that is, science outside the environments of Prague and Warsaw. Research shows that these histories have contributed to the production of global scientific knowledge.[4] The aim of this chapter is to explain narratives about local settings and how they are intertwined with the global perspective, rather than using the center–periphery paradigm or the prevalent geographical hierarchies.[5] At the same time, it attempts to move beyond the framework of mere comparison or the study of transfer, concentrating instead on the multilayered entanglements that not only co-created knowledge but also shaped the actors themselves.[6] The fates of these scholars were intertwined not only by their common fields of research, personal encounters or efforts to strengthen Czechoslovak–Polish contacts, but also by the fact that they were all in some way involved at the time when the institutionalization and professionalization of the history of science and technology was emerging. They also encountered the reality that the field was subject to ideological and political pressures. The history of science and technology was often associated with the national question and the writing of national history and therefore influenced by state ideology.

While large meetings and conferences, inter-academic cooperation agreements, and joint publications, including the aforementioned book, have secured a place in the history of Czechoslovak–Polish scientific relations, there were also (non-)historians working in the regions who played a relatively successful role during the emergence of the institutionalization of the discipline and who also attempted to establish such contacts. The ways national issues intervened in their work and careers can therefore help illuminate the role and intersection of the national and the international in the entanglements of scientists in Central European socialist countries.

Juraj Voda: An Interim Leader

Juraj Voda was born in Zagreb in 1908, where his father worked as a civil engineer. In 1922 his parents moved to Bratislava, where he completed his secondary education. Later, he studied mechanical engineering at the Czech Technical University in Prague, graduating in 1933, after which he completed his compulsory military service. He began work as a designer in the Škoda factory in Pilsen and then in 1935 he joined the Military Technical and Aviation Institute in Prague, where he worked on the development and introduction of new weapon materials. He was involved in weapons research, especially new types of infantry ammunition, including mines and rockets. From January 1939 he worked at the Považská Bystrica arms factory and later at the Slovak Ministry of Defense (in the field of aviation) during the war.[7]

In 1948 Voda was transferred to the military reserve force. At the beginning of the 1950s, he helped organize and introduce the publication of technical literature in Slovakia (among other things, he was involved in the creation of the State Publishing House of Technical Literature).[8] In 1954 he found himself in the vanguard of the institutionalization of the history of (natural) sciences and technology in Slovakia, when he became the head of the newly established Cabinet for the History of Natural and Technical Sciences within the SAV.[9] In his own words, he had been given the task of "building from the ground up a scientific workplace for the history of the technical sciences."[10]

Two other cabinets were established at the same time as Voda's: the Cabinet for the History of Biological and Medical Sciences and the Cabinet for the History of Agriculture. They all had the advantage that they did not fall under the social sciences section, but under the medical, agricultural, and technical sections of the ČSAV. Thus, unlike in the humanities, there was less pressure from Marxist ideology in these areas.[11] However, they were very small departments with only one to three employees,[12] most of whom had no academic background in history, including Voda, who had a degree in mechanical engineering.

The history of science and technology in Slovakia was mainly focused on the study of typical Slovak industries, such as mining and metalworking. Along with the study of the impact of technological development on the status of the peasantry, these were the subjects that the state supported most.[13] Juraj Voda dealt with such topics in the 1950s, focusing on biographies of important figures in the history of science and technology who lived in or came from the territory of present-day Slovakia. These were biographies of the engineer and physicist Aurel Stodola (1859–1942),[14] the inventor, architect, and botanist Jozef Murgaš (1864–1929),[15] and the engineer and inventor Jozef Karol Hell (József Károly Hell) (1713–1789).[16] The focus on important personalities in the history of science was in line with the contemporary assumption that the history of science was important "for the consolidation and development of national pride and socialist patriotism."[17]

In the case of Voda's biographies of "Slovak" figures from the 1950s, contemporary reviews focused more on the subject than on the writing. In other words, Voda received recognition primarily due to who he wrote about rather than for his writing. An example of this is an article on Voda's biography of Stodola written by Jan Barica in the daily newspaper *Večerník* in May 1957. In his introduction, Barica described Voda's monograph as "a meritorious work in the popularization of our greats." He referred to Stodola as "ours" and expressed his indignation at the fact that the world's encyclopedias referred to him as Swiss (he spent most of his life in Switzerland) or even Hungarian. However, according to Barica, the fault lay with the Slovaks themselves, as it had taken them so long to publish a work on Stodola and other great Slovaks.[18]

Voda's most controversial book was probably his biography of Jozef Karol Hell, who is generally considered a Hungarian. It was also the case with this book that the subject was appreciated for promoting "the glory of the past of Slovak mining and engineering," as Róbert Vlach stated in his review of the book.[19] The reviewer also lauded the book's "political and educational character," as it "dispels the not yet completely overcome harmful superstition of bourgeois historiography that Slovakia has always been backward in this respect and has given nothing of value to the world." He also praised the fact that Voda had referred to Hell as "the first Slovak mechanical engineer."[20] Vlach concluded his review by emphasizing how necessary Voda's book was for Slovakia, and that the country was still in need of books "about dozens of other important figures from Slovakia's scientific and technical past."[21]

However, what Voda was most praised for—writing books about important figures in the history of Slovak science and technology—was sometimes criticized, especially when the work was evaluated by professional historians. In his review of Voda's biography of Jozef Murgaš, Slovak historian František Bokes acknowledged Voda's contribution to "eliminating in the consciousness of the members of our nation the feelings of smallness and insignificance that still exist" and to showing that "even a small nation has had important scientists in its past who have contributed more than once to the general cultural development of mankind."[22] However, Bokes accused Voda of paying little attention to the final period of Murgaš's life, and, in particular, to his visit to Slovakia in 1920, which "left a deep mark on his life, thinking and views of his homeland."[23] According to Bokes, Voda should also have paid more attention to Murgaš's articles published in certain American Slovak newspapers in 1920, in which Murgaš had demonstrated that he was "a very sharp, astute and penetrating observer of life in Slovakia after 1918," in particular, in criticizing "the political movement and life of Slovak Catholics and their leaders."[24] Regardless of whether the Slovak aspect of Juraj Voda's work in the 1950s was appreciated or criticized by reviewers, the aforementioned reviews show how the history of science and technology was perceived at the time, as being closely associated with national history.

As head of the Cabinet for the History of Natural and Technical Sciences, Voda particularly sought to establish cooperation with Poland, and he visited Warsaw in 1956 and 1957. The trip to Warsaw in 1957 was combined with a visit to the Soviet Union, and its aim was "to obtain information on the organization of those scientific departments of the PAN and the Academy of Sciences of the USSR which deal with the historiography of technical and natural sciences" and "to establish personal contacts with the heads of these departments and their staff."[25] The Slovak delegation, which included Voda as a representative of the SAV, consisted mainly of the staff and directors of Slovak museums, whose aim was to learn about museological issues and the organization of museums in Poland and the Soviet Union. Voda and this delegation visited Warsaw, Brest, Moscow, Leningrad, and Kyiv from 24 August to 8 September 1957. In Warsaw, Voda met Polish engineer and historian of technology Eugeniusz Olszewski (1911–1994), who was at that time head of the Department of History of Technology and Technical Sciences at the Institute of History of Science and Technology of the PAN. He also met engineer Alfred Wiślicki (1913–1995), the then director of the Technical Museum in Warsaw.[26] Contacts with these important Polish figures in the history of technology continued on Slovak soil. In September 1957, Voda accompanied Wiślicki during his stay in Slovakia.[27] Less than a year later, in July 1958, Olszewski visited Bratislava as part of a tourist bus tour organized by the Polish travel agency Orbis. He also met Juraj Voda.[28]

An important milestone in the institutionalization of Slovak history of (natural) sciences, and also in the life of Juraj Voda, was the year 1960, when the Cabinet for the History of Biological and Medical Sciences and the Cabinet for the History of Natural and Technical Sciences were merged into the Department of History of Science and Technology, which became part of the Institute of History of the SAV. The resulting department was small, with only five employees (including Juraj Voda), and led by Ján Tibenský (1923–2012). It was the first academic department in Slovakia "which was to research the history of science and technology."[29] The decision was partly the result of efforts by the Technical Section of the SAV, to which Voda's cabinet was subordinated, to eliminate the Cabinet, allegedly because of his unrealistic financial and professional demands. However, the Institute of History of the SAV, to which the Cabinet for the History of Technical and Natural Sciences was attached, did not have enough employees to staff the department.[30]

Tibenský became head of the department, although his research had hitherto focused more on the study of national revival, economic history and the history of technology.[31] Thus, in 1960, Juraj Voda went from being a leading figure to a mere collaborator, and his opportunities to influence the history of science and technology as a field were significantly reduced. At the same time, his salary was lowered, something he had long fought against, but to no avail.[32] At the Institute of History he was classified as a mere "academic worker," not a researcher, given his technical training and qualifications were not relevant to social science research. This justification was given to Voda by the then director of the Institute, Ľubomír Holotík, after he had demanded a higher salary and a higher qualification classification.[33]

It is clear from Juraj Voda's letters that he perceived his lack of academic background and experience in history as a major handicap that prevented him from furthering his academic career at his new department. In January 1969, for example, he stated that even before the merger of the Cabinet into the Institute of History, he had noticed that specialists in the field of general history did not look kindly on "non-historians" who were also doing "their" science as part of their work. He also complained that they did not consider him "erudite enough in their science to be considered an equal partner."[34] Voda also complained that after his cabinet became part of the Institute of History, he was not allowed to work on research into the development of technology and engineering or publish the results of his work,[35] and his ability to travel abroad for research was restricted.[36] The reason for such restrictions on foreign contacts was most likely not the language barrier, as he spoke Serbo-Croatian and German, and had a passing knowledge of Russian, English, and French.[37] Nevertheless, despite these restrictions on his international and publishing activities, Voda was able to defend his doctoral thesis at the Institute of History of the SAV in June 1965, which was entitled "The Development of Energy and Technology of Water-Powered Machines in the Banská Štiavnica Mining Industry in the Eighteenth Century and the First Half of the Nineteenth Century."[38]

The 1960s marked a change not only in the institutional organization of the history of science and technology in Slovakia, but also in the way this history was to be written. The close association of the history of science and technology with the (legitimization of) national history, typical of the 1950s and of Voda's books, began to be problematized in the following decades, hand in hand with the professionalization of the field and the introduction of specialist standards in historical work.[39] Tibenský, for example, distinguished between the "history of science and technology in Slovakia," which before 1918 included personalities of various nationalities and was a science supported by state institutions (Austrian, Hungarian), and the "history of Slovak national science," which mainly included fields such as national history, linguistics, folklore studies, and so forth, and had nationalist goals.[40] The development of metallurgy and mining in contemporary Slovakia and the application of technical innovations and discoveries since the sixteenth century undoubtedly fell into the first category—the history of science and technology in Slovakia. Its actors were not only Slovaks, but also Germans, Hungarians, Czechs, and other "nationalities."[41] The association of these personalities with Slovak national science, as had been the case in the 1950s, proved to be unsustainable.

Although in the 1960s Juraj Voda no longer determined the shape of Slovak–Polish relations in the field of the history of science, he was still a part of them. For example, he took part in a meeting of Polish and Slovak historians of mining and metallurgical technology held in the Polish part of the Tatra Mountains in Bukowina Tatrzańska in May 1966. At this meeting, the organization of which was agreed at the 11th International Congress for the History of Science in Warsaw in October 1965,[42] he presented a paper on the mining water system in the region of Banská Štiavnica,[43] that is, a similar topic to his doctoral thesis. The 1966 meeting in Bukowina Tatrzańska was attended by ten Slovak and twelve Polish historians of science and technology. While Voda represented Slovak research, one of the Polish scholars was Stanisław Miczulski from the Institute of the History of Science of the PAN in Cracow, who wrote a report on the meeting that was later published in the journal *Kwartalnik Historii Nauki i Techniki*.[44]

The example of Juraj Voda thus shows the paradoxical consequences of participating in the institutionalization of the history of science and technology in Slovakia. What enabled Voda to become head of the cabinet, to establish scientific contacts with Polish colleagues, and to write several books on "Slovak" scientists in the 1950s, that is, the emergence of the institutional background for the history of science and technology in Slovakia in connection with the establishment of the SAV, led to his demotion from head of the department, when the further institutionalization and professionalization of the field (and its connection with the Institute of History) necessitated a stronger integration of the specialist methods of history and thus a qualified historian at the head of the department (Ján Tibenský). And although Voda continued his scientific work in the decades that followed, going on to earn a doctorate, he did not play such a significant role in the history of science and technology in Slovakia as he did in the 1950s. His three biographies were written at a time when the history of science and technology in Slovakia was still focused primarily on the life stories of prominent personalities, in which their authors attempted to combine the incompatible—the history of Slovak national science and the history of science in Slovakia. Nevertheless, Voda made efforts to become involved in international networks, and his contacts with Polish colleagues were intended to be the gateway.

Stanisław Miczulski: A Historian in the Shadow of His Master

Stanisław Miczulski was born in Żarnowiec, Silesia in 1923. During the World War II he worked as a laborer in a sawmill in Muszyna (in today's Lesser Poland Voivodeship near the border with Slovakia) and attended secret high school classes as a student. After the end of the war, he finished high school, graduating in 1946 from the grammar school in Muszyna. Around the turn of 1947, he worked for a few months at the manufacturing company H. Cegielski in Poznań, and from 1947 to 1948 he worked in Nowa Sól (*Neusalz*) and Lubin in western Poland as an administrative worker. From 1948 to 1952 he studied history at the Jagiellonian University in Cracow.[45] After graduating, he worked as an archivist at the Voivodship State Archives in Cracow, and from 1958 to 1988 at the Institute of the History of Science and Technology of the PAN in Cracow, in the section dealing with Polish mining and metallurgy.[46]

In contrast to the ČSAV (and the SAV), the PAN played a greater role in institutionalizing the history of science and technology. Soon after its foundation, the discipline had its own institute, as in November 1953 the Presidium of the PAN decided to establish the Institute of the History of Science (from 1958 the Institute of the History of Science and Technology). The Institute began its work in May of the following year. In 1957 a branch of the Institute was established in Cracow, where Miczulski later worked.[47] The section concerned with the history of mining and metallurgy was led by Mieczysław Radwan. In the mid-1960s, shortly before the aforementioned symposium in Bukowina Tatrzańska, Miczulski began to develop intensive contacts with Czech and Slovak historians of science and technology. Most of these contacts were the result of a two-week stay in Czechoslovakia in November 1965.

Miczulski was in Czechoslovakia from 6 to 20 November 1965 in the framework of a reciprocal exchange between the PAN and the ČSAV. In Prague he visited the Department of the History of Natural Sciences and Technology at the Institute of History of the ČSAV and a similar department at the Institute of History of the SAV in Bratislava. Miczulski was particularly interested in Czech and Slovak research into the history of metallurgical technology. During his stay, he studied in the State Central Archive in Prague and the Mining and Metallurgical Archive in Banská Štiavnica. He also visited the National Technical Museum in Prague, the Technical Museum in Košice, and the Institute of Archaeology of the SAV in Nitra.[48] In his report for the PAN, he stated that it would be advisable to "establish permanent contacts through meetings or exchanges between the Institute of the History of Science and Technology of the PAN and the Department of the History of Science and Technology of the ČSAV and the SAV especially."[49]

Miczulski actively sought to develop these contacts on his return to Cracow. Specifically, in November and December, he sent many letters to people he had met during his stay in Czechoslovakia, and he thanked his hosts for their hospitality and promised to send Polish publications on the history of science and technology.[50] Correspondence with Slovak colleagues shows that during Miczulski's visit to Bratislava a more concrete form of the planned Polish–Slovak meeting on the history of metallurgy in the Tatra Mountains was discussed.[51]

Like Voda, Miczulski wrote his dissertation in the 1960s, but unlike the Slovak engineer, he was not able to defend it. Miczulski's work dealt with metallurgy in the seventeenth and eighteenth centuries in the area of the towns of Bobrza and Samsonów in the present Świętokrzyskie Voivodeship.[52] The text was ready for the defense and had passed the examination process, but Miczulski was unable to defend the thesis, allegedly because of a long illness, and thus did not successfully complete his doctoral studies.[53] Miczulski's publishing activities were also rather limited. In the 1960s, several of his shorter texts on the beginnings of metallurgy in the territory of present-day Poland were published.[54] One of his later published works was a chapter in the collective volume *Dzieje i technika świętokrzyskiego górnictwa i hutnictwa kruszcowego* (History and technology of the Świętokrzyskie ore mining and smelting industry).[55] Although Miczulski was not very active in publishing, he was particularly valued at the Institute for his archival practice and his ability to navigate Polish archival materials, in which he was of great assistance to his colleagues.[56]

Although the meeting of Polish and Slovak historians of science and technology in Bukowina Tatrzańska in 1966 was certainly a milestone and was reported in several specialist journals, today it is rather forgotten in the history of science and Polish–Slovak scientific contacts. This is partly due to the fact that the papers from the symposium were never published in a comprehensive form, although it been planned and preparations had been made for it. The papers were to be published in the Polish journal *Studia z dziejów górnictwa i hutnictwa* (Studies in the history of mining and metallurgy). In a letter to the Slovak historian Pavol Hapák in January

1967, Miczulski announced that "the materials are already going to press," without the two articles whose authors did not submit the texts, one of whom was Juraj Voda.[57] This was despite the fact that half a year earlier, in August 1966, Ján Tibenský had assured Mieczysław Radwan that Voda would send him the text of his presentation "within a few days" after he had made "some minor adjustments," which apparently did not happen.[58] The definitive end to the (non-)publication of the papers in the form of a special issue was the letter dated March 1968 stating that *Studia z dziejów górnictwa i hutnictwa* would not publish the articles. The letter from the editor, Jan Pazdur, was sent back with the articles to the Institute of the History of Science and Technology. The journal was originally supposed to receive all the material by the end of 1966, as volume 13 of the journal had been reserved for this special issue. This did not happen, and in the following issues there would be "no room for these papers," according to Pazdur's justification for the return of the manuscript.[59]

Even though in the end there was no joint publication that documented and symbolized Slovak–Polish cooperation in this field, it was possible to trace this cooperation, not only through mutual correspondence and visits, but also through the mutual provision of bibliographical and archival sources. Polish historians of science such as Miczulski and Radwan, at least in the field of research, seem to have made frequent use of Czechoslovak sources. For example, as shown by Radwan's correspondence with Karol Suchanek (1929–2018), a Polish historian specializing in the history of the Polish village of Węgierska Górka, which is around 15 kilometers from the Czechoslovak (today's Slovak) border, they sent each other transcripts or photocopies of archive material, such as that held in the town of Třinec in Czech Silesia.[60]

The fields of the history of technology that Miczulski and Radwan dealt with, that is, the history of mining and metallurgy, may at first glance appear to be less ideologized, especially when they focused on a very ancient period, as the archaeologists did. However, even though there was no Polish state in the period in question, it did not mean there was no sense of patriotism or territorial claim to the technological achievements of the time, especially in popular discourse.[61] Moreover, as in the case of Voda's monographs on the pioneers of technology, (not only) ancient technical monuments on the territory of present-day Poland could serve as a "cure for various kinds of complexes, led by the complex of technical backwardness."[62]

The story of Stanisław Miczulski is also the story of an actor in the period of the emergence of the institutionalization of the history of science and technology, this time in Cracow. Although Miczulski was present when the Cracow branch of the Institute of the History of Science and Technology within the PAN originated, unlike Voda, he was not in a leading position. Therefore, he attempted to establish contacts with Czechoslovakia over a longer period, in the second half of 1960s. At that time, Mieczysław Radwan had already worked closely, for example, with the Czech archaeologist Radomír Pleiner (1929–2015).[63] Thus, participation in the early days of the Cracow branch of the Institute most likely did not bring him any particular advantages in his professional career. This was partly because the history of mining and metallurgy, in which he was involved, was also being developed by other institutions at the time, such as the AGH University in Cracow (Akademia Górniczo-Hutnicza), where Mieczysław Radwan headed the Department of the History of Technology. Miczulski's career was also marked by his loyalty to the Institute, where he worked for decades, as evidenced by the jubilee award he received for thirty years' service at the Institute in 1988.[64]

Stanisław Zahradnik: A Czechoslovak Pole Writing History of Polish Czechoslovakia

In Czechoslovakia, not all central scientific institutions existed in the capital or in Bratislava. For example, the ČSAV had a branch in Opava in the Moravian-Silesian Region.[65] This was the Silesian Institute, which had been there since May 1948,[66] even before the establishment of the ČSAV, of which it became a part in 1958.[67] As literary historian and employee of the Silesian Institute Viktor Ficek (1902–1986) noted in the late 1960s, "the Institute was spontaneously created twenty years ago out of the cultural and political needs of the Silesian and Ostrava region, which, with its complex economic, social, political and national conditions, has different prerequisites and conditions for scientific work than other regions of our country."[68] This interpretation of Ficek, namely that this part of Czechoslovakia was so specific that it required a "special approach" to scientific work, was widespread among scholars in the region.[69]

The Institute's foreign contacts were mainly directed toward Poland. In the period 1950–52, the Institute even had a department for scientific relations with Poland.[70] In particular, the Institute dealt with the bibliography of Polish Silesians, organized the exchange of books and journals, and established contacts with scientific institutions and workers in Poland. It also published two volumes of *Česko-polský sborník vědeckých prací* (Czech–Polish collection of scientific works).[71] The Institute cooperated mainly with the Silesian Scientific Institute in Katowice,[72] and, to a lesser extent, the Silesian Institute in Opole.[73] The issue of Czechoslovak–Polish relations came to the fore in the mid-1960s in connection with the celebration of the millennium of the Polish state. A scientific conference was held near Opava in September 1965, entitled "A Thousand Years of Czech–Polish Reciprocity." Among the dozens of participants was Stanisław Zahradnik, who represented the Polish Cultural and Educational Association (Polski Związek Kulturalno-Oświatowy) in Třinec, still the main official representation of the Polish minority in the Czech Republic.[74]

The characteristics of the Silesian Institute of the ČSAV in Opava thus reflected the contemporary perception of local science—most scientists from this area were expected to focus on local (transnational) issues. Thus, Silesia was perceived as a specific scientific region with two main characteristics: its industrial nature[75] and its Polish influences and relationship with Poland. The two subjects were interrelated, since a "thorough study" of the national question was seen as a prerequisite for research into the development of industrial areas and their characteristics.[76] Throughout the history of the Silesian Institute, there were attempts to approach the history of Silesia in cooperation with Polish historians and move away from a purely national perspective. In the mid-1950s, the Institute organized a conference to discuss the theses of the forthcoming history of Silesia that was being written by Polish historians in Wrocław. According to Czech historian Andělín Grobelný, the comments on their draft (which were highly critical) were intended to "give the Polish historians a helping hand" and also to "establish lasting fraternal cooperation."[77] Although the influence of the Czech contributions to the conference on the final Polish publication[78] is difficult to assess, the conference at least made an effort to address the pressing issues of shared Czech–Polish history.[79]

Particularly in relation to research on nationality (and minority) issues, the work of the Institute was also directed toward the regional representatives of the Communist Party. For example, the report on the Institute's activities in the years 1971–75 states that "the conclusions of the sociologists' research, especially the reports on contemporary national relations and on the socialist development of national and ethnic cultures in the Ostrava region, were received with great interest by the Party and state authorities of the North Moravian Region and were applied in the administrative work of these authorities."[80] Similarly, the Institute's 1983 activity report stated that the Institute's work on national and ethnic relations in the region, including issues of education, culture, and language, had been made available to the Communist Party of Czechoslovakia.[81] The actual research into ethnic issues in the region was allegedly initiated in the 1960s by the North Moravian Regional Committee of the Communist Party of Czechoslovakia, with the aim of "a deeper understanding of the ongoing national processes."[82] When, in June 1976, the Silesian Institute, in cooperation with Třinec Iron and Steel Works, organized a conference on the problems of developing industrial areas, it was attended not only by Party representatives of the North Moravian Region but also by representatives of the Government Office of the Czechoslovak Socialist Republic.[83] The study of the national question was, as the historian Karel Sommer put it in 1984, "a very complex and demanding task," because it involved ideas "with political implications" that affected the perception of the population.[84]

One of the most important and long-standing collaborators of the Silesian Institute in the field of the history of science and technology or the history of factories was Stanisław Zahradnik, the aforementioned participant in the Czechoslovak–Polish Conference. In addition to his involvement in organizations of the Polish minority in Czechoslovakia, Zahradnik was the director of the company archive of Třinec Iron and Steel Works from 1956 to 1992,[85] and in 1969 he co-founded and directed the Třinec Metallurgical Museum (now the Museum of the Třinec Iron and Steel Works and the Town of Třinec), the first of its kind in Czechoslovakia.[86] He also published some of his work through the Institute.[87]

The Silesian Institute and the Třinec Iron and Steel Works also signed a number of agreements, one of which covered the period from 1975 to 1980 concerning the preparation of a history of the steelworks. In the agreement, the Silesian Institute undertook to prepare a "clear history of the Třinec Iron and Steel Works," which was to be published in 1979 on the occasion of the 140th anniversary of its founding.[88] Among other things, Třinec Iron and Steel Works offered to put its computer center at the disposal of the Silesian Institute staff "for statistical and mathematical processing of research results and technical equipment for the reproduction of documentary works and materials of the scientific library."[89] In the context of regional science, it was probably easier to establish cooperation between research institutions and enterprises, especially when the enterprises themselves were specific subjects of research and had their own archives.[90] Although not a member of a research institute or university, Zahradnik was involved in scientific networks, and not only regional ones. In addition to his membership of the Czechoslovak Society for the History of Science and Technology, he participated, for example, in seminars on the history of mining and metallurgy held at the National Technical Museum in Prague,[91] and was member of the Coordinating Board of the Technical and Economic Research Institute of the Metallurgical Industry in Prague, which was tasked in 1970 with preparing a history of iron metallurgy in Czechoslovakia.[92]

Zahradnik also involved the Třinec archive in cross-border research networks. In October 1967, for example, the Třinec Iron and Steel Works hosted an international meeting on the problems of processing the history of metallurgy and mining after 1945, which was organized as part of the cooperation between the Silesian Institute in Opava, the Silesian Scientific Institute in Katowice and the University of Mining and Technology in Freiberg (Saxony).[93]

It is no coincidence that many of Zahradnik's contacts also went to Poland, as Zahradnik belonged to the Polish minority in the Czech part of Cieszyn Silesia (Zaolzie).[94] Born in Kojkovice near Třinec, Zahradnik studied in Český Těšín and at the Faculty of Arts of Charles University in Prague. In 1967 he was awarded the degree of Candidate of Sciences (CSc) in history.[95] He was also involved in cultural and sports activities of the Polish minority in Czechoslovakia and often wrote for its press (including the monthly *Zwrot,* the main magazine of the Polish minority in Czechoslovakia, which was founded in 1949).[96]

Zahradnik's pro-Polish commitment became a thorn in the side of State Security (*Státní bezpečnost*), especially during the first half of the 1980s, when Czechoslovakia was undergoing a nationwide "Action North" (*Akce Sever*), organized by State Security, which was focused on the events in Poland related to the Solidarity movement and the subsequent declaration of martial law in December 1981. Czechoslovak security forces and the Communist Party perceived the events in Poland as a threat to Czechoslovakia's security (it feared strikes and demands for democratization), so it increased surveillance of Polish activities in the Czechoslovak–Polish border area and kept (alleged) Solidarity supporters under surveillance. Thus, Zahradnik was investigated by State Security on suspicion of "Polish nationalism." In 1984, Zahradnik helped organize a celebration and exhibition of historical documents on the history of Polish education in Návsí (Nawsie), a Czechoslovak border village with a large Polish minority. He was also the author of a publication marking the 200th anniversary of the founding of the Polish primary school in Návsí.[97] According to State Security records, Zahradnik was allegedly collecting archival material on the activities of Poles in the Cieszyn region in order to write a book that would "prove the priority merits of Poles for the development of all industry in Karviná (Karwinna) and other adjacent territories of the Polish People's Republic."[98]

The 1985 report of State Security in Ostrava stated that Zahradnik would be removed from his position as head of the Třinec Iron and Steel Works archive on 1 January 1986 because of his "hostile" activities and thus "deprived of the possibility of access to archival materials and their misuse in literary works with a nationalist orientation."[99] However, this did not happen. According to a State Security document from January 1987, Zahradnik was kept on as head of the company archive despite the efforts of State Security thanks to his friends from the Polish national minority.[100] This is therefore an interesting case in which the omnipotence of State Security was called into question, as it failed to achieve its goals due to the strength of the Polish national minority in Czechoslovakia. Nevertheless, State Security continued to monitor Zahradnik. For example, secret agents recorded Zahradnik's participation in the "Zaolzie Day" discussion at the Cieszyn branch of the Silesian University in Katowice in October 1987. The discussion was organized by Tadeusz Wantuła,[101] a Czechoslovak citizen working with the Silesian University. The State Security report assessed Zahradnik's involvement negatively, mainly because of the alleged anti-Czechoslovak orientation of the event. The discussion was described as "nationalistically oriented against the Czechoslovak Socialist Republic with reference to the oppression of the Polish minority living in the Czechoslovak Republic."[102]

Documents produced by the Czechoslovak security forces give us a rough idea of how Zahradnik was viewed by the security services. But how was he perceived by his fellow historians? We can reconstruct a picture of this by means of contemporary reviews of Zahradnik's books. Zahradnik's 1969 book *Třinecké železárny* (Třinec Iron and Steel Works) was reviewed by historian Ludomír Česenek in the pages of *Slezský sborník* (Silesian Collection). According to Česenek, Zahradnik's work shows "a sharply critical attitude toward the existing historical production."[103] Zahradnik criticized not only the previous treatment of the history of Třinec Iron and Steel Works, but also "the whole of Czechoslovak production in the field of factory history"[104] from a methodological point of view. In particular, Zahradnik reproached this field for its excessive orientation toward the history of the Labor movement, which could often lead to the mythologization of historical reality.[105] While Česenek appreciated Zahradnik's efforts to draw on a wealth of archival material, he criticized him in some cases for his regional approach. For example, he claimed that it was not possible to determine the contribution of Třinec Iron and Steel Works to the Polish national movement solely on the basis of regional sources, as Zahradnik had done.[106] Similarly, Česenek criticized the lack of wider use of secondary literature on Czech–Polish relations and the Těšín issue, which he said resulted in "a very unreliable picture of the significance and character of the Polish issue in Těšín."[107] According to Česenek, Zahradnik had "abandoned the view of the wider historical context" and worked with regional material "not very correctly" (he accused him, for example, of using Polish literature of the interwar period, which was strongly anti-Czechoslovak). Despite these reservations, however, Česenek rated the book as "among the best that has been written in this country on the history of factories" and described it as "an avant-garde work of its kind, inventive and flexible in its methodological approach, yet based on painstaking diligence in working with material, the use of which is in many ways quite admirable."[108] The historian Milan Myška later classified Zahradnik's book on Třinec Iron and Steel Works as a monograph "which, as far as possible, tried to follow the ideas of business history monographs developed by Western European and American historiography in the middle of the twentieth century."[109]

Other Czech historians found similar shortcomings in Zahradnik's texts. For example, historian Vladimír Škaloud, in his review of the publication *Z dějin dělnického hnutí na Ostravsku za předmnichovské republiky* (From the history of the Labor movement in the Ostrava region in the pre-Munich republic),[110] criticized Zahradnik's chapter on the Great Economic Crisis of the 1930s and the Třinec Iron and Steel Works[111] stating that "a serious flaw is a certain bias in the selection of sources (of the archival material, the records of the Třinec Iron and Steel Works company archive are used almost exclusively, and only the Polish press)." According to Škaloud, this had resulted in a "one-sided interpretation."[112] In general, Czech historians criticized Zahradnik for having a fundamentally too pro-Polish approach to sources and their interpretation.[113]

As mentioned above, Zahradnik's work was appreciated for its innovative approach to the history of factories. Although this was not an entirely new discipline, it became institutionalized in Czechoslovakia in the early 1960s. It also took on a highly politicized form, as the history of the productive forces, including the history of the Labor movement, was seen as one of the key aspects of the history of the communist movement, as in other socialist countries. In 1960, after the meeting of the Secretariat of the Communist Party of Czechoslovakia, a national commission on factory history was established, and the following year methodological guidelines were issued for writing such histories, inspired by the Soviet Union. The principle of workers' participation in writing the history of factories was also prescribed, but as the historian Milan Myška later commented, "fortunately, there was no one who took such nonsense seriously."[114] In many factories, however, halls of revolutionary and production traditions were created.[115] Similar to the history of factories, after 1948 the history of technology in general in Czechoslovakia was perceived mainly in the context of the history of the means of production within the framework of social production.[116]

Although Zahradnik's works were written in a very different context from the works of Juraj Voda discussed at the beginning of this chapter, a common feature is observable, which is how the question of nationality and minorities (for example, in the case of the origin of sources or even the choice of topic itself) shaped the reception of scholarly works on the history of science and technology. Moreover, certain areas of the history of science and technology were perceived as national, that is to say "ours," and thus, in the contemporary imagination of some historians of science and technology, it was only proper for such areas to be dealt with by "our" historians. This was true not only of Czechoslovakia but also of Poland. It is no coincidence that Radomír Pleiner, who studied the history of Wieliczka in Poland, complained to Mieczysław Radwan that "some Polish researchers do not like it when someone from abroad interferes in the issue,"[117] meaning the study and writing of the history of Polish territory.

Like Voda and Miczulski, Zahradnik was also active during the emergence of the institutionalization or professionalization of the history of science and technology, specifically the writing of the history of factories. It was most likely his Polish origins and contacts with the Polish scientific milieu that made his works on the history of factories more original and less ideological in comparison with contemporary Czechoslovak historiographical production, although, as noted above, it was quite heavily criticized (mostly in relation to the national question). Of the three main actors in this chapter, Zahradnik was most involved in Czechoslovak–Polish entanglements, not only because of his geographical proximity to Poland, but also because of his Polish roots. However, his contacts with historians from Poland were made possible not so much by his status as a scholar in Czechoslovakia as by his position as director of the company archive of Třinec Iron and Steel Works and, above all, by his public activities among the Polish minority in Czechoslovakia.[118]

Conclusion

This chapter has attempted to approach the history of science from below by looking at the history of science from the perspective of the actors (individuals) who initiated the institutionalization of the history of science and technology in the SAV, the Cracow branch of the PAN and in the Třinec Iron and Steel Works. All three historians contributed to the entanglements of Polish, Czech, and Slovak historians. What Voda, Miczulski, and Zahradnik had in common was that they had relatively limited opportunities for foreign travel and contacts with world science, such as participation in world congresses, which representatives of central scientific institutions in Prague or Warsaw had greater access to. As a result, "friendly" (socialist) neighboring countries, which were easier to travel to, were often the focus of their foreign contacts.

The stories of the three historians also reveal that the history of science and technology in Czechoslovakia and Poland after World War II was highly politicized and ideologized. This was true not only of the history of the humanities and social sciences, where ideological pressures were to be expected but also of the history of technical disciplines, in which such ideologization might have been expected to a much lesser extent. The main reason was the linking of the history of technology with the history of factories, which then became fundamental to the ideological Marxist–Leninist interpretation of the history of the workers' movement and the productive forces. Although politicization was less pronounced in Poland, and the leaders of the ruling Polish United Workers Party had a weaker influence on this discipline, it was evident that a patriotic interpretation of the history of science and technology also emerged occasionally in the Polish communist press. In this respect, politicization was to a greater extent associated with those areas of the history of science that dealt with national issues.

Voda, Miczulski, and Zahradnik were confronted with the question of nationality and its instrumentalization in the history of science, either in their works or how their work was evaluated by other historians. Stanisław Zahradnik is an example of entangled Czech–Polish identities, but this feature is also present, for example, in the Slovak–Hungarian milieu and elsewhere. It was the problem of not only the identity of the actors (historians) but also their object of study, as Voda's monograph on Jozef Karol Hell illustrates.

While Juraj Voda was traveling to Poland and writing works on important figures of Slovak science and Stanisław Miczulski was sending dozens of letters to Czech and Slovak historians of science and technology, and while Stanisław Zahradnik was organizing an exhibition on the history of the education in Návsí, international cooperation agreements were being negotiated at the highest levels of the academies of sciences and scholars were working on joint publications, which would then go on to become the most visible part of the history of Czechoslovak–Polish scientific relations. However, the history of cross-border cooperation and entanglements was not only this big story, but also the little ones, such as those discussed in this chapter. Despite their small scale and scope, such regional developments were not isolated from national and transnational frameworks.

Author Biography

Michaela Šmidrkalová is a historian at the Masaryk Institute and Archives of the Czech Academy of Sciences, specializing in twentieth-century Central European history and the history of science. She earned her PhD in Modern History from Charles University in 2015. Her research focuses on scientific institutions, international scholarly exchanges, and the cultural history of technology under socialism.

List of Abbreviations

A SAV	Archív Slovenskej akadémie vied (Archive of the Slovak Academy of Sciences)

ABS	Archiv bezpečnostních složek (Security Services Archive)
AN PAN i PAU	Archiwum Nauki Polskiej Akademii Nauk i Polskiej Akademii Umiejętności w Krakowie (Archive of Science of the Polish Academy of Sciences and Polish Academy of Arts and Sciences in Cracow)
ČSAV	Československá akademie věd (Czechoslovak Academy of Sciences)
PAN	Polska Akademia Nauk (Polish Academy of Sciences)
PZKO	Polski Związek Kulturalno-Oświatowy (Polish Cultural and Educational Association)
SAV	Slovenská akadémia vied (Slovak Academy of Sciences)

Notes

1. Nový and Stasiewicz-Jasiukowa, *Problems of Teaching the History of Science.*
2. In the period 1950–57, it was officially called the Třinec V. M. Molotov Iron and Steel Works. After that, the official name became the Třinec Iron and Steel Works of the Great October Socialist Revolution until the 1990s.
3. Miczulski, "Z pobytu w Czechosłowacji."
4. Lebow, Mazurek and Wawrzyniak, "Making Modern Social Science," 138.
5. Ibid., 138–39; Krige, "Introduction," 4–6.
6. Werner and Zimmermann, "Beyond Comparison," 38.
7. A SAV, Coll. Historický ústav SAV, box 62, inv. no. 354, "Juraj Voda: Curriculum Vitae"; Pejs, "Slovenská armáda," 29.
8. Juraj Voda: Curriculum Vitae.
9. The SAV was founded in 1953 and was partly subordinate to the ČSAV (especially in the field of foreign relations). For more information on the relationship between the ČSAV and SAV, see Hudek, "The Development of Relations."
10. A SAV, Coll. Historický ústav SAV, box 50, inv. no. 314, Letter from J. Voda to the Presidium of the SAV, 1960 March 19.
11. Hudek, "Inštitucionalizácia výskumu," 214. However, as we will see later, this did not mean that these fields were not subject to ideological or political demands.
12. Tibenský, "20 rokov Slovenskej spoločnosti pre dejiny vied a techniky," 8.
13. Hudek, "Inštitucionalizácia výskumu," 213.
14. Voda, *Aurel Stodola.*
15. Voda, *Jozef Murgaš.*
16. Voda, *Jozef Karol Hell.*
17. Bokes, "Činnost a plány kabinetov pre dejiny vied," 152.
18. Barica, "Pozná ho svet, poznáme ho my?" 4.
19. Vlach, "Juraj Voda: Karol Hell," 555.
20. Ibid.
21. Ibid., 556.
22. Bokes, "Dr. Ľ. Ivan – J. M. Nový."
23. Ibid.
24. Ibid.

25. A SAV, Coll. Kabinet dejín technických a prírodných vied, box 2, inv. no. 6, "J. Voda: Zpráva o služobnej ceste do SSSR," 1957 September 26.
26. Ibid.
27. A SAV, Coll. Kabinet dejín technických a prírodných vied, box 2, inv. no. 6, Letter from A. Turecký to Ústavy technickej sekcie SAV, Bratislava, 1957 October 14; Ibid., inv. no. 2, "Zpráva o činnosti Kabinetu dejín technických a prírodných vied za 5-ročné obdobie SAV."
28. A SAV, Coll. Kabinet dejín technických a prírodných vied, box 2, inv. no. 6, Letter from Z. Skubała to J. Voda, Warsaw, 1958 July 7; Ibid., inv. no. 3, "Zpráva o činnosti Kabinetu dejín technických a prírodných vied za 5-ročné obdobie SAV," 10.
29. Tibenský, "20 rokov Slovenskej spoločnosti," 8.
30. Hudek, "Inštitucionalizácia výskumu," 215.
31. Ibid.
32. See, for example, A SAV, Coll. Historický ústav SAV, box 52, inv. no. 323, Letter from J. Voda to the Rehabilitation Commission of the SAV, 1969 January 13.
33. A SAV, Coll. Historický ústav SAV, box 50, inv. no. 314, Letter from the director of the Institute of History of the SAV to the Presidium of the Section of Social Sciences, 1960 August 5. Voda also complained that his research at the Military Technical Aviation Institute in Prague, where he worked from 1935 to 1940, had not been recognized as scientific practice. Ibid., Letter from J. Voda to the Presidium of the SAV, 1960 March 19.
34. See, for example, A SAV, Coll. Historický ústav SAV, box 52, inv. no. 323, Letter from J. Voda to the Rehabilitation Commission of the SAV, 1969 January 13.
35. Voda never wrote another book, except for a small booklet on technology and religion. See Voda, *Technika a náboženstvo.*
36. A SAV, Coll. Historický ústav SAV, box 52, inv. no. 323, Letter from J. Voda to the Rehabilitation Commission of the SAV, 1969 January 13.
37. Juraj Voda: Curriculum Vitae.
38. Bokes, "Obhajoby kandidátskych dizertácií v HÚ SAV," 644; A SAV, Coll. Historický ústav SAV, box 55, inv. no. 342, Defence of the candidate's thesis "Development of Energy and Technology of Water-Powered Machines in Mining in Bánska Štiavnica," 1964 December 12. Voda received the title "Candidate of Science" (CSc) for defending his thesis, that is, the contemporary Czechoslovak academic equivalent of Doctor of Philosophy (PhD).
39. This was also true of Slovak historiography in general, in which, in the 1960s, there were discussions about national and territorial approaches and the search for boundaries between Slovak and Hungarian history. See Lipták, "Úloha a postavenie historiografie," 112; Ľubomír Lipták, for example, called for "overcoming the narrow national-defensive character" of Slovak historiography, which leads to "uncritical apologetics." Lipták, "Úloha a postavenie historiografie," 112.
40. Tibenský, "Národná veda a jej vzťah k svetovej vede," 39–41.
41. Ibid. Although the use of the term nationality is rather anachronistic for the period.
42. Ibid. Coll. RO SAV I, box 2652, "J. Tibenský: Zpráva o účasti na XI. Medzinárodnom kongrese pre dejiny vied vo Varšave."
43. Tibenský, "Národná veda a jej vzťah k svetovej vede," 39–41. Coll. SSDVT, box 6, inv. no. 71, "Sympózium slovenských a poľských historikov o baníctve a hutníctve

v karpatskej oblasti v Tatrách (Cestovná zpráva – J. Tibenský, J. Voda a P. Hapák)." See also Hapák, "Sympózium slovenských a poľských historikov," 664.

44. Miczulski, "Spotkanie słowackich i polskich historyków."
45. AN PAN i PAU, Coll. KIII-266, "Stanisław Miczulski: Życiorys [CV]," January 1978.
46. Owczarek, "Pracownicy Instytutu Historii Nauki PAN," 432; AN PAN i PAU, Coll. KII-8, sign. 180, 154/87; Miczulski, "Mój brat: Stanisław Miczulski," 295–96.
47. Sygocki, "Instytut Historii Nauki PAN w latach 1953–2003," 17–19. For more on the history of the Cracow branch of the Institute for the History of Science of the PAN, see Sroka, "Trzydziestolecie krakowskiej filii," 573–74.
48. Miczulski, "Z pobytu w Czechosłowacji," 192–93; AN PAN i PAU, Coll. KII-8, sign. 184, 5/87, "Sprawozdanie z podróży służbowej do Czechosłowacji odbytej w dniach 6–21. XI. br."
49. AN PAN i PAU, Coll. KII-8, sign. 184, 5/87, "Sprawozdanie z podróży służbowej do Czechosłowacji odbytej w dniach 6–21. XI. br."
50. He wrote almost twenty letters, among others to Luboš Nový, head of the Department of History of Science and Technology of the Institute of History of the ČSAV, and Ján Tibenský, head of the corresponding department at the Institute of History of the SAV in Bratislava. In his letter to Tibenský he also greeted Juraj Voda, among others. AN PAN i PAU, Coll. KII-8, sign. 184, 5/87. He wrote his letters in Polish and the addressees replied in Czech or Slovak, which was the standard way of Czechoslovak–Polish scientific communication at that time (Juraj Voda also corresponded with Poles in a similar way). Miczulski also corresponded in English with other foreign historians (for example, the historian of technology Mario Loria). He also spoke and wrote German.
51. AN PAN i PAU, Coll. KII-8, sign. 184, 5/87.
52. AN PAN i PAU, Coll. KII-8, sign. 251, 5/87, "Sprawozdanie z przebiegu prac nad rozprawą doktorską pt. 'Bobrzańsko-Samsonowski zespół hutniczy XVII–XVIII w.' za okres lipiec 65 r. – maj 66 r., 23 May 1966." Originally, Miczulski's work concerned metallurgy in Samsonów in the first half of the nineteenthth century (in the early 1960s, he changed the subject of his doctoral thesis).
53. This is the reason given by Miczulski's brother Bartołomiej. Miczulski, "Mój brat – Stanisław Miczulski." Several documents from the Institute of the History of Science of the PAN in Cracow from the 1960s and 1970s refer to Miczulski's doctoral thesis as "almost finished." AN PAN i PAU, Coll. KIII-266.
54. See Miczulski, "Archeologia—Metalurgia—Historia"; Miczulski, "Początki i rozwój pierwszego ośrodka."
55. Miczulski, "Początki rozwoju górnictwa i hutnictwa."
56. AN PAN i PAU, Coll. KII-8, sign. 180, 154/87, "M. Radwan: Opinia o pracy mgr. Stanisława Miczulskiego, 1960 May 15"; Ibid., Letter from M. Radwan to the Institute of the History of Science and Technology of the PAN in Cracow, 1965 February 21.
57. AN PAN i PAU, Coll. KII-8, sign. 251, 5/87, Letter from S. Miczulski to P. Hapák, 1967 January 12.
58. AN PAN i PAU, Coll. KII-8, 154/87/P.30, Letter from J. Tibenský to M. Radwan, 1966 August 22.
59. AN PAN i PAU, Coll. KII-8, 154/87/P.56, Letter from J. Pazdur to the Institute of the History of Science and Technology of the PAN, 1968 March 28.

60. AN PAN i PAU, Coll. KII-8, sign. 183, 5/87.
61. See, for example, Wanacka, "Gdy świętokrzyskie żelazo wspierało."
62. Ibid. Ewa Wanacka interprets the achievements of ancient technicians from the period of the Roman Empire on the territory of contemporary Poland as part of Polish mining tradition.
63. Malina, *Vzpomínky na minulost,* 272; see also Radwan and Pleiner, "Polnisch-tschechoslowakische Schmelzversuche."
64. AN PAN i PAU, Coll. KIII-266.
65. This was not an exception; other ČSAV/SAV workplaces were located, for example, in Brno (Institute of Archaeology of the ČSAV, Institute of Biophysics of the ČSAV and others), in the Slovak Tatras (Institute of Astronomy of the SAV), and in Košice in Eastern Slovakia (Institute of Experimental Biology of the SAV). However, most of the institutes were located in Prague (ČSAV) and Bratislava (SAV).
66. Before that, the Institute existed in the form of an association.
67. For more on the institute, see Ficek, *Slezský ústav*; Knapík, *Slezský studijní ústav.*
68. Ficek, *Slezský ústav*, 6; see also Knapík, *Slezský studijní ústav*, 44.
69. Bohumil Sobotík, director of the Silesian Museum in Opava and former director of the Silesian Study Institute, the predecessor of the Silesian Institute of the ČSAV, held a similar view. As early as 1947, he stated that "Silesian scientific activities should be supported directly in Silesia," thus emphasizing the need for research to be carried out directly in the region. Knapík, "Slezský ústav v Praze," 509.
70. Ficek, *Slezský ústav*, 20–21.
71. See Kudělka et al., *Česko-polský sborník vědeckých prací.*
72. Ficek, *Slezský ústav*, 61.
73. Kolejka, "Další rozvoj spolupráce Slezského ústavu," 14.
74. Grobelný and Bakala, *Tisíc let česko-polské vzájemnosti*, 285.
75. At the national level in the 1960s, the Silesian Institute was the coordinator of the research topic "the emergence and development of industrial areas in Czechoslovakia". ČSAV, *Státní plán badatelského výzkumu*, 83.
76. Sommer, "Diskusně o národnostní otázce," 7.
77. Grobelný, *K otázkám dějin Slzeska*, 6.
78. See Maleczyński et al., *Historia Śląska*, vol. 1; Długoborski et al., *Historia Śląska*, vol. 2, part 1; Michalkiewicz et al., *Historia Śląska*, vol. 2, part 2; Michalkiewicz et al., *Historia Śląska*, vol. 3.
79. In his memoirs, Andělín Grobelný later noted that "the Polish side welcomed our attitude with appreciation." Grobelný, *Vzpomínky*, 73. Criticism from the Czech side mainly concerned the portrayal of the national situation in Silesia, especially in Cieszyn, and Czechoslovak-Polish relations in the interwar period. Grobelný, *Vzpomínky*, 73.
80. Kocích, "Nejdůležitější výsledky výzkumné činnosti," 4.
81. Kocích, "K výsledkům činnosti Slezského ústavu," 3.
82. Zpravodaj, "Výsledky výzkumu soudobé národnostní problematiky," 1; "ongoing national processes" in this context referred specifically to the issue of national minorities and politics in bilingual regions.
83. Kolejka, "Zpráva o konferenci," 15.
84. Sommer, "Diskusně o národnostní otázce," 7.
85. The archive was founded in 1953. Bystrický and Hrubý, *Přehled archivů ČSR*, 184.

86. Jaworski, "Stanisław Zahradnik," 28–30.
87. See Zahradnik, *Třinecké železárny v 19. století.* On the cover, Zahradnik's first name is written in its Czech form as "Stanislav."
88. Vytiska, "Dohoda o spolupráci," 2; see also Vytiska et al., *Dějiny Třineckých železáren.*
89. Vytiska, "Dohoda o spolupráci," 2.
90. The most important company archives include those of ČKD, Škoda Works (V. I. Lenin Works), Tatra, J. Šverma Works (Zbrojovka Brno) and others.
91. See, for example, Jílek, "VI. pracovní seminář," 67; Rasl, "XVIII. Seminář z dějin hutní výroby."
92. Pleiner et al., *Dějiny hutnictví železa v Československu. 1*; see also Danihelka et al., *Dějiny hutnictví železa v Československu. 2*; Dostál et al., *Dějiny hutnictví železa v Československu. 3.*
93. Hříbek, "Mezinárodní porada v Třinci," 132.
94. The Polish term *Zaolzie* (Trans-Olza, also Trans-Olza Silesia, that is, from the Polish point of view, the lands beyond the river Olza) refers to the part of Cieszyn-Silesia that fell to the former Czechoslovakia after World War I (and is now part of the Czech Republic). The territory was claimed by Poland in the interwar period and annexed in October 1938. It then became part of Germany and, after 1945, returned to Czechoslovakia. It still has a large Polish minority. In Czech, the area is usually referred to as České Těšínsko or Těšínské Slezsko (Cieszyn Silesia).
95. The contemporary Czechoslovak academic equivalent of Doctor of Philosophy (PhD).
96. Jaworski, "Stanisław Zahradnik," 28–30.
97. See SRPSz, *Stój mocno złocista nadziejo.*
98. ABS, Coll. AKR 36, inv. no. 837, "Vyhodnocení plánu práce 2. Odboru S StB Ostrava za rok 1984 – roční + 1.–4. čtvrtletí (zprac. Kpt. Karel Kudla), Hodnocení bezpečnosti práce a dosažených výsledků ve služební činnosti za III. čtvrtletí 1984 po linii vnitřního zpravodajství – Krajská správa sboru národní bezpečnosti, Správa státní bezpečnosti Ostrava," 19.
99. ABS, Coll. AKR 36, inv. no. 838, "Vyhodnocení plánu práce 2. odboru S StB Ostrava za rok 1985 – roční + 1.–4. čtvrtletí (kpt. Karel Kudla), Hodnocení bezpečnostní práce a dosažených výsledků ve služební činnosti za IV. čtvrtletí 1985 po linii vnitřního zpravodajství," 28.
100. ABS, Coll. AKR 36, inv. no. 1000, "Denní informace ze S StB Ostrava (leden-prosinec 1987)."
101. Tadeusz Wantuła (1950–2024) was born in Bystřice, Czechoslovakia, to Polish parents. He studied philology and film studies at the Jagiellonian University in Cracow and then taught at the University of Silesia. He was a member of Polish–Czechoslovak Solidarity and Civic Forum. In the years 1990–92, he was elected to the Czech National Council for this movement. In 1990 he became the first chairman of the Council of Poles in Czechoslovakia. For his memoirs, see "Mgr. Tadeusz Wantuła." Paměť národa. Retrieved 4 April 2026 from https://www.pametnaroda.cz/cs/wantula-tadeusz-1950.
102. ABS, Coll. AKR36, inv. no. 962, "Informace o bezpečností situaci a aktuálních otázkách bezpečnosti na úseku boje proti vnitřnímu nepříteli – X. správa, 1., 3.–5. odbor, 2. odbor XII. správy, Správy StB v Krajích ČSR a SSR 4. Čtvrtletí 1987,

Krajská správa sboru národní bezpečnosti – Správa státní bezpečnosti Ostrava: Hodnocení bezpečnostní situace a dosažených výsledků ve služební činnosti za IV. Čtvrtletí 1987 po linii vnitřního zpravodajství," 15. The discussion probably referred to the persecution of Poles living in Czechoslovakia who supported Polish Solidarity in connection with the above-mentioned "Action North."

103. Česenek, "Stanisław Zahradnik: Třinecké železárny," 202.
104. Ibid.
105. Ibid.
106. Ibid., 203.
107. Ibid.
108. Ibid., 204.
109. Myška, *Problémy a metody hospodářských dějin*, 138.
110. Dohnal, Hájková, and Mariánek, *Z dějin dělnického hnutí*.
111. Zahradnik, "Projevy světové hospodářské krize."
112. Škaloud, "Z dějin dělnického hnutí," 432.
113. See, for example, Steiner and Borák, "Stanisław Zahradnik: Zaolziańskie ofiary."
114. Myška, *Problémy a metody hospodářských dějin*, 137; Jíša, "Dějiny závodů," 25; see also Komise, "Metodické zásady"; ÚD KSČ, *Regionální dějiny*; activities in this area were conducted by the Institute for the History of the Communist Party of Czechoslovakia in cooperation with the Cabinet for Trade Union History at the Central Council of Trade Unions.
115. Jíša, "Dějiny závodů," 30.
116. See, for example Havránek, *Úvod do studia dějepisu*, 24.
117. AN PAN i PAU, Coll. KII-8, sign. 196, Letter from R. Pleiner to M. Radwan, Březno, 1965 July 19. Pleiner was referring to his dispute with the Polish archaeologist Jerzy Piaskowski (1922–2013) over the method of identifying iron in archaeological finds.
118. Among his many activities, he sang in a (Polish) vocal quartet and church choir, he was an active sportsman, and he was one of the leaders of the local Polish Cultural and Educational Association.

Bibliography

Barica, Ján A. 1957. "Pozná ho svet, poznáme ho my?" [The world knows him, we know him?] *Večerník* 110, no. 11: 4.

Bokes, František. 1955. "Činnost a plány kabinetov pre dejiny vied v Slovenskej akadémii vied" [Activities and plans of the departments for the history of science at the Slovak Academy of Sciences]. *Historický časopis* 3, no. 1: 152.

———. 1956. "Dr. Ľ. Ivan – J. M. Nový, K. G. Zechenter [...]; Juraj Voda, Jozef Murgaš, priekopník radiotelegrafie [...]" [Jozef Murgaš: a pioneer of radiotelegraphy]. *Historický časopis* 4, no. 1: 124.

———. 1965. "Obhajoby kandidátskych dizertácií v HÚ SAV" [Defenses of doctoral dissertations at the Institute of History of the SAV]. *Historický časopis* 13, no. 4: 641–44.

Bystrický, Vladimír, and Václav Hrubý. 1984. *Přehled archivů ČSR* [Overview of Czechoslovak archives]. Prague: TEPS místního hospodářství.

Česenek, Ludomír. 1970. "Stanisław Zahradnik: Třinecké železárny [review]." *Slezský sborník* 28, no. 2: 202–5.

ČSAV. 1967. *Státní plán badatelského výzkumu v oblasti společenských věd na léta 1966–1970* [State plan for research in the social sciences for 1966–1970]. Prague: Ústav plánování vědy ČSAV.

Danihelka, Antonín, Zdeněk Deyl, Jozef Faltus, Pavla Horská, Jan Josif, Jiří Matějček, Václav Průcha, Mária Šarudyová, and Ladislav Tajták. 1986. *Dějiny hutnictví železa v Československu. 2, Od průmyslové revoluce do konce druhé světové války* [The history of iron metallurgy in Czechoslovakia. Vol. 2: From the industrial revolution to the end of World War II]. Prague: Academia.

Długoborski, Wacław, Józef Chlebowczyk, Julian Janczak, Tadeusz Ładogórski, Stanisław Michałkiewicz, and Kazimierz Orzechowski. 1966. *Historia Śląska* [The History of Silesia], vol. 2, part 1. Wrocław: Zakład Narodowy im. Ossolińskich.

Dohnal, Miloň, Eva Hájková, and Vladimír Mariánek, eds. 1966. *Z dějin dělnického hnutí na Ostravsku za předmnichovské republiky* [From the history of the labor movement in the Ostrava region in the pre-Munich Republic]. Ostrava: Krajské nakladatelství Ostrava.

Dostál, Vladimír, Zdeněk Jadrný, Jan Josif, Václav Průcha, and Zdeněk Wolf. 1988. *Dějiny hutnictví železa v Československu. 3, Od znárodnění k současnosti (1945–1980)* [The history of iron metallurgy in Czechoslovakia. Vol. 3: From nationalization to the present (1945–1980)]. Prague: Academia.

Ficek, Viktor. 1969. *Slezský ústav ČSAV v Opavě 1948–1968* [The Silesian Institute of the *ČSAV* in Opava, 1948–1968]. Ostrava: Profil.

Grobelný, Andělín, ed. 1956. *K otázkám dějin Slezska: diskuse a materiály z konference v Opavě ve dnech 3. a 4. listopadu 1955* [On questions of Silesian history: discussions and materials from the conference held in Opava on November 3 and 4, 1955]. Ostrava: KNV.

———. 2017. *Vzpomínky* [Memories]. Opava: Slezské zemské muzeum.

Grobelný, Andělín, and Jaroslav Bakala, eds. 1966. *Tisíc let česko-polské vzájemnosti. Materiály z vědecké konference, pořádané na Hradci u Opavy ve dnech 9.–10. září 1965* [A thousand years of Czech-Polish mutuality. Proceedings from the academic conference held in Hradec near Opava on September 9–10, 1965]. Opava: Slezský ústav Československé akademie věd.

Hapák, Pavel. 1966. "Sympózium slovenských a poľských historikov o starom hutníctve a baníctve v Karpatskej oblasti na Slovensku a Poľsku" [Symposium of Slovak and Polish historians on early metallurgy and mining in the Carpathian region of Slovakia and Poland]. *Historický časopis* 14, no. 4: 664.

Havránek, Jan. 1967. *Úvod do studia dějepisu* [Introduction to the study of history]. Prague: SPN.

Hříbek, Jan. 1968. "Mezinárodní porada v Třinci o dějinách hornictví hutnictví po roce 1945" [International meeting in Třinec on the history of mining and metallurgy after 1945]. *Dějiny věd a techniky* 1, no. 2: 132.

Hudek, Adam. 2008. "Inštitucionalizácia výskumu v oblasti dejín vied a techniky na Slovensku od počiatkov po rok 1989" [The institutionalization of research in the history of science and technology in Slovakia from its beginnings to 1989]. *Studia bibliographica posoniensia* no. 1: 211–21.

———. 2022. "The Development of Relations between the Slovak and the Czechoslovak Academy of Sciences until 1970." *Práce z dějin Akademie věd* 14, no. 2: 27–47.

Jaworski, Kazimierz. 2012. "Stanisław Zahradnik: Osiemdziesięcioletni historyk w świetle swego dorobku" [Stanisław Zahradnik: An eighty-year-old historian in the light of his life's work]. *Zwrot* 62, no. 4: 28–30.

Jílek, František. 1968. "VI. pracovní seminář z dějin hornictví a hutnictví" [6th workshop on the history of mining and metallurgy]. *Dějiny věd a techniky* 1, no. 1: 67.

Jíša, Václav. 1989. "Dějiny závodů: programová vědecká disciplína" [History of factories: a programmed scientific discipline]. *Československý časopis historický* 37, no. 1: 16–43.

Knapík, Jiří. 2003. "Slezský ústav v Praze a vědecký výzkum Slezska v letech 1945–1950" [The Silesian Institute in Prague and scientific research on Silesia in 1945–1950]. In *Semináře a studie Výzkumného centra pro dějiny vědy z let 2002–2003*, 475–549. Prague: Výzkumné centrum pro dějiny vědy AV ČR.

———. 2004. *Slezský studijní ústav v Opavě 1945–1958: proměny vědeckého pracoviště v regionu* [The Silesian Research Institute in Opava, 1945–1958: The evolution of a scientific institution in the region]. Prague: Výzkumné centrum pro dějiny vědy.

Kocích, Miroslav. 1977. "Nejdůležitější výsledky výzkumné činnosti Slezského ústavu ČSAV za léta 1971–1975" [The most important results of the research activities of the Silesian Institute of the ČSAV for the years 1971–1975]. *Zpravodaj Slezského ústavu Československé akademie věd v Opavě* no. 11: 1–5.

———. 1984. "K výsledkům činnosti Slezského ústavu ČSAV v Opavě v roce 1983" [On the results of the activities of the Silesian Institute of the ČSAV in Opava in 1983]. *Zpravodaj Slezského ústavu Československé akademie věd v Opavě* no. 1: 1–7.

Kolejka, Josef. 1977. "Zpráva o konferenci" [Conference report]. *Zpravodaj Slezského ústavu Československé akademie věd v* Opavě no. 11: 15–16.

———. 1984. "Další rozvoj spolupráce Slezského ústavu ČSAV s institucemi v Polsku" [Further development of cooperation between the Silesian Institute of the ČSAV and institutions in Poland]. *Zpravodaj Slezského ústavu Československé akademie věd v Opavě* no. 6: 14.

Komise pro dějiny závodů při ÚD KSČ. 1961. "Metodické zásady pro zpracování dějin průmyslových závodů" [Methodological guidelines for researching the history of industrial plants]. *Příspěvky k dějinám KSČ* 1, no. 4: 574–85.

Krige, John. 2019. "Introduction: Writing the Transnational History of Science and Technology." In *How Knowledge Moves: Writing the Transnational History of Science and Technology*, edited by John Krige, 1–32. Chicago: The University of Chicago Press.

Kudělka, Milan, ed. 1955. *Česko-polský sborník vědeckých prací* [Czech-Polish collection of scientific papers]. Prague: Státní pedagogické nakladatelství.

Lebow, Katherine, Małgorzata Mazurek, and Joanna Wawrzyniak. 2019. "Making Modern Social Science: The Global Imagination in East Central and Southeastern Europe after Versailles." *Contemporary European History* 28, no. 2: 137–42.

Ľipták, Ľubomír. 1969. "Úloha a postavenie historiografie v našej spoločnosti" [The role and status of historiography in our society]. *Historický časopis* 17, no. 1: 98–118.

Maleczyński, Karol, Józef Andrzej Gierowski, Roman Heck, Kazimiera Wanda Heckowa, Włodzimierz Hołubowicz, Stefan Inglot, Józef Leszczyński, Ewa Maleczyńska, Kazimierz Orzechowski, and Kazimierz Piwarski. 1960–1964. *Historia Śląska* [The History of Silesia], vol. 1, part 1–4. Wrocław: Zakł. Narod. im. Ossolińskich.

Malina, Jaroslav. 1992. *Vzpomínky na minulost* [Memories of the past]. Brno: Masarykova univerzita.

Michalkiewicz, Stanisław, Józef Chlebowczyk, Juliusz Demel, Tadeusz Gospodarek, Julian Janczak, Tadeusz Ładogórski, Kazimierz Orzechowski, Jerzy Sydor, and Mieczysław Urbanowicz. 1970. *Historia Śląska*, vol. 2, part 2. Wrocław: Zakł. Narod. im. Ossolińskich.

Michalkiewicz, Stanisław, Antoni Gładysz, et al. 1976–1985. *Historia Śląska* [The History of Silesia], vol. 3, part 1–2. Wrocław: Zakład Narodowy im. Ossolińskich.

Miczulski, Bartołomiej. 2012. "Mój brat: Stanisław Miczulski" [My brother: Stanisław Miczulski]. *Almanach Muszyny*: 295–96.

Miczulski, Stanisław. 1962. "Archeologia—Metalurgia—Historia" [Archaeology—Metallurgy—History]. *Problemy* 18, no. 4: 265–71.

———. 1962. "Początki i rozwój pierwszego ośrodka wielkopiecowego na ziemiach polskich" [The origins and development of the first blast furnace complex in Poland]. *Wiadomości Hutnicze* 18, no. 7–8: 227–29.

———. 1966. "Z pobytu w Czechosłowacji" [From a stay in Czechoslovakia]. *Kwartalnik Historii Nauki i Techniki* 11, no. 1–2: 192–93.

———. 1967. "Spotkanie słowackich i polskich historyków techniki górniczej i hutniczej" [Meeting of Slovak and Polish historians of mining and metallurgical technology]. *Kwartalnik Historii Nauki i Techniki* 12, no. 1: 222–25.

———. 1972. "Początki rozwoju górnictwa i hutnictwa kruszczowego w rejonie Kielc do końca XVI. w." [The beginnings of mining and ore smelting in the Kielce region up to the end of the 16th century]. In *Dzieje i technika świętokrzyskiego górnictwa i hutnictwa kruszcowego*, edited by Zbigniew Kowalczewski, 79–99. Warsaw: Wydawnictwa Geologiczne.

Myška, Milan. 1995. *Problémy a metody hospodářských dějin. Část 1. Metodické problémy studia dějin sekundárního sektoru* [Issues and methods in economic history. Part 1. Methodological issues in the study of the history of the secondary sector]. Ostrava: Ostravská univerzita, Filozofická fakulta.

Nový, Luboš, and Irena Stasiewicz-Jasiukowa, eds. 1981. *Problems of Teaching the History of Science: Studies of Czechoslovak and Polish Historians of Science for the 16th International Congress of the History of Science*. Prague: Ústav československých a světových dějin ČSAV.

Owczarek, Janina. 2004. "Pracownicy Instytutu Historii Nauki PAN w latach 1953–2004" [Staff of the Institute for the History of Science of the PAN in the years 1953–2004]. In *Instytut Historii Nauki Polskiej Akademii Nauk w latach 1953–2003*, edited by Joanna Schiller and Leszek Zasztowt, 427–36. Warsaw: Instytut Historii Nauki PAN.

Pejs, Oldřich. 2024. "Slovenská armáda: nultá hodina" [The Slovak Army: The Zero Hour]. *Vojenská história* 28, no. 2: 19–44.

Pleiner, Radomír, Jan Kořan, Matúš Kučera, and Jozef Vozár. 1984. *Dějiny hutnictví železa v Československu: 1, Od nejstarších dob do průmyslové revoluce*. Prague: Academia.

Radwan, Mieczysław, and Radomír Pleiner. 1963. "Polnisch-tschechoslowakische Schmelzversuche in den Rennöfen der römerzeitlichen Bauarten" [Polish-Czechoslovakian smelting experiments in Roman-era kilns]. *Archeologické rozhledy* 15: 47–71.

Rasl, Zdeněk. 1979. "XVIII. Seminář z dějin hutní výroby [18th seminar on the history of metallurgical production]." *Dějiny věd a techniky* 12, no. 3: 190–91.

Škaloud, Vladimír. 1966. "Z dějin dělnického hnutí na Ostravsku za předmnichovské republiky [review] [From the history of the labor movement in the Ostrava region in the pre-Munich Republic]." *Slezský sborník* 24, no. 3: 430–32.

Sommer, Karel. 1984. "Diskusně o národnostní otázce v průmyslových oblastech" [A discussion on the national question in industrial regions]. *Zpravodaj Slezského ústavu Československé akademie věd v Opavě* no. 6: 7–13.

Sroka, Stanisław. 1988. "Trzydziestolecie krakowskiej filii IHNOiT PAN" [The 30th anniversary of the Cracow branch of the IHNOiT PAN]. *Kwartalnik Historii Nauki i Techniky* 33, no. 2: 573–74.

SRPSz. 1984. *Stój mocno złocista nadziejo: 200 lecie oświaty szkolnej w Nawsiu* [Stand firm, golden hope: The 200th anniversary of school education in Nawsie]. Nawsie: SRPSz przy PSP w Nawsiu.

Steiner, Jan, and Mečislav Borák. 1989. "Stanisław Zahradnik: Zaolziańskie ofiary okupacji Hitlerowskiej [review]" [Stanisław Zahradnik: Victims of the Hitler occupation in Zaolzie]. *Slezský sborník* 47, no. 2: 138–42.

Sygocki, Witold. 2004. "Instytut Historii Nauki PAN w latach 1953–2003: Kalendarium" [The Institute for the History of Science of the PAN, 1953–2003: A Chronology]. In *Instytut Historii Nauki Polskiej Akademii Nauk*, edited by Joanna Schiller and Leszek Zasztowt, 17–25. Warsaw: Instytut Historii Nauki PAN.

Tibenský, Ján. 1981. "Národná veda a jej vzťah k svetovej vede ako metodologický problém dejín vedy" [National science and its relationship to world science as a methodological problem in the history of science]. *Dějiny věd a techniky* 14, 2: 37–46.

———. 1985. "20 rokov Slovenskej spoločnosti pre dejiny vied a techniky pri SAV v slovenskej vede a kultúre" [20 years of the Slovak Society for the History of Science and Technology at the SAV in Slovak science and culture]. In *Veda a technika v dejinách Slovenska: zborník prác pri príležitosti 20. výročia založenia Slovenskej spoločnosti pre dejiny vied a techniky pri SAV.* 2, 5–23. Bratislava: Slovenská spoločnosť pre dejiny vied a techniky pri SAV.

ÚD KSČ. 1965. Regionální dějiny dělnického hnutí a dějiny závodů: metodická příručka [Regional history of the labor movement and history of factories: a methodological guide]. Prague: NPL.

Vlach, Róbert. 1957. "Juraj Voda: Karol Hell." *Naša veda* 4, no. 12: 555–56.

Voda, Juraj. 1955. *Jozef Murgaš: Priekopník radiotelegrafie* [Jozef Murgaš: Pioneer of Radiotelegraphy]. Martin: Osveta.

———. 1957. *Jozef Karol Hell: životné dielo strojného inžiniera štiavnických baní v 18. storočí* [Jozef Karol Hell: The life's work of a mechanical engineer at the Štiavnica mines in the 18th century]. Martin: Osveta.

———. 1959. *Aurel Stodola 1859–1942: pamiatke storočnice narodenia* [Aurel Stodola 1859–1942: in commemoration of the centennial of his birth]. Bratislava: Vyd. SAV.

———. 1961. *Technika a náboženstvo* [Technology and religion]. Bratislava: Čs. spoločnosť pre šírenie politických a vedeckých poznatkov.

Vytiska, Josef. 1975. "Dohoda o spolupráci mezi Třineckými železárnami VŘSR a Slezským ústavem ČSAV v Opavě" [Cooperation agreement between the Třinec Iron and Steel Work VŘSR and the Silesian Institute of the ČSAV in Opava]. *Zpravodaj Slezského ústavu Československé akademie věd v Opavě* no. 11: 2.

Vytiska, Josef et al. 1979. *Dějiny Třineckých železáren VŘSR: 1839–1979* [History of the Třinec Iron and Steel Work VŘSR: 1839–1979]. Prague: Práce.

Wanacka, Ewa. 1972. "Gdy świętokrzyskie żelazo wspierało Cesarstwo rzymskie" [When Świętokrzyskie iron supported the Roman Empire]. *Trybuna robotnicza* 284, no. 29: 4.

Werner, Michaela, and Bénédicte Zimmermann. 2005. "Beyond Comparison: Histoire Croisée and the Challenge of Reflexivity." *History and Theory* 45: 30–50.

Zahradnik, Stanisław. 1961. *Třinecké železárny v 19. Století* [The Třinec Iron and Steel Works in the 19th Century]. Opava: Slezský ústav Československé akademie věd v Opavě.

———. 1966. "Projevy světové hospodářské krize v 30. letech v Třineckých železárnách" [The effects of the global economic crisis of the 1930s at the Třinec Iron and Steel Works]. In *Z dějin dělnického hnutí na Ostravsku za předmnichovské republiky*, edited by Miloň Dohnal, Eva Hájková, and Vladimír Mariánek, 167–96. Ostrava: Krajské nakladatelství Ostrava.

Zpravodaj. 1979. "Výsledky výzkumu soudobé národnostní problematiky v pracech Slezského ústavu ČSAV v Opavě" [Research findings on contemporary nationality issues in the works of the Silesian Institute of the ČSAV in Opava]. *Zpravodaj Slezského ústavu Československé akademie věd v Opavě* no. 6: 1–6.

CHAPTER 2

Scientific–Technological Revolution and the Failure of Czechoslovak–Polish Cooperation

Jan Surman

> Scientific–technological revolution is a profound transformation in the domains of science, technology, and their interrelationships, along with the social functions that accompany them. This transformation has resulted in a comprehensive restructuring of societal productive forces, with a notable shift in the role of humans within this system. This shift is based on the comprehensive, technological application of science as a direct productive force, which has permeated all aspects of production and transformed the material conditions of human existence.[1]

If history is written from the perspective of the victors, entangled history is written from the perspective of successful collaborations, networks, transfers, and translations. This is particularly the case with the history of science, which was, for a long time, oriented toward demonstrating scientific achievements and successes in the name of progress. In the 1970s, historians and sociologists of science deemed it necessary to pay equal attention to both success and failure and to use the same tools to investigate them.[2] This paved the way for current cultural-science oriented investigations of science as a social system. While the history of failed collaborations has yet to be written, this chapter will examine one illustrative case study in which a potential collaboration ultimately did not materialize. By examining the vagaries of scientific–technological revolution (hereafter STR) in Czechoslovakia and Poland, this chapter will elucidate the personal and structural factors that impeded collaboration between the two states during the 1970s, a period when both STR and socialist cooperation were politically reinforced.

In one of the foundational articles of twenty-first-century history of science, James Secord posits that the central question for the discipline is "How and why does knowledge circulate?"[3] While this also addressed obstacles to the circulation of knowledge, primarily in the context of cultural bias toward alternative forms of knowledge (such as traditional, non-academic knowledge), it was the study of ignorance that placed the issue of barriers to knowledge exchange and various forms of unknowing at the core of the research agenda. In his writings on agnotology, Robert Proctor drew attention to the concept of "ignorance as a lost realm, or selective choice (or passive construct)," arguing that any inquiry is inherently selective. Consequently, each stage of the research process is associated with a duality of attention and inattention, knowledge and ignorance, the mobilization of specific resources and the neglect of others.[4] Recently, there have been calls for a global and longitudinal history of ignorance that would include an investigation of the various obstacles to knowledge, including those of a physical, mental, social, and political nature.[5] Some of these fields are more well-known than others. For instance, the influence of colonialism, racism, nationalism, and the exclusion of women and their knowledge has been widely discussed in recent decades. Other fields, such as the impact of the privatization of medical research and the secrecy connected with it on medical knowledge, are emerging topics.[6]

The issue of hurdles to knowledge circulation during the Cold War has been frequently highlighted by historians, yet it has not been a prominent focus within the discipline of the history of science. Instead, there has been a tendency to concentrate on instances where the Iron Curtain was breached.[7] Nevertheless, the Cold War was a period characterized by significant challenges to the circulation of knowledge, including the primary obstacle of the Iron Curtain. This is evidenced by the existence of numerous institutions that were designed to facilitate exchange, including initiatives such as cover-to-cover translations, the International Institute for Applied Systems Analysis (in Laxenburg near Vienna), the European Coordination Center for Social Sciences in Vienna, and various UN/UNESCO programs. Other entities, such as the Council on Mutual Economic Assistance (COMECON), sought to foster collaboration within the respective blocs. These endeavors were responses to perceived global or regional deficiencies in the circulation of knowledge caused by the politicization of scholarship, as well as recognition of the need for political projects that could facilitate the coordination of programs addressing transregional issues. The majority of inquiries focus on problems concerning contacts with the respective opposing side of the Cold War divide. These inquiries tend to prioritize the politicization of research and restrictions on scholarly contacts, particularly in the context of the Eastern Bloc.[8] While such problems were not exclusive to this region, other instances may be less prominent in the historical record. McCarthyism in the United States and the potential consequences for scholars who engaged with East European scholars is a noteworthy exception.[9] Consequences of Cold War knowledge ignorance could even be fatal, as evidenced by the cases of mushroom poisoning in the United States during the 1970s and the lack of awareness among medical professionals of Chinese publications on avian influenza in the first decade of the twenty-first century, as observed by Christopher Hollings.[10]

The investigation of contacts that did not happen and knowledge that was not known in a locality other than the one in which it was produced should not be guided by an assumption that science is per se global and relies on the free flow of scholars, data, knowledge, and so forth. The historiography of science has demonstrated that this early twentieth-century utopia was not a realistic expectation, as contacts were always established and maintained through various practices, and could also easily disintegrate if not supported. In order to travel from one space to another, in all possible meanings of the word "space," knowledge had to be mobilized and appropriated, always also changing in the process. Therefore, the investigation of failed knowledge movement should also include questions of networks and translations, which would lead us to a more refined understanding of what makes knowledge move and what makes it stagnant.

When writing about a connection that did not happen, it is necessary to substantiate that it actually could have or should have done. In the case of Czechoslovak and Polish STR, there are numerous reasons why this cooperation should have occurred. As will be discussed in greater detail below, STR was a socialist prestige endeavor that was presented as the key discipline of the Eastern Bloc from the late 1960s. Transsocialist cooperation was also a central aspect of STR, occurring on both bilateral and multilateral levels (COMECON). Moreover, scholars from both countries engaged in frequent communication with their counterparts abroad, both within the Eastern Bloc and beyond. More importantly and significantly, one of the key players of Polish STR, Lech Zacher (1942–2022), stated that he had wanted to cooperate with his Czechoslovak counterparts, but could not for political reasons.[11] Thus, while there were indications that cooperation was inevitable, not only it did it not happen, but also the reception of the leading Czechoslovak STR author, Radovan Richta (1924–1983), who achieved global renown, was so insignificant in Poland that some historians even record the lack of reaction to his ideas.[12]

The following part of this chapter will analyze this "failed" cooperation in STR within the context of Czechoslovak–Polish collaboration in the social sciences following the suppression of the Prague Spring in August 1968. It can be argued that the obstacles that arose were not exclusively the result of the political factors that Zacher highlighted, but also due to social and epistemological considerations. Doubravka Olšáková (1977–) posited that the international position of socialist (non-Soviet) scholars could be described as "restricted internationalism."[13] I have also pointed out elsewhere that Richta, a prominent figure in this chapter, was acutely aware of his ambiguous position as both an internationally renowned scholar and one whose political views placed him in a precarious position. This led him to navigate a complex landscape of pressures and interests with a cautious and strategic approach.[14] Political restrictions affected the transfer of his writings, which also had an influence on the Czechoslovak–Polish entangled history. At the same time, Richta became a very cautious and statist writer, at odds with the development of the social sciences in Poland, which from the 1960s became increasingly internationalized. Zacher's assessment of his unsuccessful collaboration with Richta must also be considered in this context, even though Zacher himself did not reflect on this particular aspect.

The following part will commence with an overview of socialist STR, with a particular focus on its variants that have been the subject of investigation in Czechoslovakia and Poland. Subsequently, I will delineate the patterns of international cooperation of Central European STR, with particular attention to their intersection with politics. The events of 1968 and the subsequent repression in Czechoslovakia and to a lesser extent in Poland had a profound impact on the region of Central Europe. It disrupted established individual-based international cooperation and replaced it with formalized exchanges, as well as strengthening the role of the Soviet Union as the Eastern Bloc's hegemonic power. The final part of the chapter discusses the lack of collaboration between Richta and Zacher. While "politics" is identified as the primary factor, it is argued that the reasons are manifold and may extend beyond bilateral relations.

Richta and Czechoslovak STR

In 1920, Vladimir Lenin (1870–1924) famously declared that the electrification of the Soviet Union was an essential step toward the establishment of communism.[15] Although electricity did not play that role in the second half of the twentieth century, socialist modernization was connected with techno-optimism and presented predominantly a technical, science, and later increasingly (scientific) expertise-driven process.[16] In other states of the Eastern Bloc, this concept blended seamlessly with the respective traditions of science-led development, which had been promoted since they gained independence following the Great War, and in some cases even earlier.[17] This movement reached its zenith in 1961 when Soviet leaders proclaimed science to be a pivotal direct productive force, a notion that was echoed by political leaders across the Eastern Bloc.[18] Despite the discrepancy between these declarations and their actual implementation, they elevated scientific expertise to a central position in the pursuit of an ideal socialist future.

The declaration of science as a key productive force was something of a revolutionary act. While it was to be found in the writings of Karl Marx (1818–1883), it was not the canonical reading practiced during Stalinism and early post-Stalinism, and it mostly appeared in Marx's early writings, particularly the *Grundrisse der Kritik der Politischen Ökonomie* (Foundations of a critique of political economy), which remained unpublished until 1939 (in Russian translation; in German in 1953). Although the subject was discussed in Moscow throughout the early 1950s, it could not gain traction in the Stalinist and early post-Stalinist period.[19] However, when the first writings postulating this change were published in Moscow and Berlin in 1957, the new idea of science began to gain ground rapidly.[20] It also reached Czechoslovakia. Gerhard Kosel (1909–2003), author of the 1957 German publication *Produktivkraft Wissenschaft* (Science as a productive force) that famously proposed treating science as a direct productive force, spoke in Prague in 1959 at the Symposium on Planning in Science.[21] The invitation to Kosel by the vice-chairman of the Association of Workers of Education, Science, Arts, and Print, Jan Štěpánek, noted that his book had been "very well received in Czechoslovakia,"[22] and indeed it can be found in many libraries throughout the country, although I was not able to find any translations of it in Czech or Slovak and only one review.[23] While Kosel's ideas were initially controversial, he was also supported by the First Secretary of the Socialist Unity Party of the GDR, Walter Ulbricht (1893–1973), and GDR Prime Minister Otto Grotewohl (1894–1964), which also well illustrates the connection between the scientific and political Marxist elites. While the new recognition of science in both theory and politics was complete, the social implications remained to be worked out.

The term that proved useful in this regard was scientific–technological revolution (STR), most commonly associated with the writings of the British communist biologist and historian J. D. Bernal (1901–1971).[24] Bernal coined the term to describe what he saw as a second scientific revolution, a period beginning around the 1930s in which science assumed the role of a transformative force in society. His books *The Social Function of Science* (1939) and *Science in History* (1954) became international bestsellers and were translated into the languages of the emerging Eastern Bloc.[25] After World War II, Bernal was one of the key players in science internationalism, with a decisively pro-Soviet stance. He was awarded the Soviet Peace Prize in 1953 and even interviewed the First Secretary of the Communist Party of the Soviet Union, Nikita Khrushchev (1894–1971), during the post-Stalinist power struggle.[26] The term "STR" was used in socialist speeches from the mid-1950s as a synonym for "scientific–technological progress" but became popular in the 1960s as the Soviet answer to Rostow's modernization theory.[27] At the 24th Party Congress in 1971, the General Secretary of the Communist Party of the Soviet Union, Leonid Brezhnev (1906–1982), issued a call to "to combine organically the achievements of the scientific–technological revolution with the advantages of the socialist economic system, and to develop more broadly our own, intrinsically socialist, forms of combining science with production."[28]

Although Brezhnev's appeal fueled STR research, at the same time it may have dealt it a fatal blow. While STR research was already flourishing in 1971, it mushroomed after that, entering the lexicon of more and more disciplines and fields, and frequently used in semi-academic and non-academic settings. Initially, it was a field dominated by philosophers, sociologists, and economists, but it soon spread beyond these groups. Since STR was both an academic area and a period analyzed, the number of articles referring to STR skyrocketed. However, there was no well-articulated definition of what STR actually was.[29] While the history of STR is yet to be written, the late 1970s was more of a time of transition from STR as an academic field in the making to STR as a reference to be included in any publication that touched on socialist society.[30] Significantly, even the early academic pioneers of STR seemed to have abandoned the idea of developing STR into a coherent research program and concentrated their efforts on researching how STR affected individual areas of society. Tellingly, the discipline that could be considered the successor to STR was futurology/prognostics, marking a shift of interest from the present to the future.[31]

This is not to suggest that STR constituted at any point a coherent field or discipline. During its brief heyday, it developed into what could be called a transdisciplinary area, consisting of scholars from philosophy, sociology, economics, and science studies. It also lacked a proper disciplinary belonging and was housed in institutes of philosophy and sociology (often merged into one) and institutes of the history of science and technology, rarely within specialized departments. Unlike futurology, whose proponents had been working on socialist disciplinary dictionaries since the early 1970s,[32] STR never developed a unified terminology. Over the years, relatively few international projects emerged, although scholars from the respective countries frequently met at various conferences.

One of the centers of STR from the late 1960s to the late 1970s was Prague, more precisely the Institute for Philosophy and Sociology of the Czechoslovak Academy of Sciences (Ústav pro filosofii a sociologii Československé akademie věd, hereafter ÚFS ČSAV), which was led by Radovan Richta. Richta became well-known in Czechoslovakia in 1963, when he postulated the need to study the socioeconomic impact of science—which he described as a direct productive force. His later writings developed this idea and explored the consequences of this reorientation. In the late 1960s, he achieved worldwide recognition with *Civilization at the Crossroads*, a book prepared by an interdisciplinary team that described the impact of STR on twentieth-century society.[33] Following the book's agenda, the Institute developed a wide range of interdisciplinary inquiries into the impact of STR on twentieth-century people, from philosophy—Richta's own field—to leisure studies, science studies and prognostics. In the decade following the publication of *Civilization* in 1966, Richta was one of the preferred socialist speakers at international congresses and a leader of several prestigious projects. At the same time, his prominent position during the Prague Spring led to problems within the Eastern Bloc after the reform process was halted by the Warsaw Pact invasion. Ironically, after 1969, Richta transformed from a representative of liberal communism into a representative of "normalization," shying away from statements that could be seen as anti-socialist and generally adopting a very cautious and statist position that surprised even his staunchest supporters.[34]

Importantly, Czechoslovak STR was a broad and very diverse movement, even around Richta. The study of scientific social change required researchers dedicated to the study of science (for example, Ladislav Tondl [1924–2015], Karel Berka [1923–2004], Karel Müller 1936–]), social structure (František Charvát [1940–2007]), culture (Marie Huláková [1923–]), lifestyle and leisure (Blanka Filipcová [1928–), Jindřich Filipec [1926–2013]), social planning/prognostics (František Kutta [1923–2004], Ota Šulc [1932–?]), and many other areas. In fact, the Institute often published in teams or as a group, reflecting Richta's conviction that complex research could not be carried out by individuals alone. In the 1970s, the Institute also participated in numerous bilateral and COMECON projects, adding an international dimension to this community.

Polish STR did not have such a central figure or center of STR research as Richta and ÚFS ČSAV. The term itself appeared quite early in the writings of sociologist Kazimierz Owoc (1914–1961), who died leaving his announced STR-related publications unpublished.[35] In 1972, one of the Polish reviewers of the Polish translation of Richta's *Civilization* spoke about Poland's "intellectual backwardness" in STR.[36] However, by the mid-1970s, there were STR courses and a Research Team on the Scientific–Technical Revolution (Zespół Badań nad Zagadnieniami Rewolucji Naukowo-Technicznej), first at the Institute of the History of Science of the Polish Academy of Sciences (Institut Historii Nauki PAN, hereafter IHN PAN) and from 1974 at the Institute of Philosophy and Sociology of PAN (Instytut Filozofii i Socjologii PAN, hereafter IFiS PAN), and the main actors of early Polish STR were Zacher and Eugeniusz Olszewski (1911–1994). The head of IFiS PAN in the early 1970s, Jan Szczepański (1913–2004), was a renowned sociologist with good contacts in Czechoslovakia, but he did not embrace STR and did not have close contacts with Richta.[37] In contrast, his successor from 1974, Tadeusz Jaroszewski (1930–1988), focused on STR and its effects on individuals and individualities and was on friendly terms with Richta.[38] Moreover, they addressed each other on a first-name basis, which was rare among socialist scientists. Jaroszewski was also the most internationally renowned Polish STR scholar, although both Zacher and Olszewski were also internationally active. Olszewski is considered one of the founders of the International Committee for the History of Technology, along with, among others, Semyon Shukhardin (1917–1980), one of the most eminent Soviet STR researchers.[39]

Importantly, Polish scholars had their own agendas that were closely aligned with, yet distinct from, STR. On the one hand, there was praxeology, the science of (good) human action (not directly related to post-Bourdieu practice theory),[40] and, on the other, there was the science of science, a general (and specifically Polish) research program of broadly understood science, including its social ramifications.[41] Only a select few scholars, such as the Pole Wojciech Gasparski (1936–2022) (and, to a lesser extent, the Czech Ladislav Tondl), were engaged in both of these fields. Although there were Czechoslovak–Polish contacts in both fields, Richta did not appear to be particularly engaged with the field of praxeology, unlike some of his collaborators.[42] He was undoubtedly more interested in the science of science, since the final major projects he was involved in were the collective publications *Socialism and Science* (1983) and *Introduction to Science of Science* (1983).[43] He was also a member of the editorial board of the Polish journal *Science of Science* (along with Karel Berka). Ironically, the collaboration with Polish scholars engaged in the study of science was, from a structural standpoint, fraught with challenges. These scholars were situated within an interdisciplinary framework, namely the Committee of the Science of Science (Komitet Naukoznawstwa) of the PAN. Institutes such as ÚFS ČSAV should only formally cooperate with institutes,[44] although in time this problem was solved.

Of course, STR was not solely a Czechoslovak and Polish undertaking. Soviet scholars played a significant role in this quite international field—from the abovementioned Sukhardin, through Victor Afanasiev (1922–1994), Dzhermen Gvishiani (1928–2003) (both writing extensively on planning), to Bonifaty Kedrov (1903–1895) and Semjon Mikulinskij (1919–1991), who were prominent in STR science of science. The latter two scholars frequently engaged in collaborative research with Richta. Bulgarian and Hungarian scholars also authored notable publications, although I was not able to identify any notable bilateral collaborations with Czechoslovakia or Poland. Scholars from the German Democratic Republic (GDR) such as economist Jürgen Kuczynski (1904–1997) and science of science scholars Günther Kröber (1933–2012) and Hubert Laitko (1935–2024) achieved more prominence, particularly in Czechoslovakia. In addition, there were a few non-socialist researchers of STR, such as the Japanese Marxist philosopher Shingo Shibata (1930–2001) (who was frequently published in Poland) and numerous other studies with varying names, beginning with Daniel Bell's (1919–2011) postindustrial society and concluding with various forms of *Wissensgesellschaft* (the German name for knowledge-based society).

As numerous STR initiatives were undertaken at the multilateral and bilateral levels, the following part will examine the internationalism that was embraced within the Eastern Bloc, particularly with regard to STR. It will be demonstrated that, after 1965, multilateralism began to supersede bilateral collaboration, prioritizing larger projects. However, this did not address the deficiencies of the exchange system that favored senior scholars and cooperation with Moscow, ultimately leading to a situation in which a few scholars were responsible for maintaining all the cooperation.

Socialist Cooperations

The Cold War period saw significant shifts in the vision of internationalism in science.[45] Scholars in non-Soviet countries of the Eastern Bloc found themselves situated between the East and West. On the one hand, their traditional, pre–World War II, channels of exchange were with the West. By the late 1960s, these were outweighed by the politically prescribed orientation to the East, which was not symmetrical, since the countries of Central and Eastern Europe were expected to follow the Soviet example.[46]

The practice was evidently more complicated than the prescription. For scholars from countries such as Czechoslovakia and Poland, socialism entailed navigating between disparate avenues of international collaboration and between diverse institutions within a single country, each of which had its own notions of cooperation.[47] Furthermore, this dynamic shifted over time. While Czechoslovakia was pursuing a more Western-oriented foreign policy in the 1960s, the subsequent invasion of the Warsaw Pact and the advent of normalization compelled those who remained to align with Moscow's directives. In contrast, during the 1970s, Poland pursued a policy of greater independence from the Soviet Union.[48] Polish academics were at the vanguard of the call for greater openness and diversity of contacts, incorporating scholarly arguments into their rationale.[49] It is noteworthy that, while normalization in Czechoslovakia is often regarded as the period when Soviet dependence was most pronounced, within academia the demand for an expansion of contacts with capitalist countries was a prominent feature of the discourse, even among those who were staunchly statist like Richta.[50]

The evolution of science policy was concomitant with the decision of the leaders of socialist academies of sciences—the primary research institutions of the Eastern Bloc—to engage more in international endeavors. In 1965, the heads of academies of sciences resolved that socialist scholars should be more actively engaged in international organizations, including those dominated by scholars from capitalist countries. Concurrently, intersocialist collaboration should be founded upon multilateralism, rather than the preceding bilateral agreements.[51] This resulted in the reinforcement of COMECON, which subsequently became a significant entity in the coordination of research initiatives. Cooperation within COMECON was structured according to specific objectives and partial objectives, which required the collaboration of two or more institutes under the guidance of a lead institution. By the 1980s, many long-term multilateral projects had been established, involving dozens of research institutes.[52] This was in addition to, rather than in lieu of, bilateral cooperation. Furthermore, both levels were subject to renegotiation every five years, with some elements being required to align with state research programs. In the period between 1971 and 1975, STR was the subject of one bilateral Czechoslovak–Polish project (another project in which both institutes participated was general works in philosophy and sociology; in total twenty-eight topics were listed),[53] and ÚFS ČSAV and IFiS PAN cooperated on multilateral levels. Nevertheless, at the conclusion of the period, the Polish side expressed reservations about the efficacy of the STR cooperation.[54] The absence of more detailed projects and travel plans was identified as a major problem. However, Richta asserted that the issue stemmed from the lack of clarity in distinguishing between bilateral and multilateral projects. While this may have been a valid observation, given the evident overlap in projects focusing on STR, three additional structural considerations merit attention.

The first issue is the tendency to prioritize exchange activities by senior scholars, a practice that was repeatedly criticized in the internal documents of the institutes. Consequently, travel was a privilege reserved for a few select scholars, frequently for those proficient in languages. For example, the most frequent travelers from STR part of ÚFS ČSAV were Richta, Šulc, Kutta, Tondl, and Berka, as well as Müller in the 1970s. All of these scholars were renowned experts who collaborated with a variety of partners. From the Polish side, the most frequently occurring names were those of Zacher, Gasparski, and Jaroszewski. It is evident that there was a gender bias, despite the participation of a few women, such as Salomea Kowalewska (1924–2004), a work sociologist, and Czesława Skrzypek, a historian of science, both from Poland, and Filipcová, a Czechoslovak leisure sociologist. Additionally, there are individuals whose names appeared only once or twice in the travel records, typically for a week or two, with travel aims that are often not clearly defined.[55] It is likely that these scholars were sent abroad to satisfy a prescribed quota.

The second issue is that, while Richta was calling for international collaboration, the majority of his international partners were based in Moscow. These were particularly Kedrov and, following his removal from directorial functions in 1973, Mikulinskij. This might have had something to do with the Soviet scholars allegedly helping Richta out when he was facing the consequences of his participation in the Prague Spring, but it might also have been his strategy to secure his position by allying himself with those in power in the Eastern Bloc. This was beneficial for the Soviet scholars as well, as Richta was esteemed and admired in the West, and Kedrov, at least, purportedly enjoyed spending time in Prague and Czechoslovak health and spa resorts.[56] This prominent collaboration was also noteworthy because, as directors of pivotal institutes, they both had a vested interest in STR and could oversee research teams. Indeed, the seminal publications resulting from this partnership were published as collective volumes.

The final reason why Czechoslovak–Polish STR cooperation in the 1970s experienced difficulties is the elephant in the room of many socialist scholarly accounts—politics. As STR was designed as a political endeavor, the entanglement was more pronounced than in other disciplines. However, it would be difficult to identify a place in the Eastern Bloc where political influence was absent.[57] The next part of this chapter will examine the political involvement of STR and its impact on Czechoslovak–Polish relations within the field of science studies, which was STR's primary area of expertise. After that, I will examine Richta's connections with Polish scholars. It will be demonstrated that the process of normalization in Czechoslovakia disrupted long-standing personal collaborations. While it created structural conditions conducive to the formation of new interconnections, these were overshadowed by politically based mistrust.

STR and Politics in Central and Eastern Europe

As a general science of the modernizing socialist society, STR was eminently political. This was especially true in Czechoslovakia, where, in the 1960s, Richta led one of four Academy teams charged with preparing the knowledge that would lead Czechoslovakia into a better socialist future. The result of this was the renowned work *Civilization*, and the political implications of the project were obvious to the participants from the beginning. In addition, in July 1968, the main Czechoslovak daily newspaper *Rudé právo* (Red right) published a three-part article by Richta's team entitled "Before a Decision: On the New Czechoslovak Model of Socialism," which outlined a proposed reform of socialism, based on democracy and humanism and oriented toward modernity.[58] From 1968 Richta was actively engaged in the political process as one of the authors of the program for the extraordinary 14th Congress of the Communist Party of Czechoslovakia (CPCz), which was hindered by the Warsaw Pact invasion.[59] During normalization, the team—or those who remained at the ČSAV after 1969—was tasked with preparing "Czechoslovakia 2000," one of many socialist projects to forecast the country's future.[60] Richta and STR were also omnipresent, in television and radio programs and popular magazines, especially before 1968.

Although Richta's popularity declined after the suppression of the Prague Spring, his contacts with politicians did not suffer. He—and the Institute—continued to write internal reports for the government.[61] Richta intervened in various issues by writing directly to high officials (although I have not found any letters in return), informing them of favorable reviews of his work and sending them copies of published books—for example, *Man–Science–Technology* (1973), the result of a joint Czechoslovak–Soviet project, was sent to the First Secretary of the Communist Party of Czechoslovakia, Gustáv Husák (1913–1991), and Prime Minister Lubomír Štrougal (1924–2023).[62]

Since STR was a socialist project, it was also promoted in the principal pan-socialist media, including the pan-socialist journal *Otázky míru a socialismu* (Problems of peace and socialism). There it was translated into more than forty languages, including Polish, and Richta was a frequent contributor in the 1970s. Publications that Richta prepared with Soviet counterparts were subsequently translated into multiple languages, including English, and disseminated widely. However, preprints also circulated, including among politicians. For instance, the First Secretary of the Polish United Workers' Party, Władysław Gomułka (1905–1982), and the First Secretary of the German Socialist Unity Party, Walter Ulbricht had unofficial translations of *Civilization* on their desks before the official translations were published.[63]

Polish STR did not attain the same level of significance. While the concept was disseminated through various media outlets, it lacked a prominent and charismatic proponent, akin to Richta in Czechoslovakia. In the 1970s, STR was a common theme in political speeches and popular articles. However, there are no indications that scholars benefited from its political prominence in the same way that Richta did. STR was frequently discussed in specialist articles, but no book emerged that would make it intellectually fashionable. In his review of the Polish translation of *Civilization*, Zacher (along with other contemporary authors) emphasized that it could contribute to the popular recognition of STR, although this did not fully materialize.[64]

One possible explanation for this outcome is the impact of the suppression of the Prague Spring. While a detailed account of these consequences will be provided below, here I would like to mention the consequences for Richta. He was one of only a few scholars involved in the Prague Spring who were later able to remain at the Academy and advance in their careers, eventually going on to become the director of ÚFS ČSAV. The price he paid for it was twofold. First, his institute was subjected to close scrutiny and expected to conform to the demands of the new authorities.[65] This subjugation and cleansing of politically problematic scholars did not happen overnight—for instance, Tondl, one of the signatories of "The Two Thousand Words" manifesto, was initially expected to be reassigned to a non-academic position but was permitted to remain at the institute until the mid-1970s. Second, the political change undoubtedly influenced the content of the written output of Richta, whose supporters observed a certain degree of caution in his works in the 1970s, which differed from the straightforward approach he had taken in *Civilization*. This shift in style may have been influenced by the changing international landscape. While Richta was exonerated in Czechoslovakia following his self-criticism, he received criticism from German colleagues and encountered obstacles in cooperation with the Soviet Union.[66]

Despite its status as a component of the normalization landscape, STR was significantly impacted by the aftermath of the invasion. Among the ramifications were problems of the STR-oriented science of the aforementioned Tondl. This also had repercussions for Czechoslovak–Polish collaboration, as will be discussed below.

Bilateralism and Politics

The establishment of ÚFS ČSAV was a consequence of the reforms within the ČSAV that were introduced in 1968. These reforms resulted in the centralization of STR expertise within a single institution, while simultaneously enhancing the capacity for control. ÚFS ČSAV was established by uniting three existing institutions: the Institute of Sociology, the Institute of Philosophy, and the Center for the Theory and Methodology of Science (Kabinet pro teorii a metodologii vědy), which was led by Tondl. All of the aforementioned institutions had contacts with Poland, although to varying degrees. Most of them did not survive after 1968 because of personnel changes on both sides. In the spring of 1968, Leszek Kołakowski (1927–2009) and Bronisław Baczko (1924–2016), two prominent Polish philosophers, were unable to attend a significant STR conference in Mariánské Lázně due to the repression they faced following the Polish political crisis in March. It is noteworthy that Richta interceded with the presidium of ČSAV, requesting the provision of long-term scholarships for both of them. While the outcome of this appeal remains unknown, neither philosopher relocated to Prague; instead, Baczko and Kołakowski sought refuge in France and Canada respectively.[67]

The primary factor impeding collaboration within STR was related to Tondl, who had been an open supporter of the Prague Spring. His standing at the Academy diminished significantly after 1968, and he finally had to leave academia to work in Prague's Design Institute PUDIS. However, Tondl was allowed to work in ČSAV until at least 1973, and it appears that Richta was not inclined to terminate his employment. In an obituary written by Gasparski, it was noted that international pressure, including that exerted by Polish praxeologists, played a role in limiting the political consequences for the Czech philosopher.[68]

Before he was compelled to leave academia, Tondl had established an impressive network of connections on both sides of the Iron Curtain and was the Czechoslovak representative within the International Union of History and Philosophy of Science. Additionally, he held a prominent position at the UNESCO science policy division. While his primary expertise was in logic and semantics, during the 1960s he was engaged in the development of science policy and science planning, which were of significant political importance in Czechoslovakia and other socialist countries. In Richta's team he was also tasked with investigating changes to the scientific landscape, a subject that gained prominence in the 1970s, during which he alleged that the institute had misappropriated his work after his departure.

Tondl's scholarly interests aligned closely with the tenets of Polish science studies, and it is no surprise that Polish scholars were among his close international cooperation partners. In the late 1960s and early 1970s, he visited Warsaw and Kraków on a regular basis, approximately twice a year. During this period, he initiated collaboration with other academic institutions, including Nicolaus Copernicus University in Toruń, where the logician Tadeusz Czeżowski (1889–1981) was teaching. He was primarily in contact with scholars engaged in the study of science, but also with praxeologists, including the founder of the field, Tadeusz Kotarbiński (1886–1981). He was a member of the inaugural editorial board of the Warsaw-based journal *Problems of the Science of Science*, which served as a primary medium for COMECON scholars in this discipline. In 1969, despite the historically strained relations between Czechoslovakia and Poland, Tondl reported a friendly demeanor among his Polish colleagues upon his return from Warsaw.[69] Although Tondl's departure from ČSAV constrained the scope of international exchange, he continued to maintain connections with colleagues based abroad, who also visited him in Prague. He himself was able to travel abroad to a limited extent and to publish. For example, in 1977 he contributed to a volume celebrating Kotarbiński's ninetieth birthday, as one of a small number of foreign scholars invited to participate.[70] He also maintained contact with Gasparski, with whom he was friends, although the two men were not able to engage in structured cooperation.[71]

Tondl's institutional contacts were adopted by his close collaborator and friend, Berka.[72] Additionally, Berka assumed Tondl's role on the editorial boards of Polish scientific journals. Although there was a certain degree of continuity, Berka's concentration on logic and translations made his work more parallel to than part of STR research. Another member of ÚFS ČSAV, Müller, who worked on questions of scientific cadres, cooperated with the Institute of Science Policy and Higher Education (Instytut Polityki Naukowej i Szkolnictwa Wyższego; a joint institute of the Ministry of Science and PAN). Despite numerous visits and the exchange of data, no long-lasting cooperation developed.[73]

At the same time, a number of proposed collaborative initiatives within science-of-science-leaning STR failed, most likely due to lack of funding. A joint publication, which was to be completed in accordance with the interacademy agreement, appears not to have been even discussed. The *Small Encyclopaedia of Logics* (Mała encyklopedia logiki) by logician Witold Marciszewski (1930–) and the *Basis of Marxist Philosophy of Sciences* (Zarys marksistowskiej filozofii nauki) by Poznań philosopher Leszek Nowak (1943–2009) were considered for translation into Czech, but neither appeared in print.[74] Most Czech STR scholars' travels to Poland were contingent upon reciprocity and were primarily related to their involvement on editorial boards.

While Richta himself moved in the direction of the science of science in the second half of the 1970s, Polish scholars were not among his primary cooperation partners, despite his inclusion on the editorial board of the Warsaw-based journal *Science of Science.* His main cooperation partners were in Moscow and East Germany. The latter partnership was facilitated by the fact that one of the primary collaborative partners, Laitko, spoke Czech, while Richta was fluent in German. Moreover, epistemic commonality was a significant factor. The 1983 volume *Introduction to the Science of Science* (Osnovy Naukovedenija), included Ignacy Malecki (1912–2004) as a representative of the Polish side, one of few Polish scholars who espoused a Marxist analysis of science. Similarly, Kröber, a German representative in the volume, was prone to Marxist theory and also helped to promote Richta's ideas in German during the mid-1970s. He was also named as the editor of the German translation of *Man–Science–Technology*, and was probably instrumental in the book appearing at all, since the names of the editors of the original publication were removed for political reasons.[75]

Richta in Poland

Although Tondl was the most connected Czechoslovak STR scholar in Poland, Richta undoubtedly was the most prominent, particularly following the publication of *Civilization*. However, he was undoubtedly known prior to the Polish publication of the book, even among those outside the specialized audience. In 1970, a year before the Polish publication of *Civilization*, Richta delivered a speech on STR on Polish radio as part of an international university series.[76] The journal *Literary Monthly* (*Miesięcznik Literacki*), which normally printed excerpts of fiction, published selected passages from the Polish translation of *Civilization* in 1971.[77] As previously stated, Gomułka had a copy of the translation dated 1967 on his desk. It was likely sent in preparation for the 1968 conference in Mariánské Lázně, which, as it happened, the Polish delegation did not attend. Sending translated preprints was not an uncommon practice. Preparing significant conferences often involved sharing materials, often in translation, although primarily in the form of articles.[78]

Following the publication of *Civilization* in Polish in 1971, the popularity of STR and Richta increased significantly. The Polish version was one of the last socialist bloc translations to be published. It is reasonable to conclude that the reason for the delay of the Polish translation was political, given that both the book and Richta's personality were becoming a problem after the suppression of the Prague Spring. In his obituary of Richta, Zacher reported that it required two meetings of a high-level political body to approve the translation.[79] Regardless of the veracity of this claim, in socialist discourse, rumors often assumed a tangible form. By 1971, however, Richta was already actively involved in the networks of socialist directors of institutes of sociology and philosophy, and had been chosen to lead the socialist delegations for the 15th World Congress of Philosophy in Varna in 1973 and the 8th World Congress of Sociology in Toronto in 1974, which demonstrated that he remained a leading socialist scholar. The limited availability of printed copies of *Civilization* (3,000) and the relatively high price of the book (50 Polish zlotys), can be regarded as a strategy to restrict the book's circulation.[80]

The manner in which *Civilization* was presented in the foreword indicates that Richta's past had not been overlooked. Kazimierz Secomski (1910–2002), an eminent economist and specialist in planning and prognostics, emphasized that STR was a fundamental element of the emerging political lexicon, citing the perspectives of both Polish and Soviet politicians, and acknowledged the pioneering nature of the Czechoslovak research team's contributions. Not only did it raise plenty of "questions worthy of discussion," but also because of its interdisciplinary nature it could reframe old topics. However, the majority of the introduction was dedicated to highlighting the shortcomings of the book compared to "current scholarly work in Poland."[81] This pertained to the subjects of the position of the individual and the conditions conducive to its actualization. The statist perspective excluded COMECON as a factor influencing development and also neglected to address the "third world" or the lack of long-term perspectives. In short, according to Secomski, the book was a noteworthy contribution to a pertinent topic, but by the time of its publication, it had already been superseded, particularly by work conducted within the field of prognostics.[82]

Secomski's assessment of the book is not unusual, as similar arguments were made in other reviews. However, its placement at the beginning of the book is somewhat perplexing. It was normally the case that, within the socialist camp, introductions (or afterwords) were more oriented toward discussion of the authors and their theses. Polemics, such as that of Secomski, were typically reserved for texts that framed capitalist publications, often with the aim of demonstrating the superiority of socialist knowledge. However, in this instance the dividing factor was not so much ideology but more the nationality of the author. Indeed, this was not the only occasion on which the argument that Polish contribution was superior was employed in relation to books with Richta's involvement.[83]

Civilization was reviewed in numerous academic journals and periodicals, the majority of which reviewed it positively. Recurring points of criticism echoed Secomski, occasionally addressing unfortunate translation choices.[84] It is noteworthy that *Civilization* was reviewed in journals from a range of academic disciplines, reflecting its in-between status. The author who provided the most comprehensive analysis of the book, across various media, from the scholarly *Zagadnienia Naukoznawstwa* (Problems of the science of science) to the official publication of the Central Committee of the Polish United Workers' Party, *Nowe Drogi* (New ways),[85] was none other than Zacher.

Zacher was twenty-nine years old when the translation of *Civilization* was published. He had recently completed his doctoral studies in economics and commenced employment at the IHN PAN, where he worked alongside Olszewski, who directed the STR research program. His reviews were not only extremely positive but also described *Civilization* as *wybitny* (eminent), a word that was rarely used in reviews at the time. Although he identified flaws in the book—chief among them the lack of consideration of the third world, a topic that was of interest to him at the time—there was no doubt in his mind that it represented a new quality of research.[86] However, he did not stop at reviews. In the issue that followed his review of *Civilization* in *Zagadnienia Naukoznawstwa,* Zacher published an overview of STR research, once again emphasizing the originality of the book and arguing that only large, multidisciplinary teams could effectively address the complexity of research.[87] A year later, he contributed a positive review of Richta and his Soviet colleagues' volume *Man–Science–Technology,* also calling for the book to be translated into Polish.[88] This review was published in *Trybuna Ludu* (People's tribune), the main Polish daily newspaper, which illustrates the integration of Czechoslovak STR into the ideological mainstream.

Zacher also served as an envoy for IHN PAN, facilitating the establishment of more enduring collaborative relationships. His contributions were certainly appreciated by his Czech colleagues. For instance, during his visit to the ÚFS ČSAV in July 1972, he was invited to deliver several presentations and engaged in discussions regarding the future of the collaboration. The institute arranged for him to attend a concert and a theater performance, and facilitated a visit to Karlštejn Castle (probably a half-day excursion) and a tour of Prague. All these activities took place in a period of three and a half days during his stay in Czechoslovakia.[89]

It should be noted that the institution that represented PAN in cooperation with ÚFS ČSAV in STR was neither IHN PAN nor IFiS PAN (which also dealt with prognostics),[90] but rather the Committee of the Science of Science. It was a unit directly subordinated to a presidium and constituted a forum for eminent scholars in the field of science of science, with functions analogous to those of an advisory body. However, it did not constitute a separate entity with its own budget or research mandate. A contract was signed in 1971 for a period of five years and included a number of specific provisions, such as personal exchange, two joint workshops, the submission of articles to each other's journals, and the preparation of a joint volume (which never materialized).[91] This dispersion of STR on the Polish side resulted in a lack of clarity regarding the respective areas of expertise and responsibilities. In the early years, IHN PAN demonstrated the greatest intellectual engagement. For instance, in 1973, its members held three separate meetings to discuss *Man–Science–Technology,* with each chapter referred to by one expert. Additionally, IHN PAN held a special meeting to discuss Olszewski's and Zacher's responses to the book, which was to be subsequently presented in a special panel at the 15th World Congress of Philosophy in Varna in 1973.[92]

Polish engagement in STR was not a one-way interest; Richta visited Warsaw on a number of occasions. In June 1972, he and the vice-director of ÚFS ČSAV and member of Richta's team Jindřich Filipec presented and discussed *Civilization* at a meeting of IHN PAN, after being introduced by Zacher.[93] Richta also visited IFIS PAN and participated in an interview for *Człowiek i światopogląd*.[94] In 1972 he was keen on the idea of a dictionary or encyclopedia of the science of science. Although there were some intermittent updates during that year, the project appears to have remained in its conceptual phase.[95] In the following year, he and Filipec visited IFIS PAN and formalized their collaboration through a cooperation agreement.[96] *Człowiek i światopogląd* published a special issue on STR, with Richta's opening article (and another Czech contribution included). However, the majority of these interactions were of an official nature and lacked the depth and dynamism that the Polish side may have desired.

The disharmony between Richta and PAN is exemplified by a grievance filed by the latter in 1974 at the assembly of international officers of socialist academies of sciences. The complaint addressed problems in "STR research, philosophy and sociology," which were under Richta's purview. The issues pertained to the precise planning of joint research and the failure to satisfy the exchange quotas.[97] In his response, Richta provided a detailed account of the joint endeavors undertaken by the two parties, while also acknowledging a slight imbalance between the expectations of Polish colleagues and the capabilities of ÚFS ČSAV.[98] Nevertheless, records from the subsequent years demonstrate a marked increase in travel, although it was predominantly accounted for by younger scholars. Even so, the level of cooperation in STR remained relatively limited, with the majority of the quota taken up by the science of science (Berka) and prognostics (Šulc).

Although STR never attained the same degree of influence in Poland as it did in Czechoslovakia, the prerequisites for fruitful collaboration were present. Despite initial challenges, in the early 1970s Richta's contributions were acknowledged and appreciated by his Polish colleagues and STR was becoming a household concept. This raises the question of what went wrong. A number of reasons can be identified, some of which were structural in nature, such as the challenge of accurately allocating corresponding research units between ČSAV and PAN. Others were more social in nature, such as the apparent lack of a particularly explicit interest in cooperating with scholars from PAN on Richta's part. Epistemic disparities also played a role, as the generalizing approach of STR stood in contrast to the growing trend of specialization in the Polish humanities. It is also possible, however, that this was a combination of social and epistemic factors. Richta seldom cited Polish research, instead favoring works from the Soviet Union, East Germany, Hungary, and Bulgaria, as well as those from Western countries. Nevertheless, given that Zacher presented himself as an astute admirer of Richta and that there was pressure for STR cooperation, a collaboration between the two of them would have been the more likely outcome. The next part will examine the reasons why such an outcome did not materialize and situate it within the broader political context of the post–Prague Spring Eastern Bloc.

Mistrust

The failure of cooperation with Richta, which provides the context for this chapter, was repeatedly described by Zacher. His students also claimed he mentioned it on several occasions.[99] Zacher recalled the following incident:

> the Czech–Russian book [*Man–Science–Technology*] was criticised by us during the 1973 Varna Congress . . . and such a review was published then in *Studia Historyczne* (Study of History), Polish leading journal; I paid the price for that for three years I could not go abroad from my Academy's Institute. This was rather political and personal restriction imposed by the Czech and Russian "colleagues" (or comrades). However, nobody of the Academy's scientific leaders did blame my theoretical standpoint. Of course, I was never included into any cooperative research programs in spite of the friendly personal relations with Richta. I was treated as enfant terrible, not in opposition, but inconvenient for the opportunists and "brotherly" relations within COMECON's science. It was an evident victory of *homo sovieticus* mentality of my elder colleagues.[100]

The quotation above is, of course, a personal reminiscence, which cannot be fully trusted. For instance, memory may omit structural factors that may have been known to the respective actors but were not deemed relevant to convey.[101] In any case, there was no bilateral cooperation between Zacher and Richta in the aforementioned years. Zacher was not registered as an official guest in Prague for several years after Varna. However, it might not have been Zacher who was sanctioned, but Richta, whose position as lead STR scholar was undermined in Varna. Alternatively, it could also have been Richta's idea not to cooperate with Zacher. As director of the institute, Richta was expected to collaborate with other prominent scholars, whereas Zacher was still in the early stages of his career. However, this would not have excluded cooperation with Richta's team, but, again, it did not happen. The lack of available sources makes it difficult to test such a hypothesis, several of which were discarded while researching this chapter.

A review of the archival materials reveals no evidence of any tensions concerning the Varna commentaries by Zacher and Olszewski. Furthermore, Richta did not protest in writing, either to his Polish counterparts at the Academy or to his Soviet collaborators, at least, there is no a paper trail of any such complaint. However, the *Man–Science–Technology* project turned out to be a highly emotional undertaking for Richta and his coeditor Kedrov, which may have led to a reluctance to engage with scholars who had made critical comments. Moreover, the controversy surrounding the volume may have deterred Polish scholars or administrators from pursuing collaboration with Richta, who was Zacher's preferred partner. A brief overview of the history of the book will aid understanding of emotions around it.

Man–Science–Technology constituted one of three socialist contributions to the 15th World Congress of Philosophy in Varna in 1973. This conference was regarded as a crucial event in the new strategy of socialist internationalism.[102] The text was prepared by key institutes in STR, and edited by its directors—Pavel Kopnin (1922–1971) (Institute of Philosophy of the Academy of Sciences of the Soviet Union), Kedrov (Institute for the History of Science and Technology), and Richta. Correspondingly, the preparation of the book, which appeared in Czech, English, and Russian for the congress, was well funded. The team convened regularly to discuss the various chapters, and the entire manuscript was subjected to rigorous debate, as was the case at the meetings held in Warsaw. However, shortly before the book's publication, pressure from the Science and Education Department of the Central Committee of the Communist Party of the Soviet Union obliged Kedrov to remove Richta's name from the list of editors and delete all references to his works and those of other scholars involved in the Prague Spring (the latter was done only in the Russian version, as the other versions had already been printed). In solidarity Kedrov decided that the institutes should be named as editors. While Richta protested and also wrote to his political contacts about having his name reinstated in the following editions, his pleas fell on deaf ears. Additionally, Richta's name was removed from another book with Kedrov in Russian at the same time. While both deletions might have been a consequence of a campaign targeting Kedrov in Moscow, who faced accusations of revisionism and was dismissed from his positions, Richta's credibility was also damaged. Even though the book received favorable reviews and was translated into Polish (in addition to German, Hungarian, and Japanese), it was translated from the Russian "original," from which Richta had been airbrushed.[103]

The removal of Richta's name from the cover of a book that had been prepared publicly was a significant act, and it is highly probable that Polish scholars were aware of it. As a director of an institute of the Academy of

Sciences of an allied country, Richta could not be ignored. However, more extensive cooperation could have been perceived as problematic, particularly in relation to the fall from grace of Richta's Moscow allies. The criticism of Richta at the meeting of representatives of the international officers of socialist academies of sciences may be viewed in this way. Although not explicitly personal, it explicitly targeted Richta. Moreover, such a high-profile meeting was an unusual venue for such a discussion, given that institute directors communicated frequently and convened in person at least once a year. The intensification of cooperation that resulted from this incident occurred without Richta's personal involvement. It is also noteworthy that no representative of PAN attended Richta's funeral in July 1983 and only two Polish historians of science who had recently collaborated with him on *Introduction to the Science of Science* sent their condolences.[104] However, the funeral occurred only shortly after the lifting of martial law in Poland, and thus opportunities to travel may have been restricted.

After his initial success after the publication of *Civilization*, Richta was scarcely present in Polish-language media. His articles were featured in *Otázky míru a socialismu* and via the journal's cover-to-cover translations they also reached Polish audiences.[105] However, these were infrequent instances of his presence in the Polish media. In 1979, he was featured in one article each in the Polish English-language journals *Problems of Science of Science* and *Dialectic and Humanism*. While several other articles by Richta were translated between Czech, English, German, and Russian, none were translated into Polish.[106] It is evident that the initial interest of Polish scholars in Richta and STR had diminished. Nevertheless, prominent figures such as Zacher and Jaroszewski continued to cite Richta in their works.

It is also noteworthy that this occurred at the same time as the shift in Richta's writings toward a more conformist and ideological stance. While *Civilization* can be seen as a liberal voice reflective of the Prague Spring atmosphere, Richta's subsequent writings can be considered representative of the process of normalization. To a significant extent, they were polemical works that engaged in ideological debates and offered critiques of Western developments in science, aligning with the orthodoxy of the Communist Party. In retrospect, Zacher identified these works as "politically correct, full of ideological declarations and quotations of 'classics.' They marked the decline of Richta's intellectual power and his growing ideological subordination."[107] During a period when Polish philosophy and the science of science were increasingly exposed to international influences from both sides of the Iron Curtain, Richta's embrace of Soviet ideology diminished his appeal as a collaborative partner.

The likely cooperation between Zacher and Richta did not happen, probably for political reasons. Nevertheless, it is not clear whether Zacher's and

Olszewski's criticism of Richta and Kedrov and the subsequent prohibition on Zacher's travel were the primary cause of this outcome or whether it was merely a coincidence that occurred at the same time as the increased skepticism of Polish academics toward Richta following indications from Moscow that he should not be entirely trusted. It is also possible that other personal factors were involved.

Conclusions

This chapter utilized the lack of successful or even non-existent cooperation as a point of departure to discuss the entanglements between Czechoslovakia and Poland in the social sciences of the late 1960s and 1970s. It aimed to uncover structural and personal factors that are not typically discussed in historiography, drawing on narratives laid out in archival materials. The research for this text was intentionally "messy" and "disorganized," pursuing leads that were initially obscured. Much of the research conducted is not visible in this text, as it proved to be unproductive. This included a multitude of resources consulted for indications of helpful or hindering factors in the context of STR cooperation, in alignment with the historical understanding of how socialist institutions functioned. The process of researching a "failure" proved to be more time-consuming and methodologically challenging than that of researching a "success." However, it allowed for the posing of questions that were initially unforeseen.

This chapter posits that, even in a highly structured field of intersocialist cooperation, personal connections play a pivotal role, outweighing the formal requirements. It is crucial to note that these personal connections do not readily translate into successful institutional cooperation when the individual who facilitated them is no longer available, as evidenced by the Tondl case. While Tondl was responsible for establishing connections and ensuring their continuity through the appointment of Berka, the latter lacked the capacity to maintain and nurture personal relationships. The case of Richta illustrates the importance of person-based trust in intersocialist cooperation. Without a trustworthy individual to facilitate collaboration, the potential for successful institutional cooperation was significantly diminished. In Richta's case, it was unclear whether it was he or Zacher who lacked trustworthiness. However, it is evident that both individuals were needed to lay the foundations for STR cooperation. Richta's prominence and Zacher's enthusiasm for the new field and Czechoslovak STR achievements made them crucial figures in this endeavor, even though it did not bring long-lasting results, at least, not those that Zacher anticipated. Once it was not possible to establish the requisite basis for cooperation—and in this particular context, it was necessary to include both individuals, given their shared epistemic and theoretical affinities—the process stalled.

A second point of significance is the geographical context. The Eastern Bloc constituted a space in which a variety of hierarchies were at play, sometimes in conflict and sometimes intersecting. The most crucial was the existence and function of the Soviet Union as a hegemonic power. Richta's apparent lack of interest in collaborating with Polish scholars may have been motivated by his focus on engaging with those in positions of power, namely Kedrov and Mikulinskij. Moreover, Polish skepticism toward Richta may have been linked to Moscow's mistrust of Richta. Zacher suggested that the prohibition on international collaboration placed on him was associated with Soviet scholars. The fact that he maintained cordial relations with Richta suggests that it was likely the other editors of *Man–Science–Technology* who were responsible. It is worth noting that STR may be an anomalous case among socialist disciplines, as the scholars involved were predominantly pro-Soviet and more conformist, and, on account of its connections to power, STR was subjected to greater scrutiny than other disciplines.

However, the relationship between Czechoslovak and Polish scholars was also not entirely symmetrical. In the context of COMECON, which espoused the ideal of intellectual camaraderie, Polish scholars frequently likened Richta's approach to that of Polish scholarship, which they regarded as inherently superior. In this context, the long-standing superiority complex of Polish scholars toward their Czechoslovak counterparts was not the only factor at play. The Polish intellectual focus, which was more oriented toward non-socialist science, was also a factor. This geographical distribution of scholarly attention was also highly disciplinary and field-specific, or even personal. Furthermore, it is likely that this distribution was not stable but rather strengthened because of the mistrust toward Richta that emerged after 1971.

Finally, this chapter argues that even bilateral cooperation—or the failure to implement it —was located in translocal networks, which have a strong influence on it. In this particular case, the presence of the Soviet Union and the role of COMECON as an intermediary agent were crucial. The lack of trust expressed toward Richta in Moscow, most likely reduced the willingness of Polish scholars to invest in deeper and official collaboration with him in the early 1970s. At the same time, because of role of STR in COMECON research, some forms of collaboration between both academies persisted in this field, although they largely exhibited traits of official exchanges to satisfy the prescribed quotas, especially after the issue raised at the highest forum of socialist academies by representatives of PAN.

The interplay of the different scales—from interpersonal to socialist (and potentially global, which was not explored here)—underscores the importance of seeing the history of bilateral connections as a history of multiple entanglements. These are structured by both formal and informal asymmetries, which should be taken into account and carefully weighed against each other. As this chapter has argued, simple ideas like the Sovietization thesis do not account for the complexity of socialist internationalism, which is equally valid for all composite units (blocs, unions, and more).

Author Biography

Jan Surman is a historian specializing in the intellectual and scientific history of Central and Eastern Europe. He earned his PhD in History at the University of Vienna and has held fellowships in Vienna, Moscow, Princeton, and Marburg. His research explores the intersections of science, language, and politics. He has published widely on multilingualism in academia, the history of positivism, and knowledge circulation, and serves on editorial boards of several international journals in the history of science.

List of Abbreviations

AAN	Archiwum Akt Nowych (Central Archives of Modern Records)
COMECON	Council on Mutual Economic Assistance
CPCz	Communist Party of Czechoslovakia
IHN PAN	Institut Historii Nauki Polskiej Akademii Nauk (Institute for the History of Science of the Polish Academy of Sciences)
IFiS PAN	Instytut Filozofii i Socjologii Polskiej Akademii Nauk (Institute of Philosophy and Sociology of the Polish Academy of Sciences)
GDR	German Democratic Republic
MÚA AV ČR	Masarykův ústav a Archiv Akademie věd České republiky (Masaryk Institute and Archives of the Czech Academy of Sciences)
PAN	Polska Akademia Nauk (Polish Academy of Sciences)
STR	scientific–technological revolution
ÚFS ČSAV	Ústav pro filosofii a sociologii Československé akademie věd (Institute for Philosophy and Sociology of the Czechoslovak Academy of Sciences)
UNESCO	United Nations Educational, Scientific and Cultural Organization

Notes

1. Palonka, "Rewolucja naukowo-techniczna." Unless otherwise stated, all translations by the author.
2. Bloor, *Knowledge and Social Imagery*.
3. Secord, "Knowledge in Transit," 655.
4. Proctor, "Agnotology," 6–8.
5. Burke, *Ignorance*.
6. Gajski, "Privatization of Science."
7. Péteri, "Nylon Curtain"; Nickelsen and Krämer, "Introduction."
8. For example, Pleskot, "Jak wyjechać na Zachód?"; Nowaczewska, "Penetrating the Iron Curtain."
9. Wang, "Science, Security, and the Cold War"; Krige, "Regulating International Knowledge Exchange"; Daniels, "Controlling Knowledge, Controlling People"; Wheeler, *Letters from Prague*.
10. Hollings, *Scientific Communication*, 89. The latter example can be found in Montgomery, *Does Science Need a Global Language ?*, 109.
11. Zacher, "The Unrealized Vision of Future," 101; see also Elzinga, "The Rise and Demise of the International Council," 290-1.
12. Kilias, "Socjologia polska i czeska," 295.
13. Olšáková, "Expert Cultures and Infrastructural Globalism."
14. Surman, "Internationalism, Normalisation."
15. Banerjee, "Electric Origins."
16. Calic, Neutatz, and Obertreis, "The Crisis of Socialist Modernity," 9–14; Kangaspuro and Smith, *Modernisation in Russia*; Musiał, *Modernizacja Polski*; Spurný, *Making the Most of Tomorrow*; Kopeček, *Architekti dlouhé změny*; Sommer, "Towards the Expert Governance"; Polák, "'We Want Light!'"; Sommer, Spurný, and Mrňka, *Řídit socialismus jako firmu*.
17. Dybiec, *Nie tylko szablą*; Míšková et al., *Bohemia Docta*.
18. Laitko, "Der lange Weg zum Kröber-Institut," 37; Josephson, "Science and Ideology," 180–82.
19. Kosel, *Unternehmen Wissenschaft*, 93–156.
20. Laitko, "Der lange Weg zum Kröber-Institut," 34; Josephson, "Science and Ideology," 177.
21. Hermann et al., "Dokumenty o rozvoji věd v Československu," 11–98; Olšáková, *Věda jde k lidu!*.
22. Kosel, *Unternehmen Wissenschaft*, 259.
23. Křížek, "K otázce 'duševní výroby.'"
24. Teich, "J. D. Bernal the Historian."
25. Steiner, *The Social Function of Science*.
26. "Beseda tov. N. S. Khrushcheva." This interview was translated into other languages of the Eastern Bloc.
27. Rindzevičiūtė, *The Power of Systems*, 27–33; Laitko, "Der lange Weg zum Kröber-Institut," 37, 68–69.
28. Hoffmann, "Soviet Views,'" 628.
29. Zacher, "Rewolucja naukowo-techniczna niektóre kontrowersje metodologiczne."

30. Interviews with Dieter Hoffmann, 16 December 2022, conducted by Jan Surman, stored at MÚA AV ČR; Interview with Leszek Zasztowt, 23 June 2022, conducted by Jan Surman, stored at MÚA AV ČR. Retrieved 1 February 2026 from https://entanglements.ihpan.edu.pl/interview/historia-nauki-miala-tu-dosyc-mocna-pozycje-wywiad-z-historykiem-leszkiem-zasztowtem/.
31. Andersson, *The Future of the World*; Sommer, "Vom sozialistischen Postindustrialismus zur Marktgesellschaft," 55–81; Rindzevičiūtė, "A Struggle for the Soviet Future," 52–76. See also the chapter by Luboš Studený and Lukas Becht in this volume.
32. Kosiński et al., *Słownik podstawowych terminów prognostycznych*. This multilingual dictionary was one of the results of a COMECON project led by the Center for Prognostics of the Wrocław Politechnics.
33. Richta, *Civilization at the Crossroads*.
34. Zacher, "The Unrealized Vision of Future," for example, 106. On Richta generally, see Sommer, "Are We Still Behaving as Revolutionaries?" 93–110; Sommer, "Towards the Expert Governance"; Smíšek, "Chováme se jako revolucionáři?"
35. Owoc, *Automatyzacja i jej skutki społeczne w kapitalizmie*.
36. Kowalik, "[Rec:] Radovan Richta i zespół," 615.
37. Nešpor, "Jan Szczepański," 52–61. See also the chapter by Kornelia Kończal in this volume.
38. Jaroszewski, "Perspektywy człowieka," 38–80; Jaroszewski, *Osobnost a společnost*; Jaroszewski, *Aktivnost' lichnosti*.
39. Łotysz, "Eugeniusz Olszewski," 42.
40. For an assessment of the overlap between Czechoslovak STR and Polish praxeology, see Dzida, "[rec:] F. Kutta, M. Soukup i zespoł."
41. Kokowski, "The Science of Science (Naukoznawstwo) in Poland."
42. Šmidrkalová, "Czechoslovak Praxeology," 394.
43. Mikulinskij and Richta, *Socialism and Science*. Czech version appeared 1982. The second book was published first in Russian (Mikulinskyj, *Osnovy Naukovedenija*.), then Czech (Kačaunov, *Základy teorie vědy*) and German (Bradinov and Kröber, *Grundlagen der Wissenschaftsforschung*.).
44. MÚA AV ČR, Coll. ÚFS-VTR, box 2, inv.no. 35, sig. 62, Hodnocení dvoustranných smluv a perspektiva jejich využití při zabezpečení 6. 5LP. [Evaluation of bilateral contracts and prospects for their use in securing the 6th 5LP] (1971?).
45. Doel, "Internationalism in Science after 1940," 60–74; Lalli, "Crafting Europe from CERN"; Turchetti, "From the National Laboratory to International Collaborations."
46. Franc, *Úderná skupina?*; Connelly, *Captive University*; Szobi and Stellner, "Sovětizace českého ekonomického vysokého školství."
47. Szumski, *Polityka a historia*; Avramchuk, *A History of the Fulbright Program*.
48. Romano and Romero, *European Socialist Regimes' Fateful Engagement*; Komornicka, *Poland and European East-West Cooperation*; McDermott and Stibbe, *Czechoslovakia and Eastern Europe*.
49. Infeld, "O kontaktach naukowych z zagranicą"; Kilias, *Goście ze wschodu*.
50. Surman, "Internationalism." For the official version of internationalism, see the speech by the head of the ČSAV, Jaroslav Kožešník in Kožešník, "Současné otázky naší vědy a její výhledy."

51. 1u soveshchanie predstavitelej akademij nauk socialiticheskich stran. Rabocha stenograma, 1–7 December 1965, Archive of the Polish Academy of Sciences, W.264.40, j. 204.
52. MÚA AV ČR, Coll. ÚFS-VTR, box 2, inv.no. 35, sig. 62, MÚA AV ČR, Coll. ÚFS-sociologická sekce, sig. 62, Dolgosrochnaya programma mnogostronnego sotrudnichestva sotsialistycheskych stran v oblasti obshchestvennych nauk i proekty ee realizatsii na 1981–1985 gg [The long-term program of multilateral cooperation of socialist countries within social sciences and the projects of its realisation for the years 1981–1985].
53. MÚA AV ČR, Coll. ÚFS-VTR – sekretariát ředitele, box 6, sig. 655, Tématický plán vědecké spolupráce mezi Československou akademií věd a Polskou akademií věd na léta 1971–1975 [Thematic Plan of Scientific Cooperation between the Czechoslovak Academy of Sciences and the Polish Academy of Sciences for 1971–1975], 7 April 1971.
54. MÚA AV ČR, Coll. ÚFS-VTR – sekretariát ředitele 5f, box 6, sig. 655, e.g. Příloha 2: Informace [Attachment 2: Information)], 29 May 1974.
55. See the materials stored in MÚA AV ČR, Ústav pro filosofii a sociologii, and Archives of the Institute of Philosophy and Sociology of the Polish Academy of Sciences, Warsaw. I am thankful to Oleksandr Pestrykov and Izabela Kwiatkowska for their help with accessing the latter materials.
56. Mitrokhin, "O B.M. Kedrove"; Laitko, "Der lange Weg zum Kröber-Institut," 89.
57. Even spaces designed to be apolitical, such as private meetings in socialist kitchens, depended on politics, see Gerovitch, "The Kitchen and the Dacha."
58. Hoppe and Sommer, "How the 'Richta Team' Was Born," 516.
59. Richta was responsible for long-term planning, see Pelikán, *Panzer überrollen den Parteitag*, 139, 187.
60. See the chapter by Luboš Studený and Lukas Becht in this volume.
61. MÚA AV ČR, Coll. Presidium ČSAV, Koncepce práce Ústavu pro filosofii a sociologii [Concept of the Institute for philosophy and sociology], discussed at the meeting of the presidium on 25 February 1970.
62. See the correspondence in MÚA AV ČR, Coll. ÚFS-VTR—sekretariát ředitele 6, box 6, sig. 73.
63. On Ulbricht, see Steltner, "Als Ulbricht 1967 Intertext bemühte." On Gomułka, "Cywilizacja na rozdrożu. Konsekwencje rewolucji naukowo-technicznej dla społeczeństw i człowiek" – opracowanie Radovana Richty z zespołem na Konferencję międzynarodową w Mariańskich Łaźniach, AAN, Warsaw, sign. 2/1354/0/2.1.12.3/XIA/330. I thank Lukas Becht for this information.
64. Zacher, "[Rec:] Cywilizacja na rozdrożu."
65. Míšková, Barvíková, and Šmidák, *Československá akademia věd*, 140–41.
66. Surman, "Internationalism."
67. MÚA AV ČR, Coll. Vědecké kolegium filosofie (a sociologie) ČSAV 1963-9, Radovan Richta to Jaroslav Pluhař, 12 April 1968.
68. Gasparski, "Ladislav Tondl," 429.
69. MÚA AV ČR, Coll. ÚFS-VTR, box 3, inv.no. 655, sig. 47, Ladislav Tondl: Zpráva o návštěvě a jednáni pracoviště Polské akademie věd ve Varšavě ve dnech 1.-5. prosince 1969 [Ladislav Tondl: Report on the visit and meeting of the Polish Academy of Sciences in Warsaw, 1–5 December 1969] 18 December 1969, esp. page 4.

70. Tondl, "O ocenie graficznych form danych w projektowaniu."
71. Interview with Wojciech Gasparski, 4 February 2022, conducted by Jan Surman, stored at MÚA AV ČR. Retrieved 1 February 2026 from https://entanglements.ihpan.edu.pl/interview/kontakty-z-czeskimi-kolegami-byly-intensywne-wywiad-z-prakseologiem-wojciechem-gasparskim/.
72. Tondl, *Racionální činnost a svět artefaktů*.
73. MÚA AV ČR, Coll. ÚFS-VTR – sekretariát ředitele 62, correspondence between Janusz Radziejowski and Karel Müller.
74. MÚA AV ČR, Coll. ÚFS-VTR, box 3, inv.no. 655, sig. 44, letter of Witold Marciszewski to Karel Berka, Warsaw 22. June 1974; letter of Karel Berka to Leszek Nowak, Prague 30. September 1974.
75. (see below)
76. "Radio."
77. Richta, "Cywilizacja na rozdrożu," 98–102.
78. Czechoslovak contributions were circulated in Czech as a series "Konference 'Člověk a společnost ve vědeckotechnické revoluci.'"
79. Zacher, "Wizja nowej cywilizacji," 11.
80. On the price, see "Warto."
81. Secomski, "Przedmowa do wydania polskiego," 9.
82. Secomski, "Przedmowa do wydania polskiego," 12.
83. Białoń, "On the Book"; Pawlak, "[Rec:] Osnovy naukowiedenija [...]."
84. Kowalik, "[Rec:] Radovan Richta i zespół: Cywilizacja na rozdrożu."
85. Zacher, "[Rec:] Cywilizacja na rozdrożu," *Zagadnienia naukoznawstwa*; Zacher, "U progu rewolucji naukowo-technicznej."
86. Zacher, "[Rec:] Cywilizacja na rozdrożu," *Zagadnienia naukoznawstwa*; Zacher, "[Rec:] Cywilizacja na rozdrożu," *Humanizm Pracy*. The first review was reprinted in IFIS PAN's journal, *Człowiek i światopogląd*.
87. Zacher, "Rewolucja naukowo-techniczna." Similarly, in his talks in seminars of IHN PAN, he also referred to Richta, as he himself stressed in printed reports: Zacher, "Zebrania Zespołu Badań," (1972).
88. Zacher, "Z Pozycji Marksizmu."
89. MÚA AV ČR, Coll. ÚFS-VTR, box 3, inv.no. 655, sig. 44, Záznam o návštěvě zahraničního hosta dr. L. Zachera ze Zakładu Historii Nauki i Techniki Polské akademie věd v Ústavu pro filosofii a sociologii ČSAV ve dnech 6.–9. července 1972 [Record of the visit of a foreign guest, Dr. L. Zacher from Zakład Historii Nauki i Techniki of the Polish Academy of Sciences to the Institute for Philosophy and Sociology of the Czechoslovak Academy of Sciences on 6–9 July 1972]; APAN, coll. BWZ, sign. II-2.6, inv. no. W. 472.77, travel documents Dr. Lech Zacher 1972.
90. MÚA AV ČR, Coll. ÚFS-VTR – sekretariát ředitele 5f, box 6, sig. 655, Návrh pracovního plánu spolupráce mezi Ústavem filosofie a sociologie Polské akademie věd a Ústavem pro filosofii a sociologii ČSAV a léta 1972–1975 [A draft work plan for cooperation between the Institute of Philosophy and Sociology of the Polish Academy of Sciences and the Institute for Philosophy and Sociology of the ČSAV for the years 1972–1975], June 1972.
91. MÚA AV ČR, Coll. ÚFS-VTR, box 3, inv.no. 655, sig. 44, Zpráva o návštěvě zástupců Komitétu pro vědu o vědě Polské akademie věd v Ústavu pro filosofii a sociologii ČSAV [Report on the visit of representatives of the Committee for Science

of the Polish Academy of Sciences to the Institute of Philosophy and Sociology of the Czechoslovak Academy of Sciences] (14 June 1971); MÚA AV ČR, Coll. ÚFS-VTR, box 3, inv.no. 655, sig. 44, Dohoda o spolupráci mezi Československou akademií věd a Polskou akademií věd v oblasti věd o vědě v letech 1971–1975 (návrh) [Agreement on cooperation between the Czechoslovak Academy of Sciences and the Polish Academy of Sciences in the field of science in 1971–1975 (draft)] (1971?).

92. Hadyniak and Zacher, "Zebranie Zespołu Badań"; Zacher, "Zebrania Zespołu Badań."(1973)
93. Zacher, "Zebrania Zespołu Badań."
94. MÚA AV ČR, Coll. ÚFS-VTR – sekretariát ředitele 5f, box 6, sig. 655, [Radovan Richta, Jindřich Filipec:] Zpráva ze služební cesty do Polska [Report from a business travel to Poland], 11. July 1972.
95. MÚA AV ČR, Coll. ÚFS-VTR, box 3, inv.no. 655, sig. 44, letter of Radovan Richta to Zdzisław Kowalewski, 13 January 1972; MÚA AV ČR, Coll. ÚFS-VTR, box 3, inv.no. 655, sig. 44, letter of Zdzisław Kowalewski to Radovan Richta and the answer of Zdzisław Kowalewski with handwritten note by Richta, undated. Kowalewski, "Sprawozdanie z działalności Zakładu Historii," 631.
96. APAN, coll. BWZ, sign. II-2, inv. no. W. 472, fold. 76, t.1 Przyjazdy 1973 A – G [vol. 1. Incoming A-G], Filipec Jindřich; MÚA AV ČR, Coll. ÚFS-VTR – sekretariát ředitele 5f, box 6, sig. 655, [Radovan Richta, Jindřich Filipec:] Zpráva ze služební cesty do Polska [Report from a business trip to Poland], 11 July 1972.
97. MÚA AV ČR, Coll. ÚFS-VTR – sekretariát ředitele 5f, box 6, sig. 655, Příloha 2: Informace [Attachment 2: Information], 29 May 1974.
98. MÚA AV ČR, Coll. ÚFS-VTR – sekretariát ředitele 5f, box 6, sig. 655, letter of Radovan Richta to J. Koželník from 29 May 1974.
99. I am thankful to Alina Betlej for this information.
100. Zacher, "The Unrealized Vision of Future," 100.
101. Surman and Pavlíček, "The Formal and the Informal."
102. A longer version of this story can be found in Surman, "Internationalism."
103. *Człowiek—Nauka—Technika.*
104. *Památce Radovana Richty (1924–1983).*
105. From 1974 to 1981 at least four of his articles were published.
106. See the bibliography in *Památce Radovana Richty (1924–1983)*; information was verified with the Polish Bibliography of Periodical's Contents (Bibliografia Zawartości Czasopism).
107. Zacher, "The Unrealized Vision of Future," 100.

Bibliography

Andersson, Jenny. 2018. *The Future of the World: Futurology, Futurists, and the Struggle for the Post-Cold War Imagination.* Oxford: Oxford University Press.

Avramchuk, Oleksandr. 2026. *A History of the Fulbright Program in Poland, 1945-2025: To Build a Republic of Spirit.* Manchester: Manchester University Press.

Banerjee, Anindita. 2011. "Electric Origins: From Modernist Myth to Bolshevik Utopia." In *L'ère Électrique: The Electric Age*, edited by Olivier Asselin, 289–304. Ottawa: University of Ottawa Press.

"Beseda tov. N. S. Khrushcheva s angliiskim uchenym i obshchestvennym deiatelem Dzhonom Bernalom" [Conversation between Comrade N. S. Khrushchev and the English scientist and public figure John Bernal]. 1954. *Znamia kommunizma*, 26 December 1954, p. 1.

Białoń, Lidia. 1987. "On the Book *Osnovy naukovedeniya*, Nauka, Moskva 1985, 421 pp." *Science of Science* 7, no. 1–2: 175–80.

Bloor, David. 1991. *Knowledge and Social Imagery*, 2nd ed. Chicago: University of Chicago Press.

Bradinov, Boris, and Günter Kröber, eds. 1988. *Grundlagen der Wissenschaftsforschung* [Fundamentals of science studies]. Berlin: Akademie Verlag.

Burke, Peter. 2023. *Ignorance: A Global History*. New Haven, CT: Yale University Press.

Calic, Marie-Janine, Dietmar Neutatz, and Julia Obertreis. 2011. "The Crisis of Socialist Modernity: The Soviet Union and Yugoslavia in the 1970s. Introduction." In *The Crisis of Socialist Modernity: The Soviet Union and Yugoslavia in the 1970s*, edited by Marie-Janine Calic, Dietmar Neutatz, and Julia Obertreis, 7–27. Göttingen: Vandenhoeck & Ruprecht.

Connelly, John. 2000. *Captive University: The Sovietization of East German, Czech and Polish Higher Education, 1945–1956*. Chapel Hill: University of North Carolina Press.

Człowiek—Nauka—Technika: Próba marksistowskiej analizy rewolucji naukowo-technicznej [Man—Science—Technology: An Attempt at a Marxist Analysis of the Scientific and Technological Revolution]. 1976. Warsaw: Książka i Wiedza.

Daniels, Mario. 2019. "Controlling Knowledge, Controlling People: Travel Restrictions of U.S. Scientists and National Security." *Diplomatic History* 43, no. 1 (1 January): 57–82. https://doi.org/10.1093/dh/dhy068.

Doel, Ronald E. 2020. "Internationalism in Science after 1940." In *The Cambridge History of Science*, edited by Hugh Richard Slotten, Ronald L. Numbers, and David N. Livingstone, 60–74. Cambridge: Cambridge University Press. https://doi.org/10.1017/9781139044301.006.

Dybiec, Julian. 2021. *Nie tylko szablą: nauka i kultura polska w walce o utrzymanie tożsamości narodowej 1795–1918* [Not just by the sword: Polish science and culture in the struggle to preserve national identity, 1795–1918]. Kraków: Instytut Literatury: Wydawnictwo Księgarnia Akademicka.

Dzida, Jan. 1975. "[rec:] F. Kutta, M. Soukup i zespoł: Řízení v období vědeckotechnické revoluce [...]" [[rev:] F. Kutta, M. Soukup, and team: Management in the Era of the Scientific and Technological Revolution [...].]. *Prakseologia*, no. 1: 241–45.

Elzinga, Aant. 2012. "The Rise and Demise of the International Council for Science Policy Studies (ICSPS) as a Cold War Bridging Organization." *Minerva* 50, no. 3: 277–305.

Franc, Martin. 2009. *Úderná skupina? Výprava českých lékařů a přírodovědců do SSSR v roce 1950 ve světle dopisů Ivana Málka* [A Strike Force? The 1950 Expedition of Czech Doctors and Natural Scientists to the USSR in the Light of Ivan Málek's Letters]. Prague: Masarykův ústav a Archiv AV ČR.

Gajski, Lidija. 2019. "Privatization of Science: A Silent Phenomenon." *Pannoniana* 3, no. 1–2 (1 December): 239–54. https://doi.org/10.2478/pannonia-2019-0012.

Gasparski, Wojciech W. 2016. "Ladislav Tondl." *Prakseologia* 158, no. 1: 428–30.

Gerovitch, Slava. 2024. "The Kitchen and the Dacha: Productive Spaces of Soviet Mathematics." *Studia Historiae Scientiarum* 23 (11 September): 657–83. https://doi.org/10.4467/2543702XSHS.24.018.19591.

Hadyniak, Bogusław, and Lech Zacher. 1973. "Zebranie Zespołu Badań nad Zagadnieniami Rewolucji Naukowo-Technicznej" [Meeting of the Research Group on Issues of the Scientific and Technological Revolution]. *Kwartalnik Historii Nauki i Techniki* 18, no. 3: 622–24.

Hermann, Tomáš, Doubravka Olšáková, Helena Durnová, Miloš Hořejš, and Alena Míšková. 2013. "Dokumenty o rozvoji věd v Československu z roku 1960 a jejich historický kontext" [Documents on the development of science in Czechoslovakia from 1960 and their historical context]. In *Plánování socialistické vědy: dokumenty z roku 1960 ke stavu a rozvoji přírodních a technických věd v Československu*, edited by Tomáš Hermann and Doubravka Olšáková, 11–98. Červený Kostelec: Pavel Mervart.

Hoffmann, Erik P. 1978. "Soviet Views of 'The Scientific–Technological Revolution.'" *World Politics* 30, no. 4 (July): 615–44. https://doi.org/10.2307/2009989.

Hollings, Christopher. 2016. *Scientific Communication across the Iron Curtain*. New York: Springer.

Hoppe, Jiří, and Vítězslav Sommer. 2020. "How the 'Richta Team' Was Born: The Scientific and Technological Revolution and Political Decision-Making in Czechoslovak Reform Communism." *Zeitschrift Für Ostmitteleuropa-Forschung / Journal of East Central European Studies* 69, no. 4: 495–518.

Infeld, Leopold. 1967. "O kontaktach naukowych z zagranicą" [On scholarly contacts with foreign cuntries]. *Zagadnienia naukoznawstwa* no. 3–4: 15–21.

Jaroszewski, Tadeusz Maciej. 1976. *Aktivnost' lichnosti v socialisticheskom obshestve* [The activity of the individual in socialist society]. Moscow: Nauka.

———. 1975. *Osobnost a společnost: problémy osobnosti v soudobé filosofické antropologii—marxismus, strukturalismus, existencionalismus, křesťanský personalismus* [Personality and society: problems of personality in contemporary philosophical anthropology—Marxism, structuralism, existentialism, christian personalism], trans. Jan Sedláček. Prague: Svoboda.

———. 1974. "Perspektywy człowieka w rewolucji naukowo-technicznej" [Humanist perspectives in the scientific-technical revolution]. In *Człowiek, socjalizm, rewolucja naukowo-techniczna*, edited by Joachim Liszka and Janusz Kolczyński, 38–80. Ideologia a Współczesność. Warsaw: Książka i Wiedza.

Josephson, Paul R. 1981. "Science and Ideology in the Soviet Union: The Transformation of Science into a Direct Productive Force." *The Soviet and Post-Soviet Review* 8, no. 1: 159–85. https://doi.org/10.1163/187633281X00137.

Kačaunov, S., ed. 1988. *Základy teorie vědy* [Fundamentals of the Philosophy of Science]. Prague: Academia.

Kangaspuro, Markku, and Jeremy Smith, eds. 2006. *Modernisation in Russia since 1900*. Helsinki: Finnish Literature Society.

Kilias, Jarosław. 2017. *Goście ze wschodu: socjologia polska lat sześćdziesiątych XX wieku a nauka światowa* [Guests from the East: Polish sociology of the 1960s and global scholarship]. Kraków: Nomos.

———. 2016. "Socjologia polska i czeska: wzajemne stosunki, ich charakter i konteksty" [Polish and Czech sociology: Mutual relations, their nature, and contexts]. *Stan Rzeczy* 1, no. 10 (1 April): 283–315. https://doi.org/10.51196/srz.10.15.

Kokowski, Michał. 2015. "The Science of Science (Naukoznawstwo) in Poland: The Changing Theoretical Perspectives and Political Contexts: A Historical Sketch from the 1910s to 1993." *Organon* 47: 147–237.

Komornicka, Aleksandra. 2024. *Poland and European East-West Cooperation in the 1970s: The Opening Up*. New York: Routledge.

Kopeček, Michal, ed. 2019. *Architekti dlouhé změny: expertní kořeny postsocialismu v Československu* [Architects of long change: The intellectual roots of Post-Socialism in Czechoslovakia]. Prague: Argo.

Kosel, Gerhard. 1989. *Unternehmen Wissenschaft: Die Wiederentdeckung Einer Idee: Erinnerungen* [The enterprise of science: The rediscovery of an idea]. Berlin: Henschelverlag.

Kosiński, Józef Adam, Lech Ignaszewski, Bolesław Rek, and Czesław Sztutowski. 1975. *Słownik podstawowych terminów prognostycznych* [Dictionary of basic prognostic terms]. Warsaw: Zakład Informacji i Wydawnictw. Centrum Informacji Naukowej, Technicznej i Ekonomicznej.

Kowalewski, Zdzisław. 1972. "Sprawozdanie z działalności Zakładu Historii Nauki i Techniki PAN w roku 1971" [Report on the activities of the Institute of the History of Science and Technology of the Polish Academy of Sciences in 1971]. *Kwartalnik Historii Nauki i Techniki* 17, no. 3: 624–32.

Kowalik, Tadeusz. 1972. "[Rec:] Radovan Richta i zespół: Cywilizacja na rozdrożu" [[Rev:] Radovan Richta and team: Civilization at a Crossroads.]. *Ekonomista*, no. 3: 615–17.

Kožešník, Jaroslav. 1971. "Současné otázky naší vědy a její výhledy" [Current issues in our science and its prospects.]. *Věstník Československé akademie věd* 80, no. 3: 177–81.

Krige, John. 2019. "Regulating International Knowledge Exchange: The National Security State and the American Research University from the 1950s to Today." *Technology and Culture* 60, no. 1: 252–77.

Křížek, Miroslav. 1959. "K otázce 'duševní výroby.' [G. Kosel 'Věda jako výrobní síla,' orig. něm. kniha]" [On the question of 'intellectual production.' [G. Kosel, "Science as a Productive Force," original German book]. *Politická ekonomie* 7, no. 4: 375–83.

Laitko, Hubert. 2018. "Der lange Weg zum Kröber-Institut" [The long road to Kröber-Institute]. In *Wissenschaftsforschung in Deutschland: die 1970er und 1980er Jahre*, edited by Wolfgang Girnus and Klaus Meier, 13–154. Leipzig: Leipziger Universitätsverlag GmbH.

Lalli, Roberto. 2021. "Crafting Europe from CERN to Dubna: Physics as Diplomacy in the Foundation of the European Physical Society." *Centaurus* 63, no. 1 (February): 103–31. https://doi.org/10.1111/1600-0498.12304.

Łotysz, Sławomir. 2009. "Eugeniusz Olszewski: Engineer, Historian and Cofounder of the Committee." In *ICOHTEC—International Committee for the History of Technology: 1968–2008*, edited by International Committee for the History of Technology, 2 enl. ed., 36–43. Bochum: European University Press.

McDermott, Kevin, and Matthew Stibbe. 2022. *Czechoslovakia and Eastern Europe in the Era of Normalisation, 1969–1989*. Cham: Palgrave Macmillan.

Mikulinskyj, Semyon, ed. 1983. *Osnovy Naukovedenija* [Foundations of science of science]. Moscow: Nauka.

Mikulinskij, Semen Romanovič, and Radovan Richta, eds. 1983. *Socialism and Science.* Prague: Academia.

Míšková, Alena, Hana Barvíková, and Miroslav Šmidák. 1998. *Československá akademie věd, 1969–1972: restaurace kommunistické moci ve vědě* [The Czechoslovak Academy of Sciences, 1969–1972: The restoration of Communist power in science]. Prague: Ústav pro soudobé dějiny AV ČR.

Míšková, Alena, Martin Franc, Antonín Kostlán, Melvyn Clarke, and Hana Jirkalová, eds. 2018. *Bohemia Docta: The Historical Roots of Science and Scholarship in the Czech Lands.* Historie. Prague: Academia.

Mitrokhin, L. N. 2010. "O B.M. Kedrove" [On B.M. Kedrov]. In *Bonifatiĭ Mikhaĭlovich Kedrov*, edited by V. A. Lektorskiĭ, 29–41. Filosofiia Rossii vtoroĭ poloviny XX veka. Moscow: ROSSPĖN.

Montgomery, Scott L. 2013. *Does Science Need a Global Language? English and the Future of Research.* Chicago: University of Chicago Press.

Musiał, Wojciech. 2013. *Modernizacja Polski: polityki rządowe w latach 1918–2004* [The modernisation of Poland: Governmental policies 1918–2004]. Toruń: Wydawnictwo Naukowe Uniwersytetu Mikołaja Kopernika.

Nešpor, Zdeněk R. 2014. "Jan Szczepański v dějinách české sociologie" [Jan Szczepański in the history of Czech sociology]. In *Profesorowi w hołdzie: publikacja poświęcona setnej rocznicy urodzin profesora Jana Szczepańskiego*, edited by Karol Daniel Kadłubiec, 52–61. Cieszyn: Oficyna Drukarsko-Wydawnicza Akant.

Nickelsen, Kärin, and Fabian Krämer. 2016. "Introduction: Cooperation and Competition in the Sciences." *NTM Zeitschrift Für Geschichte Der Wissenschaften, Technik Und Medizin* 24, no. 2 (June): 119–23. https://doi.org/10.1007/s00048-016-0145-4.

Nowaczewska, Renata. 2023. "Penetrating the Iron Curtain: The Rockefeller and Ford Foundations' Exchange Programs Directed to 'Friendly' Poles, 1948–1962/1968." *Pamięć i Sprawiedliwość* 41, no. 1: 401–19.

Olšáková, Doubravka. 2018. "Expert Cultures and Infrastructural Globalism: Socialist Experts under Restricted Internationalism." In *Experts and Expertise in Science and Technology in Europe since the 1960s: Organized Civil Society, Democracy and Political Decision-Making*, edited by Christine Bouneau and David Burigana, 53–68. Brussels: Peter Lang.

———. 2014. *Věda jde k lidu! Československá společnost pro šíření politických a vědeckých znalostí a popularizace věd v Československu ve 20. století* [Science goes to the people! The Czechoslovak Society for the Dissemination of Political and Scientific Knowledge and the popularization of science in Czechoslovakia in the 20th century]. Prague: Academia.

Owoc, Kazimierz. 1964. *Automatyzacja i jej skutki społeczne w kapitalizmie: wpływ postępu technicznego na sytuację klasy robotniczej w Stanach Zjednoczonych* [Automation and its social consequences in capitalism: The impact of technological rogres on the situation of the working class in the United States]. Warsaw: Książka i Wiedza.

Památce Radovana Richty (1924–1983) [In memory of Radovan Richta (1924-1983)]. 1983. Prague: Ústav pro filosofii a sociologii ČSAV.

Palonka Krystyna. 1982. "Rewolucja naukowo-techniczna" [The scientific-technical revolution]. In *Słownik filozofii marksistowskiej*, ed. by Tadeusz Maciej Jaroszewski, 303-304. Warsaw: Wiedza Powszechna.

Pawlak, Adam. 1987. "[Rec:] Osnovy naukowiedenija [...]" [[Rev:] Foundations of science of science]. *Zagadnienia naukoznawstwa* 23, no. 2: 275–77.

Pelikán, Jiří. 1969. *Panzer überrollen den Parteitag: Protokoll und Dokumente des 14. Parteitags der KPTsch am 22. August 1968* [Tanks roll over the party congress: Minutes and documents of the 14th Congress of the Communist Party of Czechoslovakia on August 22, 1968]. Vienna: Europa Verlag.

Péteri, György. 2004. "Nylon Curtain: Transnational and Transsystemic Tendencies in the Cultural Life of State-Socialist Russia and East-Central Europe." *Slavonica* 10, no. 2 (November): 113–23. https://doi.org/10.1179/sla.2004.10.2.113.

Pleskot, Patryk. 2010. "Jak wyjechać na Zachód? Procedury wyjazdów polskich uczonych do państw kapitalistycznych z ramienia uczelni wyższych i PAN w latach 1955–1989" [How to go West? Procedures for Polish scholars' trips to capitalist countries on behalf of universities and the Polish Academy of Sciences in 1955–1989.]. In *Naukowcy władzy, władza naukowcom. Studia*, edited by Piotr Franaszek, 292–325. Warsaw: Instytut Pamięci Narodowej.

Polák, Michael. 2022. "'We Want Light!' Prague Students and the Failing Scientific-Technological Revolution in the Post-Stalinist Era (1956–1968)." *Journal of Modern European History* 20, no. 1 (February): 127–45. https://doi.org/10.1177/16118944211072648.

Proctor, Robert. 2008. "Agnotology: A Missing Term to Describe the Cultural Production of Ignorance (and Its Study)." In *Agnotology: The Making and Unmaking of Ignorance*, edited by Robert Proctor and Londa L. Schiebinger, 1–33. Stanford, CA: Stanford University Press.

"Radio" [The radio]. 1970. *Trybuna Ludu*, 27 April.

Richta, Radovan. 2018. *Civilization at the Crossroads: Social and Human Implications of the Scientific and Technological Revolution*, trans. Marian Šlingová. 3rd exp. ed. Routledge Revivals. London: Routledge.

———. 1971. "Cywilizacja na rozdrożu" [Civilization at the Crossroads]. *Miesięcznik Literacki* no. 3: 98–102.

Rindzevičiūtė, Eglė. 2016. "A Struggle for the Soviet Future: The Birth of Scientific Forecasting in the Soviet Union." *Slavic Review* 75, no. 1: 52–76. https://doi.org/10.5612/slavicreview.75.1.52.

———. 2016. *The Power of Systems: How Policy Sciences Opened Up the Cold War World*. Ithaca, NY: Cornell University Press.

Romano, Angela, and Federico Romero, eds. 2021. *European Socialist Regimes' Fateful Engagement with the West: National Strategies in the Long 1970s*. London: Routledge.

Secomski, Kazimierz. 1971. "Przedmowa do wydania polskiego" [Foreword to the Polish edition]. In *Cywilizacja na rozdrożu: konsekwencje rewolucji naukowo-technicznej dla społeczeństwa i dla człowieka*, Radovan Richta et al., 7–13. Warsaw: Książka i Wiedza.

Secord, James A. 2004. "Knowledge in Transit." *Isis* 95, no. 4: 654–72. https://doi.org/10.1086/430657.

Šmidrkalová, Michaela. 2025. "Czechoslovak Praxeology: A Discipline That Did Not Exist?" *Studies in East European Thought* 77, no. 2 (April): 391–409. https://doi.org/10.1007/s11212-024-09638-1.

Smíšek, Jan. 1997. "'Chováme se jako revolucionáři?' R. Richta a příběh mezioborového týmu" [Are We Still Behaving as Revolutionaries? R. Richta and the story of an interdisciplinary team]. *Marathon*, no. 4: 26–31.

Sommer, Vítězslav. 2017. "'Are We Still Behaving as Revolutionaries?': Radovan Richta, Theory of Revolution and Dilemmas of Reform Communism in Czechoslovakia." *Studies in East European Thought* 69, no. 1 (March): 93–110. https://doi.org/10.1007/s11212-017-9280-2.

———. 2016. "Towards the Expert Governance: Social Scientific Expertise and the Socialist State in Czechoslovakia, 1950s–1980s." *Serendipities: Journal for the Sociology and History of the Social Sciences* 1, no. 2 (November 1). https://doi.org/10.7146/serendipities.v1i2.122851.

———. 2017. "Vom sozialistischen Postindustrialismus zur Marktgesellschaft: Zukunftsforschung in der Tschechoslowakei (1960er-1980er Jahre)" [From Socialist Postindustrialism to Market Society: Futures Studies in Czechoslovakia (1960s–1980s)]. *Bohemia*, 13 November, 55–81. https://doi.org/10.18447/BOZ-2017-4220.

Sommer, Vítězslav, Matěj Spurný, and Jaromír Mrňka. 2019. *Řídit socialismus jako firmu: technokratické vládnutí v Československu, 1956–1989* [Running Socialism Like a Business. Technocratic Governance in Czechoslovakia, 1956–1989]. Prague: Ústav pro soudobé dějiny AV ČR, v.v.i.

Spurný, Matěj. 2019. *Making the Most of Tomorrow: A Laboratory of Socialist Modernity in Czechoslovakia*. Prague: Karolinum.

Steiner, Helmut, ed. 1989. *J. D. Bernal's The Social Function of Science, 1939–1989*. Berlin: Akademie-Verlag.

Steltner, Günter. 1992. "Als Ulbricht 1967 Intertext bemühte" [When Ulbricht used Intertext in 1967]. *Neues Deutschland*, 15 September.

Surman, Jan. Forthcoming. "Internationalism, normalisation and the socialist social sciences: Radovan Richta and the Scientific–Technological Revolution after the Prague Spring."

Surman, Jan, and Tomáš W. Pavlíček. 2024. "The Formal and the Informal in the History of Socialist Scholarly Interconnectedness in East Central Europe." *NTM Zeitschrift für Geschichte der Wissenschaften, Technik und Medizin* 32, no. 2 (June): 137–66. https://doi.org/10.1007/s00048-024-00387-0.

Szobi, Pavel, and František Stellner. 2021. "Sovětizace českého ekonomického vysokého školství v 50. letech 20. století" [The sovietisation of Czech economic high schools in the 1950s]. In *Ne-svoboda, despocie a totalitarismus v kultuře a kulturních dějinách: Oppression, despotism and totalitarianism in culture and cultural history*, edited by Radomír Vlček, 296–311. Prague: Česká společnost pro slavistická, balkanistická a byzantologická studia, z.s.

Szumski, Jan. 2016. *Polityka a historia: ZSRR wobec nauki historycznej w Polsce w latach 1945–1964* [Politics and history: The USSR's and the historical scholarship in Poland, 1945–1964]. Warsaw: Oficyna Wydawnicza ASPRA-JR.

Teich, Mikuláš. 2008. "J. D. Bernal the Historian and the Scientific–Technical Revolution." *Interdisciplinary Science Reviews* 33, no. 2 (June): 135–39. https://doi.org/10.1179/030801808X259790.

Tondl, Ladislav. 1977. "O ocenie graficznych form danych w projektowaniu" [On the evaluation of graphical data forms in design]. *Prakseologia* no. 1–2: 477–84.

———. 2005. *Racionální činnost a svět artefaktů: studie věnovaná památce mého přítele a spolupracovníka profesora Karla Berky* [Rational activity and the world of artifacts: A study dedicated to the memory of my friend and colleague professor Karel Berka]. Prague: Filosofia.

Turchetti, Simone. 2023. "From the National Laboratory to International Collaborations (Framing Big Science for the Age of Science Diplomacy)." In *Big Science in the 21st Century: Economic and Societal Impacts*, edited by Panagiotis Charitos, Theodore Arabatzis, Harry Cliff, Günther Dissertori, Juliette Forneris and Jason Li-Ying, 58-1–58-18. Bristol: IOP Publishing. https://doi.org/10.1088/978-0-7503-3631-4ch58.

Wang, Jessica. 1992. "Science, Security, and the Cold War: The Case of E. U. Condon." *Isis* 83, no. 2 (June): 238–69. https://doi.org/10.1086/356112.

"Warto" [Recommended]. 1971. *Życie i Nowoczesność. Dodatek do dziennika "Życie Warszawy,"* 28 October.

Wheeler, Elisabeth. 2022. *Letters from Prague: 1947–1957*, edited by Doubravka Olšáková. Berlin: De Gruyter Oldenbourg.

Zacher, Lech. 1995. "The Unrealized Vision of Future: The Case of Radowan Richta Predictions." *Dialogue and Universalism* 5, no. 11–12: 100–7.

———. 1972. "[Rec:] Cywilizacja na rozdrożu" [Civilization at the Crossroads]. *Zagadnienia naukoznawstwa* 9, no. 3: 435–39.

———. 1972. "[Rec:] Cywilizacja na rozdrożu" [Civilization at the Crossroads]. *Humanizm Pracy* no. 1/2, 141–45.

———. 1972. "Rewolucja naukowo-techniczna" [Scientific-technical revolution]. *Zagadnienia naukoznawstwa* 9, no. 4: 506–17.

———. 1976. "Rewolucja naukowo-techniczna niektóre kontrowersje metodologiczne" [Scientific-technical revolution, some methodological controversies]. *Studia Filozoficzne* no. 7: 137–50.

———. 1972. "U progu rewolucji naukowo-technicznej" [At the threshhold of a scientific-technical revolution]. *Nowe Drogi* no. 11: 181–84.

———. 1983. "Wizja nowej cywilizacji" [A vision of a new civilisation]. *Polityka* no. 37: 11.

———. 1972. "Zebrania Zespołu Badań nad Zagadnieniami Rewolucji Naukowo-Technicznej" [Meeting of the Research Group on Issues of the Scientific and Technological Revolution]. *Kwartalnik Historii Nauki i Techniki* 17, no. 3: 605–612.

———. 1973. "Zebrania Zespołu Badań nad Zagadnieniami Rewolucji Naukowo-Technicznej" Meeting of the Research Group on Issues of the Scientific and Technological Revolution. *Kwartalnik Historii Nauki i Techniki* 18, no. 4: 798–801.

———. 1973. "Z Pozycji Marksizmu" [From the Marxist position]. *Trybuna Ludu*, 12 November.

CHAPTER 3

Networking the Future? Czech–Polish Futures Studies in the 1970s

Luboš Studený and Lukas Becht

Before the first Summer School of Forecasting for Participants from Socialist Countries in Rynia, Poland, in September 1972, the Czech economist Ota Šulc (1932–?) exchanged several letters in English with one of the organizers, the Polish sociologist Andrzej Siciński (1924–2006), whom he informed about his planned trip to Poland. In addition to the visit, Šulc and Siciński touched on mutual affairs: "Dear Andrzej, . . . Your article in *Sociological Journal* appeared in No 3 . . . . Your fee is deposited at 'DILIA' organization [Theatre, Literary, Audiovisual Agency] from which you can draw it when you are in Prague. Anyway, do give me a ring when you turn up here and I would very gladly be of some help to you if you need it."[1] Šulc was the head of the Forecasting Department at the Institute of Philosophy and Sociology (Ústav filosofie a sociologie, ÚFS) of the Czechoslovak Academy of Sciences (ČSAV). Siciński in turn, led the Social Prognoses Division of the Institute of Philosophy and Sociology (Instytut Filozofii i Socjologii, IFiS), which had been created at the Polish Academy of Sciences (Polska Akademia Nauk, PAN) in 1969. In addition to finalizing a cooperation agreement between the institutes, signed in 1971,[2] their correspondence was part of a series of encounters that contributed to the establishment of a Czech–Polish[3] network in socialist futures studies, which emerged in the late 1960s. It included institutionalized cooperation between the Czechoslovak and Polish Academies of Sciences, and exchanges of people, publications, and working materials. The actors supported each other with information exchange and publishing and collaborated at international events. This also involved private contacts and practical help during visits to the other country, as suggested by Šulc's letter to Siciński.

This chapter argues that through such practices of exchange a Czech–Polish network in the interdisciplinary field of futures studies emerged in the 1970s, which below the high hopes of jointly shaping the socialism of tomorrow through international forecasts, served first and foremost strategic and pragmatic purposes. This Czech–Polish futures studies network provided scholars with spaces to enhance their careers and develop vocabularies and ideas that went beyond existing disciplinary and political limitations. The aim of this chapter is not to evaluate the conceptual content, academic value, and intellectual influence of these ideas. Instead, it reconstructs the constellations and forms of exchange that developed between Czech and Polish scholars in the futures studies field in order to provide insights into the nature of scholarly entanglements between socialist Czechoslovakia and Poland. At the same time, this example of entanglement highlights the significant role played by transnational connections in general and futures studies in particular in socialist science, in the media and in political demands for expert knowledge from the late 1960s onward.

During the 1960s, futures studies became a highly transnational field of scientific and ideological competition and cooperation across the Iron Curtain.[4] Against this background, socialist contacts and collaboration developed vigorously in various spaces ranging from local events and bilateral meetings to multilateral socialist and international conferences, reaching its peak in the 1970s. The reconstruction of the most intense connections of the Czech–Polish part of this network is based on the premise that an interplay of geographical and linguistic proximity with pragmatic and ideological motivations furthered contacts and collaboration until in the early 1980s political developments linked to the socioeconomic crisis in Poland constrained the functioning of these networks.[5]

By approaching contacts and collaboration in the field of futures studies as a network, the chapter provides a perspective that includes both official institutional contacts and personal encounters, as they supplemented each other in establishing and practicing connections of varying degrees of intensity. It also takes into consideration the fact that contacts not only had purely scientific aims but also personal and/or political agendas. The conceptualization of a transnational network allows us to overcome the boundaries of national contexts. Although contacts were generally most intensive within these boundaries, many contacts and collaborations in the field of futures studies were related to, or even intensified by, socialist international and/or global contexts.[6]

The study of the Czech–Polish network in the field of futures studies allows for a more differentiated understanding of this field's historical role and evolution in socialism and beyond. It shows that the distinct shape of national institutions, actors, and paths of development in Czechoslovakia and Poland was affected by the contacts, mobility, and circulation of entities between the two countries. They were not detached national "cases" but entangled.

Arguably, these entanglements of the Czech–Polish futures studies network also helped advance the positions of the scholars involved in the global network of this field, whose history has recently become the subject of growing attention, although with a strong focus on futurology in the global Northwest.[7] The emerging literature on socialist futures studies has seldom elaborated further on transnational contacts and collaboration between scholars from different socialist states.[8] The pioneering works of Eglė Rindzevičiūtė focus on relations with the Soviet Union as well as knowledge circulation across the Iron Curtain.[9] Apart from in the first half of the 1970s, when large socialist delegations attended international futures studies' conferences, intra-socialist entanglements were largely invisible to Western eyes at these events. Participants from the Eastern Bloc were usually regarded as mouthpieces of their home institutions and national regimes, while the Western futurists claimed to speak for themselves as independent scholars.[10] This chapter attempts to challenge this perception by highlighting the fact that through the transnational Czech–Polish futures studies network individual scholars gained pragmatic and intellectual resources to enhance their academic positions and move more independently within existing institutional, ideological, and political constraints.[11]

After introducing the histories of the field of futures studies in both Czechoslovakia and Poland, the main part of the chapter describes three constellations[12] within the Czech–Polish futures studies network. The first was the multilateral socialist collaboration on the critique of bourgeois futurology, which went beyond the involved academies of sciences. It was based on a joint ideological adversary, but—as the chapter shows—did not lead to a joint socialist vision of tomorrow. Second, Czech–Polish cooperation did not necessarily have to occur within only the framework of socialist cooperation, as the case of Šulc and Siciński shows. Their contacts were linked both to their national institutes and engagement with the globally operating World Futures Studies Federation (WFSF). In fact, assisting each other in this international arena was a crucial driver of their mutual support and exchange. However, the third constellation, and very likely the most intensively entangled one, emerged as an institutional cooperation that was not primarily organized at the level of central scientific institutes. Wrocław Technical University's (Politechnika Wrocławska, PWR) Center for Prognostic Research (Ośrodek Badań Prognostycznych, OBP) was a highly active nodal point in the futures studies network, which established cooperation agreements and regular forms of exchange with institutions in Prague and Brno. For several years in the 1970s and early 1980s it even became the long-term academic home to the Czech futures studies scholar František Petrášek (1934–2014). All three constellations were sites of the most profound exchange in the Czech–Polish futures studies network.

Czech and Polish Futures Studies in the 1970s

In both countries, the designated term for the subject of this research has undergone various changes over the years. In Czechoslovakia, "futures studies" was predominantly known as "futurology" in the 1960s. While forecasting was developed in an effort later identified with *prognostika* as early as 1963,[13] it only became popular after futurology was denounced as a bourgeois science in the aftermath of the invasion of the armies of the Warsaw Pact.[14] This allowed futures studies scholars to continue their work using the same methods (while attempting to depoliticize them) even in the changed political context, although they had to accommodate new demands stemming from the new power relations.[15] The official distinction between *prognostika* and futurology lasted until 1989. While several terms were in use in Poland to describe forecasting-related research, "futurology" became omnipresent in scholarly debates and the media throughout the 1960s, during which Siciński noticed an "explosion of futurology."[16] The political leadership of Poland remained highly skeptical of futurology, although this changed slightly with the politically approved label *prognostyka*, under which the new party and state leadership of Edward Gierek began to promote and develop prognostic research actively in December 1971.[17] Thus, in this chapter we use the term "futures studies," which, in accordance with the field's transnational self-description, implies the notion of multiple futures and is a commonly used label in contemporary research.[18]

Scholars who became interested in forecasting the future in Czechoslovakia and Poland in the 1960s related not only to prewar socialist thinking and debates on economic planning in the Soviet Union, but also to contemporary intellectual and practical demands. Although each trajectory was differently affected by national political developments, they mutually influenced each other. In both countries, economic policy makers sought new methods of central planning that would be able to generate sources of economic growth beyond mass-scale industrial production.[19] At the same time, the concept of "Scientific-Technical Revolution" (STR)[20] became a major element of the post-Stalinist ideological discourse, highlighting the role of scientific and technological development as major resources of economic advancement. This concept and political idea generated a formalized and state-supported research agenda,[21] which held a particularly strong position in Czechoslovakia. From 1965, the philosopher and sociologist Radovan Richta (1924–1983) headed an interdisciplinary team that was supposed to deliver an analysis of social development and changes on the basis of STR. Both he and his theory became influential in the reform movement of the Prague Spring and the main point of reference in relating to the future in Czechoslovakia in a scientific way. Richta was interested in futures studies[22] and attempted to support it when he later became director of the ÚFS ČSAV in 1970.[23] Richta's work was regularly referred to in a Czech futures studies context as part of the field.[24]

In Poland, the concept of STR furthered the development of science planning and science studies, which was rooted in the pre–World War II philosophy of science, logics and praxeology.[25] At the PAN, this was connected to the government task of devising a long-term plan of scientific research, since in the 1960s science policy and planning on both sides of the Iron Curtain were intensified as major arenas of exchange and systemic competition.[26] Against this background, the research planning department of PAN began to collect and translate foreign works on new forecasting techniques and materials on science planning, which were then made available within the Academy. Thus, major foundational works of US and French futurology found their way to Poland and became an important source of (methodological) inspiration.[27]

Moreover, both STR and futurology grew out of the conviction that rapid and revolutionary scientific and technological developments would lead to important *social* changes, which called for more complex scientific planning and provision.[28] In the Polish media, the growing interest in long-term forecasts for the year 2000 in the West as opposed to socialist long-term planning could be observed as early as 1963.[29] In Czechoslovakia, futurology as such did not receive much attention until the late 1960s. Individual scholars such as Šulc—a future leading figure in the field—became invested earlier. In general, the future became an important topic for many authors in this decade, ranging from scholars to science-fiction writers.[30]

Thus, the efforts of social and technological scientists to establish futures studies as a field with its own methodological debates, defined (sub) fields of interest and publication formats can be found encountered in the late 1960s, with 1967 as a watershed year. In Czechoslovakia this resembled a civic initiative. Individuals who were interested in futures studies took advantage of the opportunity of the Prague Spring, which allowed for more independent research in contrast to the more restrictive and centralized conditions that had been in effect up until then. This resulted in the founding of a student association at the University of Economics, Prague (Vysoká škola ekonomická v Praze) called the Futurological Club (Futurologický klub) and Futurological Society (Futurologická společnost; they emerged from the activities of students at the university and later became part of the political and scientific educational organization known as the Socialist Academy [Socialistická akademie]) in 1968. This new interest was affirmed by the founding of the journal *Trend: Prognostické informace* (Trend: Forecasting information) in 1969. Unlike the Club and Society, the journal survived "normalization" and remained a steadfast source of information in the field until 1989. A number of translations of Western futurologists were published in Czechoslovakia, and Czechoslovak scholars contributed their own works in the attempt to establish futurology as an independent scientific field.[31]

There is much to indicate that the developments of the national fields of futures studies were interconnected. In Poland, the interdisciplinary Group for the Study of Contemporary Culture at the PAN, where Siciński was employed, channeled earlier efforts in science planning into a proposal to establish a futurological advisory body to the government, which also occurred in 1967. Declared to be a response to the call by COMECON to establish long-term forecasting in all socialist countries and similar expert groups in France and the United States, this group organized the first Polish futurological conference in May 1967.[32] Both the congress and the work of Siciński's Group for the Study of Contemporary Culture were known and discussed in Czechoslovakia, showing that scholars in socialist countries followed the development of futures studies in other countries closely.[33]

Both the invasion of Czechoslovakia in 1968 and the antisemitic and anti-intellectual campaigns following the protests by Polish students and intellectuals in March of the same year triggered dramatic changes in academic life, affecting the emerging communities of futures studies in both countries. The subsequent effort to quell civic initiatives and oppositional voices led to the disbandment of the futurological groups in Czechoslovakia. Even so, several official organs of futures studies were created, especially at the ČSAV. Leading figures in futures studies such as Radovan Richta and Ota Šulc attempted to maintain the field in the face of the subsequent repression and made concessions they felt were necessary to achieve this. Richta became the director of the newly established ÚFS at the ČSAV and cooperated with the demands of the new political leadership, while Šulc was appointed head of the forecasting department at Richta's institute. Others, such as František Petrášek, were not so fortunate and had to transfer to research institutes outside academia. Šulc made a tactical move in denouncing futurology as a bourgeois science and making a case for prognostics (*prognostika*) as a distinctively socialist science. This allowed him and others to continue their work in the field. However, the separation from futurology was mainly rhetorical and the methods and approaches remained unchanged, as they were deemed apolitical. Šulc constructed the difference mainly in terms of its goals and social framework.[34] While this development allowed work to continue, it subordinated the field to control from above, which was ideological and also affected personnel.

In Poland, many participants in the futurological conference of 1967 lost their academic positions in the purges that resulted from the antisemitic and anti-intellectual campaign of the Communist Party leadership following student protests in March 1968. Around the same time, in July 1968, Siciński left for the United States for a year on a fellowship of the Ford Foundation, which he had been unable to take up earlier.[35] Upon his return in June 1969, he took over the Group for the Study of Contemporary Culture, which was rebranded the Social Prognoses Division of the IFiS of PAN and organized methodological seminars and publications, conducted survey-based research on people's opinions on the future and their changing lifestyles.[36] Siciński explicitly considered futurology an interdisciplinary study of possible variants of the future and its achievements that was useful in the socialist context as a means of reflecting on societal goals. However, its official task was to deliver useful information for central planners.[37] From 1969, prognostic research in Poland—organized, centrally coordinated, and conducted by the Committee of Research and Prognostics "Poland 2000" at the presidium of PAN—provided expertise to the government. Social scientists such as Siciński, who became the Committee's secretary in the second half of the 1970s, played a significant role until the mid-1980s, for instance, in formulating concrete proposals for education and social policy.

During the 1970s, forecasting in both countries became formalized and was conducted at many different places inside and outside the academies of sciences. Scientific forecasts were (to be) produced by branch ministries, planning commissions, the military and security apparatus, and other government research offices and scientific institutes, such as, for instance, statistical offices. However, usually they devised normative forecasts that resembled plans and omitted other variants and probabilities by taking simple extrapolations as the predicted outcome. Nevertheless, the institutions addressed in the main part of this chapter, such as the ÚFS ČSAV and IFiS PAN, were also concerned with methodological and epistemological reflections, with the aid of which they provided guidance to other actors in the network. Moreover, they adopted a perspective that aimed at integrating multiple disciplines and issues into a more general analysis.[38]

The field's institutional development in the 1970s reflected not only academic but also political influences. Given that futures studies deal with the future, they are inherently political, and as such they are potentially dangerous to the political order. This is illustrated by the crackdown on the more independent thread of futures studies in Czechoslovakia after the invasion of 1968. When Edward Gierek became First Secretary of the Communist Party in Poland in December 1970, it provided an impetus for prognostic work at PAN, as the new leadership actively sought scientific expertise and exchange with the West to support its social and economic policies. However, after a few years, both political elites and scholars—for different reasons—became disillusioned by the limited use of forecasts for concrete political decision-making and the inability to predict and prevent the increasingly visible economic and social crisis in the country at the end of the 1970s.[39]

In Czechoslovakia, the 1970s was a period in which the authorities removed ČSAV as an independent actor in planning and forecasting, because of its involvement in the Prague Spring. The responsibility for long-term forecasts shifted back to the State Planning Commission. The Commission oversaw the newly established prognostic work together with the Federal Ministry for Technical and Investment Development in the 1970s. They only seconded researchers from the ČSAV and did not involve its institutes directly. The ČSAV only regained its position in long-term planning and prognostic work at the end of the 1970s. It was then directly commissioned to deliver the forecast report *Czechoslovakia 2000*—with Richta responsible for the task. This development led to the strengthening of the position of the ČSAV and, eventually, to the establishment of the Institute of Forecasting of the ČSAV in 1984, which delivered a renowned critical forecast in 1988.[40]

In Poland, the economic crisis and emerging political opposition after 1976 initially led to an intensification of futures studies, not only because expert knowledge was requested by the government. Semi-legal oppositional discussion groups from the precarious "gray zone"[41] between political persecution and tolerated dissent, and the official Committee "Poland 2000" even worked together to devise proposals for finding a way out of the country's crisis.[42] However, the imposition of martial law in December 1981 put an end to such collaboration and severely limited the international contacts of Polish scholars.

Reconstructing the Network

National and international contacts in the field of futures studies were already established in the late 1960s. In February 1967, the Permanent Commission on the Coordination of Scientific and Technical Research of COMECON (established in 1963), organized a conference of specialists in science and technology forecasting from COMECON countries in Prague. It was one of the first multilateral socialist meetings officially devoted to futures studies. PAN sent a delegation of members of its science and technology studies group.[43] In April 1968, Richta's book *Civilization at the Crossroads* was discussed with scholars from other socialist countries during a seminar in Mariánské Lázně organized by the Institute of Philosophy (and others) of ČSAV, in which a PAN delegation including the sociologist Andrzej Siciński took part.[44] These were the first occasions on which scholars from both countries officially began to build relationships and international reputations as specialists in futures studies, which culminated in the following decade.

International contacts were crucial for the scholarly career of Andrzej Siciński, who came from a family that had owned a private business in Warsaw before the Communist takeover and had socialized politically as a member of a group of young Catholic activists, journalists, and intellectuals.[45] After graduating in the subjects of sociology at Warsaw University in 1952 and technical engineering at Warsaw Technical University a year later, like many young Polish sociologists at the time, he was fascinated by the methodology of opinion research with questionnaire-based surveys developed by the Austrian émigré Paul A. Lazarsfeld in the United States.[46] Siciński began his sociological career conducting surveys on radio and television audiences. He first specialized in questions of international affairs, war, and peace at the public opinion research institute at Polish state television, which he had helped to establish in 1957, and from 1964 he worked at IFiS. At the same time, he became the leader of the Polish research team of a UNESCO-funded Polish–Norwegian–French comparative research project on public opinions on issues of international politics. Through his Norwegian counterpart Johan Galtung (1930–2024), an influential actor in international peace research and the emerging field of futures studies, Siciński gained access to literature, conferences and contacts in these fields, which he attempted to develop further with his colleagues at PAN.

Galtung and the Austrian journalist and futurologist Robert Jungk (1913–1994), in turn, sought to establish contacts with scholars from the Eastern Bloc in advance of a conference at the Peace Research Institute in Oslo in September 1967, at which futures studies and the idea of establishing an international organization for the field was discussed for the first time by scholars from the United States, Western Europe, and the East European socialist countries.[47] Soviet scholars did not attend the conference and only sent in papers. Western observers perceived the Polish and Czechoslovak contributions as presenting a somewhat common approach of socialist futures studies, focusing on emancipatory social and cultural changes instead of technological and economic developments (the only other socialist country present was Yugoslavia).[48] However, the delegations' papers rather reveal the intention to display the strength and originality of their own national futures studies' communities.[49]

On the Czechoslovak side, Ota Šulc and František Petrášek were among the most important actors in the futures studies network both nationally and internationally. Following his studies at the University of Chemistry and Technology, Prague (Vysoká škola chemicko-technologická v Praze), Šulc worked in a state enterprise and research institute as a planner and economist. In 1965, he became a member of a group that dealt with the economics of technical development at the University of Economics, Prague, and became involved with several futurological initiatives.[50] He spent the reform period of 1968 and the invasion on a stay at the University of Manchester, and after his return and a short research stay in the Soviet Union in 1969, he moved to the newly established ÚFS ČSAV where in 1970 he became the head of the forecasting department, thus managing to avoid the purges and stay fully involved in academia. Šulc became personally acquainted with many other futures studies scholars and traveled around the world to attend futures studies conferences (for example, Norway, Romania, Japan) and to share his experience (for example, Iraq, India), thereby establishing himself as the ČSAV's main nodal point in the international futures studies network.

František Petrášek's involvement in futures studies was preceded by studies at the Faculty of Mathematics and Physics at Charles University and postgraduate studies at the University of Economics, Prague, where he began his research career in 1967. It was there that he worked with Radovan Richta and developed his interest in futures studies, but he was dismissed after the invasion in 1968. Nevertheless, in 1970, Petrášek became head of a research group at the Institute of Economics and Management of Scientific-Technical Development (ÚVTR), a branch ministry institute. Although he parted ways with Richta, whose policy of cooperation with the new post-1968 order he rejected, Petrášek secured a part-time contract at the ÚFS at the beginning of the 1970s. During the early 1970s, when he attended multilateral meetings of experts from socialist countries related to science and technology forecasting, he established contacts with Polish scholars, among them the founders of the OBP in Wrocław. He later even claimed to have been involved in its creation and moved there in the second half of the 1970s as a contractual lecturer.[51] Nevertheless, neither preserved archival sources nor eyewitnesses confirm his involvement in the foundation of the OBP.

The person who formed a relationship with Petrášek and developed OBP's cooperation with Czech institutions was Karol Pelc (1935–2022), an electronics engineer of PWR by training, where he had started his career as an electronic technician in 1957. In the second half of the 1960s, Pelc participated in a regular discussion group on science organization and technology development gathering PWR's rectors and leading teachers. According to one of them, Wacław Kasprzak (b. 1932), this group's reception of contemporary forecasts on science and technology from the United

States and Japan led to the idea of establishing a center for prognostic research. After two annual scholarships at the University of Uppsala in the 1960s, Pelc gained a PhD in electronics in 1968. He became OBP's founding director in 1971. In this function Pelc became involved in the transnational and national networks of futures studies, among others as a member of Committee "Poland 2000" from 1976 to 1980. At this time, Pelc's interests evolved toward issues of the management of research and development and he earned a PhD in economics in 1976.

Before Pelc resigned as the OBP's director in February 1982 due to a severe illness and left Poland to the United States in 1985, Petrášek's employment there was a unique form of Czech–Polish entanglement in the field of futures studies. It came to an end when, as he later claimed, he was forced to either return to Czechoslovakia or permanently relocate to Poland in 1980.[52] No longer commuting between Wrocław and Prague, he returned and took up a position at ČSAV, where he provided methodological guidance for its forecasting efforts and later transferred to the Institute of Economics, where he stayed until 1989. He nevertheless remained in contact with OBP as member of the editorial board of its journal, as did Karol Pelc, who became a professor at Michigan Technological University.

A New Momentum for Socialist Cooperation after 1968

The reform process of the Prague Spring in 1968 had an immense impact on the patterns of foreign exchange at ČSAV and on Czechoslovak science in general, whose exchange focus moved to the West. However, after the invasion, exchange with capitalist countries significantly decreased once more.[53] The new presidium of the ČSAV attempted to "normalize" international relations. The focus shifted back to "brotherly socialist states" and the presidiums reaffirmed their power to approve exchanges of staff. In response to the protests in Poland in March 1968, Polish political authorities also tightened control over the universities and the PAN, including its staff policy and foreign relationships.[54] However, soon afterward, the ascent of Edward Gierek to power in Poland in December 1971 led to new impetus for academic exchange with the West, and multilateral socialist cooperation gained momentum after the stagnation of the 1960s.[55]

With regard to futures studies, the research cooperation initiated within the COMECON framework after 1971 included a program on technological forecasting, supervised by the national branch ministries for science and technology.[56] The latter designated research institutes responsible for the national contributions and participation in a series of rotating working meetings in which the work in progress was discussed. The COMECON project I-25.2 "Methodology of Science and Technology Forecasting" was meant to provide a toolbox for joint international forecasts. Poland's participation had been delegated to the OBP under the supervision of the Ministry of Science, Higher Education and Technology.[57] In the course of the project, OBP contributed first a dictionary of basic terminology of forecasting and later a forecast of the technological development of passive electronic subsystems, while ÚVTR—together with East German and Soviet scholars—played a leading role in defining the methodology of the group's work.[58] In this regard, the overall goal of an integrated "super-macro-forecast" on COMECON's potential for developing scientific research and technological innovation was meant to determine the most efficient division of labor among COMECON countries based on a joint catalog of indicators for assessing each member country.[59] It is difficult to tell whether Czech–Polish contacts were more intense than others in the meetings of this multilateral format, which rotated between the participating COMECON countries. However, at these frequent meetings many members of OBP met Petrášek and others—in 1974 alone three successive meetings took place in Sofia, Kyiv, and Moscow.[60]

In effect, the working group devised an elaborate and complex formal procedure, named a methodology of international forecasting, which consisted of 20(!) successive stages from goal formulation via working group selection to the synthesis and formulation of proposals for implementation at the COMECON level. In this way, futures studies were meant to "accelerate the process of socialist economic integration."[61] However, in practice, each national participating institution prepared a thematic forecast, but the highly bureaucratic procedure did not culminate in a joint forecast. In 1979, Jerzy Ludwik Baworowski (1921–2008), an economist and head of OBP in the 1980s, concluded that not everything should be subject to joint forecasting, although national forecasts were not always available instead. His experience had shown that the unification of procedures and time pressure were counterproductive to the creativity of forecasters.[62] Nevertheless, OBP hosted the meetings of the COMECON working group in Wrocław at least twice (1979, 1981) and functioned as one of the (rotating) coordinating institutes of the project, which lasted until 1986.[63] This shows that the most active nodal points in the socialist futures studies network were not necessarily located at the academies of sciences or central institutions in the architecture of national science policies.

"United We Advance against Bourgeois Futurology"

The Czech–Polish futures studies network also connected scholars who did not necessarily consider themselves representatives of the field but engaged in its critique. From the perspective of the socialist critique of bourgeois ideology, it was important to counter Western futurology, which was a popular narrative in socialist countries since the 1960s.[64] Although it was possible to write positively about futurology and some of its conclusions and methodological tools, in works and activities in the field of socialist futures studies an explicit or implicit critical stance can be observed. The critique targeted futurology's roots in the US military–industrial complex, its lack of focus on class relations, its assessment of Western technological advantage, and its adherence to US hegemony. Scholarly exchange between Czechoslovakia and Poland also developed along the lines of sharpening and disseminating such arguments. The Problem Commission of Multilateral Cooperation of the Academies of Sciences of Socialist Countries, named "Issues of Ideological Struggle in Connection with the Coexistence of Two World Systems,"[65] which was coordinated by ČSAV and established in 1967,[66] became the framework for such an endeavor. It provided socialist scholars from within and outside the academies who did not necessarily identify as forecasters or futurologists with opportunities to write about and engage with foreign literature, including otherwise unavailable classified Western works. Futurology's influence on socialist future-thinking was considered a threat. The year 1968 had left a lasting legacy, as in 1972 propagators of the project from Czechoslovakia and Poland condemned Ota Šik and other Czech reformers of the Prague Spring for their alleged uncritical "incorporation" of bourgeois futurology.[67] "[T]he role of Marxists is to unmask the essence of bourgeois futurology and its influence . . . , not allowing a revision of our science, to which we have been witness in recent years in CSSR"[68] was the motif of this constellation of socialist collaboration in the field of futures studies. At the same time, this joint defense against revisionist influences and attack on Western futurology helped shield the developing socialist futures studies network against political accusations of revisionism—and thus deepen its transnational connections.

In 1971, the Polish philosopher Jan Szewczyk (1930–1975) wrote to the head of the Commission and deputy director of ÚFS Jindřich Filipec (1926–2013) to inform him about his negotiations with the Bulgarian Academy of Sciences (Bŭlgarska akademiya na naukite, BAN) regarding a joint project on bourgeois futurology. Szewczyk was the deputy head of the Center for the Study of East–West Relationships (*Ośrodek Badania Stosunków Wschód-Zachód*), a research institute at the Communist Party's publishing cooperative RSW "Prasa," which was established in March 1969. The institute, whose research agenda Szewczyk was responsible for, had been invited by PAN to participate in the project. It required broad access to Western literature and the institute's main task was to provide selected recipients in the communist elite with detailed reports and expertise on the main trends in the foreign media, intellectual debates, and bourgeois ideology. Szewczyk had lived in the UK from 1945 to 1953, where he had been introduced to Marxism during a short stay in prison. After returning to Poland, he worked as a mining and construction worker, and during that time participated in the construction of Nowa Huta, the model city of Stalinist urban planning and industrialization near Kraków.[69] In 1957, he began philosophical studies at Jagiellonian University (UJ) in Kraków, focusing on Marxism, phenomenology and the philosophy of labor, on which he wrote his habilitation thesis in the period after 1962. Szewczyk was a critic of influential revisionist party intellectuals, and in 1969, he left Kraków, where his habilitation had been rejected. In Warsaw he helped to build the center and developed its research agenda and foreign contacts. On a visit to ÚFS before the letter to Filipec, he had had an opportunity to assess the possible level of interest. From Filipec's reply, he learned that he was open to cooperation and had assigned Šulc to deal with the matter.[70] The Bulgarian team was led by Asen Kozharov (1911–1988), deputy director of the Institute of Modern Social Theory at BAN.[71]

Šulc contributed to the project together with a postgraduate student working at the department, Jiří Slíva (b. 1947). Slíva recalled a visit to Poland together with Šulc—probably in November 1971.[72] They visited not only the center for East–West research but also IFiS, where they discussed a cooperation agreement with Siciński. Šulc left earlier, leaving Slíva without enough money to return to Czechoslovakia. To help him out, Szewczyk offered to buy the papers Slíva had brought for the conference in Warsaw, even though they were in Czech. With this transaction, Slíva acquired the money to get back home.[73] This is a plausible explanation of how both Slíva's and Šulc's papers—co-authored by Jaromír Sedlák (1925–2020)—found their way into a publication of Szewczyk's institute, in which four papers on "Selected Problems of Contemporary Bourgeois Futurology" were presented in Polish as working material from the project.[74] The paper by Šulc and Sedlák had been published in the journal of ÚFS in 1971, Slíva's appeared only later in 1975.[75] As Szewczyk explained during a visit to the Czechoslovak newspaper *Tvorba* while at ÚFS, he and his colleagues wanted to achieve "united ideological progress" as "[c]omplex international cooperation on the socialist ideological front allows [us] to go on the offensive in the struggle with bourgeois ideological directions."[76] However, the Czech–Polish volume marked the Polish side's most significant involvement in the multilateral project.

The uniting cause of combating the joint enemy of bourgeois futurology was not strong enough to achieve the project's main goal of publishing a large collective volume quickly. It took many years, during which the list of contributors changed several times.[77] Even though the project originally envisioned cooperation between Bulgarian, Czechoslovak, and Polish scholars,[78] East German, Soviet, and later Hungarian scholars were to be included as well. The research team officially worked under the label "The Future: Real Problems and Bourgeois Speculations." However, as it turned out, no Polish authors were included in the project volume and only the introduction acknowledged an unspecified contribution of Polish futures studies.[79] This outcome was influenced by the fact that Szewczyk, who had co-initiated the Polish collaboration, was removed from his positions in March 1973 and much of the institute's published materials and drafts were physically destroyed.[80] Moreover, two years later Szewczyk died in a car accident, after he had been removed from two posts and had begun to engage in opposition activities. In addition, in the Czechoslovak team Ota Šulc was replaced by a junior scholar from ÚFS, Oleg Suša (b. 1947). It was only after these turbulent events that the editors of the volume were chosen in 1977.[81] The book itself did not appear until 1979, eight years after the launch of the project.[82]

It is remarkable that a project on a shared subject of interest in all countries of the socialist world took so long to complete and did not have any direct continuation. By 1979, the issue of "unmasking" bourgeois futurology might have simply lost its appeal due to other emerging scientific, intellectual, and political trends. At the same time, the global field of futures studies had become much more heterogeneous and self-critical since the publication of the *Limits to Growth* report in 1973 and the series of counter-models for the global economic and political order it provoked.[83] In addition, the only language the project publication appeared in was Russian. The Bulgarian side offered to publish an English version if all contributing parties would deliver English translations, but it never materialized.[84] It was not unusual for research under the label of multilateral cooperation of socialist academies of sciences or COMECON projects aimed at joint publications of the Eastern Bloc to be published in Russian. However, this had a potentially demobilizing effect on non-Soviet socialist scholars, whose scholarly identities were built on distinct local and national traditions, concepts of science, and scholarly work.

Together beyond the Iron Curtain

Czech and Polish futures studies' scholars also met regularly at spaces of exchange with the non-socialist world, such as at the WFSF (established in 1973), which alongside multilateral socialist collaboration further intensified bilateral connections. Scholars from both PAN and ČSAV attended international events related to futures studies, including those organized by

UNESCO working groups or the United Nations Economic Commission for Europe, which led to intensive travel activity. Political supervisors of PAN at the Central Party Committee even complained about the financial burden, which had resulted, indirectly, from the Communist leadership's strategy of opening to the West (with the aim of securing foreign credits and technology to transfer to the Peoples' Republic), pursued since 1971 under Gierek. In a handwritten note from a party official to the head of PAN's Office of International Cooperation, a travel application from the Committee "Poland 2000" in November 1972 was harshly criticized: "Don't you have any qualms about spending cash on 'futurology' like that? First Rosset, then Siciński from the Committee 'Poland 2000' . . . , and now two more!"[85] Nevertheless, the Polish involvement with non-socialist institutions such as WFSF was much more extensive, as the Czechoslovak political situation of the 1970s severely restricted contacts with the West. As an institutional member of WFSF, the Committee "Poland 2000" hosted foreign scholars relatively frequently. It also sent its members abroad many times per year until the introduction of martial law in December 1981 placed strict limitations on such activities. However, according to the Committee's reports officially documented exchanges between Poland and both the West and socialist countries such as Hungary, Romania, and the Soviet Union were more intensive than with Czechoslovakia. ČSAV decided not to apply for institutional membership of WFSF, even though ÚFS and Šulc had recommended it. ČSAV deemed it enough to stay in contact while the WFSF was being formed, with Šulc involved in its operation individually as a member of the WFSF Board, to which Siciński had also been appointed.[86]

Therefore, while both Šulc and Siciński represented the socialist futures studies network on an international level, their exchange also had a bilateral aspect. Meeting frequently helped in building professional ties. They made each other's acquaintance at the international sociological congress in Varna in 1970, at the latest. Both held similar positions in their national futures studies communities as leaders of forecasting research units at their respective academies. As early as 1970, a member of the ÚFS management received a visit by a member of the presidium of "Poland 2000" and recommended further cooperation. However, the contact with Siciński and his team was much more frequent. An official agreement on cooperation was made between IFiS and ÚFS, that is, its forecasting department. It was negotiated by Šulc and was valid from 1972.[87]

In 1971, Siciński and the Committee "Poland 2000" came up with the initiative of launching a rotating, annual or biannual, Summer School of Forecasting. The event was to serve as an occasion for exchanges between socialist futures studies scholars, establishing a workshop of forecasting methodology. Siciński organized and conceptualized the first meeting, to which he invited Šulc personally and convinced him to give a lecture on the forecasting of lifestyles. PAN hosted the seventy-five participants, twenty-five of whom arrived from the socialist academies and fifty from Poland. It was held in Rynia near Warsaw from 18 to 27 September 1972. In addition to Šulc, František Petrášek presented a paper for the Czechoslovak side, which he published in Polish in 1974.[88] Šulc's paper was included in the conference volume that appeared in 1975.[89] Richta was also invited but subsequently canceled.[90] As the meeting revealed, understandings of futures studies were all but homogeneous among the socialist countries, thus bilateral contacts that secured open or tacit support in such multilateral arenas were deemed important.

Accordingly, the organizers' reports and the published proceedings in Rynia appear to be of a rather general nature, revolving around methodologies of forecasting and its relationship to planning. The goal of a shared methodological understanding and concept of forecasting applicable in all socialist countries, which Siciński had announced in his paper, was not achieved.[91] Later reports evaluated the meeting as useful, noting that the Czech participants had presented the most elaborated (quantitative) methodological proposal. The next gathering was to be hosted by the Hungarian Academy of Sciences in 1974.[92] However, no record of continuation exists, either at ČSAV or PAN or in Czechoslovak and Polish futures studies journals.[93]

One of the reasons for this discontinuation may have been the differences that arose during the first meeting. The proceedings show that it was difficult to reach a common stance on whether instruments developed by Western futurology could be used pragmatically or should be rejected because of their ideological background. In Siciński's view, it was crucial that socioeconomic planning left space for "the existence of unplanned spheres of life."[94] Here, forecasting came into play as a reflection of multiple social goals and the long-term consequences of decisions, instead of a simple instrument for checking the probability of social goals set by socialist ideology and economic plans in advance. In a handwritten note, most probably left by Siciński himself on the back of his copy of the conference material, he concluded from a comment of "some d[octo]r. from GDR" that his understanding of futures studies was not shared by the Marxist-Leninists participating in the meeting. He noticed that they considered a distinct science of futures studies unnecessary, since they deemed goals of social development sufficiently laid out by Marxist-Leninist philosophy, which itself had a sufficient "prognostic function."[95] In fact, in the conference proceedings the use of the term "futurology" raised controversies as the Marxist-Leninists openly rejected it, while others argued for its pragmatic use to designate methodological tools.[96] Šulc's methodological paper can be interpreted as tacitly supporting Siciński's view, as it called for a better understanding "of lifestyles as a system of alternative goals and multiple ways toward its achievement"—all of these alternatives were explicitly included in the transformation from socialism to communism.[97]

Šulc and Siciński met frequently at other futures studies conferences and events, such as the IRADES conferences in Rome in 1973, the 5th World Futures Studies Conference in Dubrovnik in 1976, and the 6th WFSF Conference in Jabłonna 1977, which was hosted and co-organized by "Poland 2000." When Šulc was not able to visit a conference in Paris in 1972, he asked Siciński to relate to him its proceedings.[98] They both visited each other's workplaces. They addressed each other by their first name and tried to support each other by exchanging publications and other scientific information personally or by mail. They sometimes exchanged their own papers before conferences and subsequent publications. Šulc informed Siciński about the status of his paper during publication procedures[99] and asked him to make efforts to place his papers in Polish journals, with the offer of the same assistance in Czechoslovakia.[100] Indeed, one of Šulc's papers appeared later in Polish translation.[101] Šulc proposed Siciński as a member of the Continuing Committee of the World Futures Studies Research Conferences—a body that worked as an organization apparatus before the formal establishment of WFSF.[102]

While their professional relationship fostered the exchange of information, potentially strengthened their standing in international organizations and provided them with insight into national contexts, it did not lead to any cooperation in the form of projects or joint publications. Although the institutional structure of futures studies at ÚFS changed, Šulc remained the main international contact of ÚFS in futures studies until the fall of the Communist regime in Czechoslovakia in 1989. Siciński's research interests shifted toward qualitative studies of lifestyles. He remained interested in futures studies as far as it allowed for alternative utopian thinking, but he had become doubtful of its ability to effectively inform political decision-making.[103] Siciński took an active role in intellectual discussion groups close to the Solidarity movement in 1980–81, which published and disseminated a critical forecast regarding possible ways out of the crisis induced by the government's decision to introduce martial law.[104] The price he paid for it was restricted foreign travel. In 1985, Siciński was expelled from the Committee "Poland 2000" along with others.[105]

Dialectics of Obstacles

Not only Siciński and Šulc but deeper Czech–Polish cooperation in the field of futures studies faced institutional and political obstacles. First, the field only emerged at the end of the 1960s. As decisions and changes generally took quite some time under state socialist governance, this did not allow for its development and proper establishment before the political upheavals in both Czechoslovakia and Poland during and after 1968. Czechoslovakia experienced an academic reshuffle, persecutions and a reinvention of forecasting after 1968. At the futurological conference in Oslo in 1967, Polish sociologists had the opportunity to meet the representatives of Czechoslovakia, Bedřich Levčík[106] and Ota Klein.[107] However, Klein died in 1968 and Levčík emigrated at the beginning of the 1970s, becoming deputy director at the Vienna Institute for International Economic Studies in 1973, and therefore these contacts were unproductive for further inter-socialist cooperation in futures studies. On the other hand, the 1970s presented new opportunities for scholars who were new to the field or remained in it both in bilateral and multilateral frameworks.

One example is the OBP, which was established at the PWR in 1971 as part of a national system of prognostic research. Its local roots lay in an interdisciplinary study group on research and technology planning, which had become interested in forecasting methods. In addition to research in this field, the newly created OBP prepared didactic programs, conducted postgraduate studies, and coordinated research projects involving members of other faculties. It had grown to around forty employees by 1981 and established a rich internal scientific life with many internal publications and its own scientific journal.[108] Its interdisciplinary staff came mainly from technical, economic, and informational–mathematical disciplines, although one of its major long-term projects was the development of a matrix of major social needs, which was then to be applied in forecasting scientific and technological developments. Beyond the national futures studies network, in which it became a leading institute in science and technology forecasting, OBP organized multiple series of events with international guests: the Winter School of Forecasting (held annually from 1975 to 1981), a series of conferences on forecasting and environmental protection (1974, 1976, and 1982), and a biannual conference on the "Management of Research, Development and Education" (from 1975 to 1980). These conferences were rarely held in Wrocław, more often close to the Czechoslovak border in tourist resorts in the mountains south of the city, such as Szklarska Poręba, Międzygórze, Cieplice, and Karpacz.

These events were regularly attended by Czechoslovak scholars[109] and some gave lectures or were active there over a longer period.[110] Miloš Zeman (b. 1944), a former futures studies scholar turned politician, recalled visiting the OBP regularly and cooperating at events with Petrášek. He considered these contacts to be the most intensive with scholars from abroad.[111] The OBP's closest collaboration partner on the Czechoslovak side was the ÚVTR at the Federal Ministry for Technical and Investment Development. Petrášek had been working there since 1970. An agreement between both institutions was signed in 1975. The agreement encompassed the standard exchange of materials and experience, and plans for joint projects.[112] Both sides embarked on research on a project titled "Procedure for Creating State Plans and Programs of Science and Technology," which was meant to provide joint policy recommendations for decision-makers in both countries. However, the bilateral protocol of the project evaluation does not contain any indication of its successful finalization. The institutional collaboration was praised as fruitful, and there was a call for further development, but there were complaints that the unsynchronized work on the national contributions was impossible to merge into a synthesis.[113] Nevertheless, an outcome of the cooperation was Petrášek's employment at OBP, which was a unique Czech–Polish entanglement in the futures studies network.

The connection was rooted in both institutes engagement within the COMECON framework, in which Petrášek was the coordinator of the relevant methodological forecasting task at the level of COMECON.[114] At OBP, he worked as a contractual employee for several years, motivated by the freer environment with regard to intellectual debate and access to foreign literature.[115] Pelc had initiated the idea of offering Petrášek employment at OBP and it was facilitated by its rector for scientific affairs and co-founder, Wacław Kasprzak.[116] Petrášek and Pelc continued to collaborate over the years. For instance, they gave a joint lecture at a COMECON symposium in Sofia in 1980.[117] During his stays in Wrocław, Petrášek fully participated in the OBP's academic life, although his Polish was very rudimentary. Petrášek's presence was a significant asset for OBP's research and its external (re)presentation as it established and intensified a network not controlled by the academies, but between a regional institute and a branch ministry research institute—both of which were at least on the semi-periphery of their own respective national academic contexts.

Eventually, this constellation of the Czech–Polish futures studies network and the density of its contacts also became intertwined with the course of political developments, which initially furthered the cooperation and later severely limited it. Although political obstacles in Czechoslovakia initially induced Petrášek to search for a more unconventional space in which he could conduct research and cooperation, he was later forced to return to Czechoslovakia, when, from the point of view of the political authorities in Prague, the political situation in Poland had become too dangerous due to the Solidarity movement's opposition to the government. As it was considered a potential entrance for revisionist ideas, contacts between Czechoslovak and Polish citizens and institutions were closely monitored during that time[118]—as with Polish academia in general, the OBP employed many open or tacit supporters of the opposition movement. In 1981, its entire budget was put under control of the relevant branch ministry. After struggling with a severe illness, Pelc asked to be released from his position as its director in January 1982.[119] Baworowski, the former head of the OBP's postgraduate program, replaced him, as a younger generation of scholars, who like Baworowski had joined it only recently, began shifting the OBP's research agenda toward the use of computers in forecasting, game theory, and model simulations. During the 1980s, the OBP's funding was reduced, and it experienced financial problems. At the same time, the collaboration with COMECON ran out of steam, as the forecasting working group met rarely and was dissolved in 1988.[120] Nevertheless, Petrášek remained a member of the editorial board of the OBP's journal. And he would return to it in person at the end of the 1980s as an employee of the Institute of Economics of ČSAV.[121] The story of Pelc and Petrášek shows how obstacles can lead to close cooperation in noncentral spaces of a national part of the futures studies network and how these bilateral contacts can be even more intensive than in the center itself.

However, as in the case of the other constellations described above, institutes or departments dealing with forecasting had different foci in terms of areas. This hindered the establishment of lasting cooperation, which would have rested with senior scholars, for whom the Czech–Polish connection was only one direction among many. Established senior scholars in the Czechoslovak context after 1968 were focused on cooperation with Moscow (Richta), while the travel logs of Siciński and other Polish scholars representing the Committee "Poland 2000" testify to a much more intense exchange with the West between 1971 and 1981. Others (Petrášek, Zeman) seem to have lacked the funds, influence, and support to establish broader projects and multiple avenues of cooperation.

Despite the lack of joint projects or collaborations, there was a substantial amount of information exchange in the network. Czechoslovak journals such as *Trend* or later *Prognostika* mediated information about Polish institutions such as the OBP and "Poland 2000" and published excerpts of Polish forecasting research and reviews of Polish books.[122] OBP's journal *Prace Naukoznawcze i Prognostyczne* (Papers on science of science and forecasting) published Czechoslovak scholars' papers in Polish or English, and apart from Soviet and East German papers in Russian and German these were the most frequent foreign publications in the journal.[123] The OBP also published Šulc's work "Methods and Principles of Forecasting Technology Development" in an internal series of research reports, and other works were also translated into Polish in special supplements dedicated to forecasting.[124] In turn, Ota Šulc's "ABCs of Forecasting," which in its Czech original was available at the Polish National Library in Warsaw and the library of IFiS PAN, included an entry on forecasting studies in Poland.[125]

Finally, this points to the far from unpolitical role of language, which played out differently in each of the depicted constellations. English was the language of choice in the correspondence between Šulc and Siciński. Siciński knew only basic Russian, like his French and German, and he indicated that English was the only foreign language he was fluent in.[126] Moreover, both scholars frequently met on the occasion of international conferences in which English was the adopted working language. Zeman did not understand Polish and communicated with his counterparts in English.[127] As mentioned above, Russian was the language of the final publication from the project on bourgeois futurology, in which a Polish contribution was ultimately absent. Remarkably, a unified language was not a prerequisite for understanding. On the contrary, Šulc was content to receive materials from Siciński in Polish.[128] The anecdotal transaction between Szewczyk and Slíva, who "sold" Czech papers for publication in a volume in Polish, is a case of translation involving both material and information resources, which enhanced both human actors' capacities but without an external lingua franca. Similarly, the collaboration between the OBP and ÚVTR, which generated the closest institutional and personal entanglements in futures studies, was bilingual. Petrášek's letters in Czech were answered in Polish. Minutes of meetings were in both languages. Petrášek navigated the formalities of his one-year lectureship at OBP in 1980 in Czech. However, Pelc explicitly listed the goal of learning Polish up to a communicative level as one of the major reasons for Petrášek's continued employment at OBP, arguing in his recommendation to the rectorate that—given this goal would be accomplished—he would be a significant asset in (re)presenting the institute's scholarly record in Poland and abroad.[129] It seems that in the cases in which Czech–Polish futures studies were "only" bilateral, translations related to bilingual knowledge circulation caused less friction than a "common" language, which was associated with the two hegemonic powers of the Cold War world.

Conclusion

This chapter explored three constellations of intensive Czech–Polish contacts and cooperation. They developed as part of a broader transnational network in which the countries were not the number one priority partner for each other. Andrzej Siciński's and Ota Šulc' relationship was related to their engagement in the international field of futures studies. The Czech–Polish collaboration in the multilateral project on a critique of bourgeois futurology led to a bilateral outcome long before the multilateral project produced its projected outcome, which was produced with no substantial Polish involvement. František Petrášek's engagement with the OBP has been shown to be the most substantial and enduring collaboration, although political developments severely reduced its intensity by the early 1980s.

In many regards, Czech–Polish contacts and collaboration in the consolidating and interdisciplinary field of futures studies in the 1970s followed the logic of scientific exchange similar to well-established disciplines. At the beginning of the 1970s, scholars from both countries used the new—both bilateral and multilateral—opportunities to build on contacts originating from the 1960s as well as to form new ones. However, this did not lead to a notable increase in joint projects or some form of sustainable long-term cooperation that would last undisturbed until the end of the socialist regimes in Central and Eastern Europe. Many projects, such as the socialist Summer School of Forecasting or the working group on the critique of bourgeois futurology, brought more intensive cooperation, but they were one-off affairs and were not continued.

This can partly be ascribed to the specific position of futures studies in socialist science. Not constituting a discipline, futures studies often dissolved in other fields. This allowed individual futures studies scholars to obtain the position of unique specialists who provided others with methodological guidance. Thanks to their participation in transnational networks, they also had privileged access to literature and new ideas, with which the officials of socialist science policy and ideology could not always keep up. For the same reason, however, socialist futures studies did not provide stable institutional positions and many actors did not manage to attain positions that would have enabled them to pursue substantial independent international bilateral and multilateral projects in the 1970s and the following decade. Therefore, individual actors often moved between different projects as their interests shifted (Siciński), or were removed from their positions (Szewczyk), or were pressured to withdraw (Petrášek).

The relationship between Šulc and Siciński was characterized by the strategic use of mutual support in interinstitutional and multilateral constellations. This pragmatic intellectual friendship, however, did not culminate in joint research. Despite the fact that they exchanged materials, furthered each other's causes in different spaces, and had frequent contacts and personal meetings, they did not engage in any direct bilateral collaboration in the form of joint projects or publications.

At the time of the formation of the futures studies field, Siciński issued a called to arms in an intellectual "war over prognoses,"[130] which he saw emerging between the proponents of socialist and liberal-capitalist visions. Šulc—and Sedlák—set "significant goals for Marxist-Leninist forecasting," which would contest these ideas in Czechoslovakia.[131] The project on the critique of bourgeois futurology shows a multilateral space for cooperation, which was motivated by the joint determination to counter capitalist ideas of the future circulating in transnational debates in futures studies. It offered an opportunity to formulate a competing version of the future. Nevertheless, the project simply led to another collective volume criticizing various Western authors. As with the other depicted parts of the futures studies network, socialist collaboration and entanglements did not change the fact that the future remained a national construct.[132] At the same time, the futures studies network was a space of active involvement with global intellectual debates concerning structural changes of the international political economy, including "limits to growth" and new technologies, such as microelectronics or automated intelligence, often before such issues received national media attention.[133]

Petrášek's involvement with the OBP points to the sometimes unforeseen intertwining of political developments with geographical and institutional factors. While he had to retreat from academia to ÚVTR after the invasion in 1968, moving toward a scientific periphery in the context of Czechoslovakia, this change enabled him to build upon his earlier Polish contacts and to engage in the closest collaboration that could be found in the Czech–Polish network of futures studies. This may suggest that the more peripheral parts of individual national scientific spaces might have been sometimes more suitable for intensive contacts than institutions in the center. In conclusion, despite all the translation, necessary transactions and commuting it required, this enabled individual scholars to speak on behalf of themselves and their particular Czech or Polish identities as scholarly persons in the international field of futures studies.

Ultimately, this chapter's focus on Czech–Polish entanglements in the futures studies network demonstrates the importance of further research on the networks of socialist science. In particular institutions that were more peripheral in the national hierarchies of science organization, such as the OBP, could be lively hubs for transnational contacts and thus important sites of knowledge exchange and production in particular fields. Furthering research on socialist scientific collaboration, including its multilateral dimension, thus can reveal entanglements that are often played down or left out in popular narratives, as they tend to focus on contacts beyond the Iron Curtain.

Author Biographies

Lukas Becht is a political scientist and historian focusing on the history of future studies, political time, and knowledge production in Central and Eastern Europe. He studied political science, sociology, and philosophy, obtained a master's degree at the University of Freiburg and interned at the Polish Sejm. In a binational PhD program between LMU Munich and the University of Vienna, he researches scientific futurology in socialist Poland. He is an Associated researcher at RECET, University of Vienna, where he also coordinated the project Economic Collectivism Old and New.

Luboš Studený is a historian specializing in the late socialist and post-socialist period of Czechoslovakia, with a focus on the history of future studies, postcommunist transformation, cybernetics and historiography theory. He holds a master's degree from Charles University, where he is currently completing his PhD in modern economic and social history. He has contributed to research on socialist enterprise transformation and Cold War scientific networks.

List of Abbreviations

APAN	Archiwum Polskiej Akademii Nauk (Archive of the Polish Academy of Sciences)
APWR	Archiwum Politechniki Wrocławskiej (Archive of the Wrocław Technical University)
BAN	Bŭlgarska akademiya na naukite (Bulgarian Academy of Sciences)
BN	Biblioteka Narodowa (National Library of Poland)
BWZ	Biuro Współpracy z Zagranicą Polskiej Akademii Nauk [Office of Foreign Relations of the Polish Academy of Sciences]
COMECON	Council for Mutual Economic Assistance
ČSAV	Československá akademie věd (Czechoslovak Academy of Sciences)
IFiS	Instytut Filozofii i Socjologii (Institute of Philosophy and Sociology)
IRADES	Instituto Richerche Applicato Documentazione e Studi (Institute of Applied Research, Documentation and Studies)
MÚA AV ČR	Masarykův ústav a Archiv Akademie věd České republiky (Masaryk Institute and Archives of the Czech Academy of Sciences)
OBP	Ośrodek Badań Prognostycznych (Center of Prognostic Research)
PAN	Polska Akademia Nauk (Polish Academy of Sciences)
PWR	Politechnika Wrocławska (Wrocław Technical University)
STR	Scientific-technical revolution
UJ	Uniwersytet Jagielloński (Jagiellonian University)

UNESCO	United Nations Educational, Scientific and Cultural Organization
ÚFS	Ústav pro filosofii a sociologii (Institute of Philosophy and Sociology)
ÚVTR	Ústav pro ekonomiku a řízení vědeckotechnického rozvoje (Institute of Economics and Management of Scientific-Technical Development)
WFSF	World Futures Studies Federation

Notes

This study is a result of research funded by the Czech Science Foundation under project GA ČR GF21-45624L "Czechoslovak-Polish Scholarly Entanglements in the Cold War between High Politics and Individual Strategies." The archival research on which this chapter builds was funded by LMU Munich's Institutional Strategy "LMUexcellent" within the framework of the German Excellence Initiative and by a short-term scholarship at the German Historical Institute in Warsaw.

1. MÚA AV ČR, Coll. ÚFS-VTR, box 3, inv. no. 44, sig. 655, Letter from Ota Šulc to Andrzej Siciński, Praha, 1972-17-07, not paginated.
2. MÚA AV ČR, Coll. ÚFS-VTR, box 2, inv. no. 34, sig. 61, Report on the progress and results of contacts between the STR section of the Institute of Philosophy and Sociology and foreign countries in 1971, n.d., p. 3.
3. In studying transnational scientific connections between Czechoslovakia and Poland in the field of futures studies, the chapter is limited to Czech–Polish contacts as they were much more developed and intense than Polish–Slovak contacts, for instance, with the Slovak Academy of Sciences. Thus, instead of introducing a multilateral space of three distinct sides, this chapter does not analyze Polish–Slovak exchange.
4. Andersson, *The Future of the World*, esp. 122–50; Seefried, "Der Kurze Traum von der steuerbaren Zukunft," 179–220, esp. 184–219.
5. For Polish sociology, see Kilias, "Socjologia polska i czeska," 283–315, esp. 291, and Kornelia Kończal's chapter in this volume.
6. This chapter's approach follows the principles of Actor–Network–Theory in that it considers the relationships it describes as situated and specific, as they are constructed by variously intense practices instead of being by definition global or bilateral. This does not imply the full adoption of its methodological program but has an influence on the understanding of the nature of a network's operation. See Latour 1996, "On Actor-Network Theory," 369–381.
7. Andersson, *The Future of the World*; Eberspächer, *Das Projekt Futurologie*; Pietruska, *Looking Forward*; Seefried, *Shaping Tomorrow's World*.
8. Rocca, "'A Second Party in Our Midst,'" 199–247; Sommer, "Forecasting the Post-Socialist Future"; Sommer, "From Socialist Post-industrialism to Market Economy"; Catanus, "Official and Unofficial Futures"; Becht, "From Euphoria to Frustration"; Rindzevičiūtė, "A Struggle for the Soviet Future"; Rindzevičiūtė, "The Future as an Intellectual Technique."
9. Rindzevičiūtė, "Toward a Joint Future"; Rindzevičiūtė, *The Power of Systems*; Rindzevičiūtė, *The Will to Predict*.

10. Jim Dator, email communication with Luboš Studený, 31 October 2024.
11. For an overview of the history of interconnectedness, see Surman. "Re-connecting."
12. This term refers to a concept used in the history of ideas and knowledge to draw attention to the intellectual constellations in which scholarly knowledge is produced by concrete actors in organizational, pragmatic, and sociopolitical settings. See Henrich, "Konstellationsforschung"; Mulsow, "Zum Methodenprofil der Konstellationsforschung," 74–96.
13. Schermer, *K problematice vědeckého předvídání.*
14. Šulc, "Marxisticko-leninská prognostika."
15. Sommer, Spurný, and Mrňka, *Řídit socialismus jako firmu.*
16. Siciński, *Prognozy a nauka*, 5.
17. Secomski, *Prognostyka*, esp. 14; Secomski 1970, "Prognostic Research."
18. Andersson, *The Future of the World*; Seefried, *Shaping Tomorrow's World.*
19. Koryś and Tymiński, "Od socjalizmu do socjalizmu"; Bukowski and Maciejewski, "Between Rationality and Reality"; Gouarné, "Mandatory Planning versus Indicative Planning."
20. Richta, *Civilization at the Crossroads*. See also Guth, "One Future Only."
21. See the contribution by Jan Surman in this volume.
22. Richta, Klein, Levčík, and Dubská , "The Perspective of the Scientific and Technological Revolution."
23. Slíva, Jiří, 2024, Interviewed by Luboš Studený. 10 October. Audio.
24. Šulc, *Futurologie.*
25. See Kokowski, "The Science of Science (*Naukoznawstwo*) in Poland"; Herman and Olšáková, *Plánování socialistické vědy.*
26. Seefried and Malycha, "Planen, um aufzuschließen"; Naumann, "International Research Planning."
27. For example, Gordon and Helmer, *Prognozy wieloletnie*. Interestingly, the Polish translation of selected parts of this US book was based on a French translation.
28. See Sommer, "Scientists of the World, Unite!"; Guth, "One Future Only."
29. Pajestka, "Rok 2000."
30. Sommer, Spurný, and Mrňka, *Řídit socialismus jako firmu.*
31. For the development of futures studies in Czechoslovakia in the 1960s, see Sommer, Spurný, and Mrňka, *Řídit socialismus jako firmu*, 64–82.
32. Kurczewski and Lutyk, "Futurological Conference."
33. Filipec, "Futurologie v Polsku"; Šulc, *Futurologie*; Siciński, *Prognozy a nauka.*
34. See Šulc, "Budoucnost vědy o budoucnosti"; Šulc, "Marxisticko-leninská prognostika."
35. APAN, coll. BWZ, sig. II-2/4, inv. no. 620, box 136, Andrzej Siciński. Życiorys naukowy, Warszawa, 1975-31-10, not paginated.
36. Siciński, "Social Prognoses Division."
37. Siciński, "Przedmowa," 7–13, esp. 8.
38. Šulc, *Abeceda prognostiky.*
39. See Siciński, "Two Sociologies," esp. 5.
40. Studený, "Die Prognostik als Weg aus der Krise."
41. Šiklová, "The 'Gray Zone' and the Future of Dissent in Czechoslovakia."
42. The most intense collaboration was with the so-called Conversatory "Experience and Future" (*Konwersatorium "Doświadczenie i Przyszłość"*). It met for the first and only time on 14 November 1978. However, after its dissolution, a smaller circle of

journalists and scholars continued to meet and produce texts, unofficially aided by "Poland 2000." See Siciński, "Konwersatorium im. Bogdana Gotowskiego."

43. Rolbiecki, "Badania prognostyczne—prognozologia—futurologia"; Šulc, "Sympozium o prognózách."
44. "Zpráva o konferenci."
45. See Friszke, *Między Wojną a Więżeniem*, esp. 124–27.
46. See Sułek, "To America!"
47. Jungk and Galtung, *Mankind 2000*; Levčík and Klein, "Futurologický kongres v Oslu."
48. Tuchel, "Die Zukunft der Zukunftsforschung."
49. Richta, Klein, Levčík, and Dubská , "The Perspectives of the Scientific and Technical Revolution."
50. MÚA AV ČR, Coll. ÚFS-VTR, box 1, inv. no. 10, sig. 45, CV of Ota Šulc, n.d., p. 4.
51. Petrášek, František. 2013. Interview by Vítězslav Sommer. 1 November. Audio. He definitely helped develop the center's research during his long-term lectureships, as noted by Karol Pelc. See AWPR, Coll. "Akta osobowe," sig. 877.678/1, Letter by the director of OBP to the prorector of PWR, signed by Karol Pelc, Wrocław, 1979-28-09, not paginated.
52. Petrášek, František. 2013. Interview by Vítězslav Sommer. 1 November. Audio.
53. Brádlerová and Kmochová, "Nástin zahraničních styků ČSAV," esp. 113.
54. See Pleskot and Rutkowski, "Wstęp"; Pleskot, "Jak wyjechać na Zachód?"
55. Brádlerová and Kmochová, "Nástin zahraničních styků ČSAV," esp. 110.
56. The program number and title was I-25.2: "Methodology of forecasting and planning of science and technology development."
57. Baworowski and Kosiński, *Działalność Ośrodka Badań Prognostycznych.*
58. Sztukowski, "Sofijskie spotkania prognostów RWPG," esp. 66–67. The multilingual dictionary, which was published as a working paper, contained twenty-five terms in Russian, Polish, Bulgarian, Czech, German, Hungarian, and Romanian: Kosiński, Ignaszewski, and Sztukowski, *Terminologicessij slovar'.*
59. Baworowski, "Charakterystyka i prognozowanie," esp. 41.
60. Sztukowski, "O metodyce prognozowania," esp. 100.
61. Sztukowski, "O metodyce prognozowania," esp. 102.
62. Baworowski and Kauf, "Metodyka wspólnego prognozowania," esp. 184.
63. AWPR, Coll. "Spółki z udziałem Politechniki Wrocławskiej: Ośrodek Badań Prognostycznych 1987–2003," sig. 74/442, Sprawozdanie z działalności OBP w obszarze współpracy zagranicznej, Wrocław, August 1989, p. 18–22, esp. 22. See also Kosiński, "Dziesięciolecie działalności," esp. 172.
64. For example, Zeman, *Futurologický sborník.*
65. Later renamed "Aspects of Ideological Struggle in the Contemporary World."
66. Hrzal, "O činnosti Problémové komise."
67. Jodkowska, "Antykomunistyczne Aspekty Futurologii," esp. 32.
68. Sedlak and Szulc [*sic!*], "Powstanie futurologii burżuazyjnej," esp. 100. This was a text that originally appeared in Czechoslovakia. See Sedlák and Šulc, "Vznik buržoazní futurologie."
69. For Nowa Huta, see Lebow, *Unfinished Utopia*. For Szewczyk's biography, see Sowa, "Jan Szewczyk."
70. MÚA AV ČR, Coll. ÚFS-VTR, box 3, inv. no. 44, sig. 655, Letter from Jindřich Filipec to Jan Szewczyk, Praha, 1971-02-09, p. 1–2.

71. MÚA AV ČR, Coll. ÚFS-VTR, box 3, inv. no. 44, sig. 655, Travel report by Ota Šulc and Jiří Slíva, Praha, 1971-03-12, p. 5.
72. MÚA AV ČR, Coll. ÚFS, box 2, inv. no. 34, sig. 61, Report on the progress and results of contacts between the STR section of the Institute of Philosophy and Sociology and foreign countries in 1971, n.d., p. 3.
73. Slíva, Jiří. 2024. Interview by Luboš Studený. 10 October. Audio. 26:15.
74. Ośrodek Badania Stosunków Wschód-Zachód 1972, ed. *Wybrane problemy współczesnej futurologii burżuazyjnej.*
75. Sedlák, and Šulc, "Vznik buržoazní futurologie"; Slíva, "Světová dynamika."
76. Szewczyk, "Jednotný ideologický postup."
77. However, several contributions to the Information Bulletin of the Problem Commission were published.
78. According to the protocol signed in November 1972, see MÚA AV ČR, Coll. ÚFS-VTR, box 2, inv. no. 34, sig 61, Analysis of the course and results of scientific contacts of the Section for Research on the Scientific and Technological Revolution with foreign countries in 1972, n.d., p. 3.
79. Kozharov, *Budushchee*, 6.
80. Sowa, "Jan Szewczyk," esp. 146.
81. MÚA AV ČR, Coll. ÚFS-VTR, box 2, inv. no. 34, sig. 61, Analysis of the course and results of scientific contacts of the Section for Research on the Scientific and Technological Revolution with foreign countries in 1972, n.d., p. 3.
82. Kozharov, *Budushchee.*
83. Andersson, *The Future of the World*; Seefried, *Shaping Tomorrow's World.*
84. MÚA AV ČR, Coll. ÚFS – Sekretariát ředitele Radovana Richty, box 44, sig. 6, Travel report by Jiří Slíva and Oleg Suša, Praha, 1977-08-12, p. 2.
85. APAN, Coll. BWZ, sig. II-2/4, inv. no. 620, box 13, file 69 – Juliusz Goryński, Note to p. Obuchowicz, Warszawa, n.d., not paginated.
86. MÚA AV ČR, Coll. ÚFS – Sekretariát ředitele Radovana Richty, box 42, sig. 6, Foreign relations plan of the Section on STR for 1979, n.d., p. 2.
87. MÚA AV ČR, Coll. ÚFS-VTR, box 2, inv. no. 34, sig. 61, Report on the progress and results of contacts between the STR section of the Institute of Philosophy and Sociology and foreign countries in 1971, n.d., p. 3.
88. Petrášek, *Systemowe ujęcie prognozowania.*
89. Šulc, "Przyczynek do metodologii prognozowania stylu życia."
90. MÚA AV ČR, Coll. ÚFS, box 3, inv.no. 44, sig. 655, Travel report by Ota Šulc, Praha, 1971-03-12, p. 3.
91. BN, sig. 3.174.425A, Andrzej Siciński. 1972. Prognozowanie jako element sterowania procesami społeczno-ekonomicznymi. Materiały I. "Letniej Szkoły Prognozowania" dla uczestników z krajów socjalistycznych Rynia k/Warszawy, 18.IX.–27.IX.1972 r. (do użytku służbowego), Warszawa, September 1972, p. 2.
92. Mikiewicz, "Letnia szkoła prognozowania w Ryni," esp. 84; Jawłowska, "Szkoła prognozowania w Ryni," esp. 302–3.
93. The only mentions are in and based on the travel report of the Czech participants. MÚA AV ČR, Coll. ÚFS-VTR, box 3, inv. no. 44, sig. 655, Travel report by Ota Šulc and Jiří Slíva, Praha, 1971-03-12, p. 3.
94. BN, sig. 3.174.425A, Siciński. Prognozowanie jako element sterowania procesami społeczno-ekonomicznymi, p. 11.

95. Ibid.
96. See, for instance, the chapters by Rutkiewicz (15–30) and Jankow (21–31) in Siciński, Kovács, and Starowierow, *Problemy prognozowania*, which argue against a distinct discipline of futures studies in socialism, as opposed to Kovács's argument that "there is—or should be—Marxist futurology." Kovács, "Naukowe Przewidywanie Przyszłości," esp. 56.
97. Šulc, "Przyczynek do metodologii prognozowania stylu życia," esp. 159.
98. MÚA AV ČR, Coll. ÚFS-VTR, box 3, inv.no. 44, sig. 655, Letter from Ota Šulc to Andrzej Siciński, Praha, 1973-22-01, not paginated.
99. Siciński, "Perspektivy přeměn životního stylu v Polsku."
100. MÚA AV ČR, Coll. ÚFS-VTR, box 3, inv.no. 44, sig. 655, Letter from Ota Šulc to Andrzej Siciński, Praha, 1972-29-03, not pag.
101. MÚA AV ČR, Coll. ÚFS-VTR, box 1, inv.no. 17, sig. 12, Ota Šulc – Plan for 1973, p. 5.
102. MÚA AV ČR, Coll. ÚFS-VTR, box 3, inv. no. 49, sig. 673, Letter from Ota Šulc to Arne Sorensen, Praha, 1973-05-01, not paginated.
103. Mazurek, "Rewizyta Etnograficzna."
104 Gotowski, B., Kapuściński, Nowak, Regulski, Siciński, Trzeciakowski and Gotowski, W., *Hipotetyczne Scenariusze Przemian Ekonomicznych.*
105. "Działalność PAN w 1984 r."; Siciński, "List otwarty do A. Gieysztora."
106. Later referred to as Friedrich Levcik as well.
107. Levčík and Klein, "Futurologický kongres v Oslu."
108. Kosiński, "Dziesięciolecie działalności Ośrodka Badań Prognostycznych." Pelc, "Futures Research in a University."
109. Potůček, Martin. 2024. Interview by Luboš Studený. 5 June. Audio.
110. Petrášek and Šebelík, "Ze světa prognostiky," esp. 53, who also wrote that OBP was "the only institute of its kind in socialist countries."
111. Zeman, Miloš. 2025. Interview by Luboš Studený. 19 February. Audio.
112. AWPR, Coll. "Umowy i Porozumienia," sig. 238/177, Protokół porozumienia w sprawie dwustronnej współpracy między OBP we Wrocławiu a UVTR, Praha, 1976-27-02, p. 3–12.
113. APWR, Coll. "Umowy i Porozumienia," sig. 237, box 177, Protokół porozumienia w sprawie dwustronnej współpracy między Ośrodkiem Badań Prognostycznych a UVTR w Pradze [opracowany w czasie wizyty dr. Z. Szymanka i T. Galanca w Pradze 18.-20.10.1976], Wrocław, 1976-20-10, p. 13–18.
114. MÚA AV ČR, Coll. ÚFS-VTR, box 2, inv. no. 38, sig. 633, Travel report by Petr Kotlas and Vladimíra Machová, Praha, June 1973, p. 15. Task 1.25.2: Methodology of forecasting of development of science and technology.
115. Petrášek, František. 2013. Interview by Vítězslav Sommer. 1 November. Audio.
116. AWPR, Coll. "Akta osobowe," sig. 877.678/1, Letter from the director of OBP to the prorector of PWR, signed by Karol Pelc, Wrocław, 1979-28-09, not paginated; Kasprzak, Wacław. 2024. Interview by Lukas Becht, 11 November.
117. The title of the paper was "Problems of Organization of a System of Forecasting and Its Use for Developing State Strategy of Technical Development." See *Prognózování vědeckotechnického rozvoje v zemích RVHP.*
118. Blažek, *Akce "Sever."*

119. APWR, Coll. "Akta osobowe," sig. 877.498/89, Letter by Karol Pelc to the rector, Wrocław, 1982-27-02, not paginated.
120. AWPR, Coll. "Spółki z udziałem Politechniki Wrocławskiej," sig. 74/442, Sprawozdanie z działalności OBP w obszarze współpracy zagranicznej, Wrocław, August 1989, p. 18–22.
121. MÚA AV ČR, Coll. ZHO ZSS UP CSAV, PAV, box 99, Travel proposal of František Petrášek, n.d.
122. For example, Zawadzki, "Komise 'Polsko 2000'"; N.a., "Oddělení společenských prognóz Ústavu filosofie a sociologie"; Slíva, "Práce prognostického výboru 'Polsko 2000'"; "Delfská prognóza vývoje železniční dopravy v PLR"; Petrášek and Šebelík, "Ze světa prognostiky"; Baworowski, "Prognosticka studie svetoveho technickeho pokroku."
123. Machová, "Projekt kierowania programami rozwoju nauki i techniki"; Petrášek, "On the Interpretation of the Forecasting Object"; Zeman, "Three Generations of Forecasting Methods."
124. Šulc, *Metody i zasady prognozowania rozwoju techniki*; Petrášek, *Systemowe ujęcie prognozowania*.
125. Šulc, *Abeceda prognostiky*.
126. APAN, Coll. BWZ, sig. II-2/4, inv. no. 620, file 136, Andrzej Siciński – Kwestionariusz osobowy, Warszawa, 1975-31-10, not paginated.
127. Zeman, Miloš. 2025. Interview by Luboš Studený. 19 February. Audio.
128. MÚA AV ČR, Coll. ÚFS-VTR, box 3, inv. no. 44, sig. 655, Letter from Ota Šulc to Andrzej Siciński, Praha, 1973-22-01, not paginated.
129. AWPR, Coll. "Akta osobowe," sig. 678/1, Pismo Dyrektora OBP do prorektora Politechniki Wrocławskiej, signed by Karol Pelc, Wrocław, 1979-28-09, not paginated.
130. Siciński, *Prognozy a nauka*, esp. 203.
131. Sedlák and Šulc, "Vznik buržoazní futurologie a její vliv v Československu."
132. Efforts to develop a socialist contribution to global models and global studies are mainly from the 1980s, even though they emerged during the 1970s. They were mostly connected to the Institute for Systems Research of the All-Union Soviet Academy of Sciences in Moscow, to which global models in other socialist countries contributed with their regional data, but these were only contributions to a Soviet endeavor. See Rindzevičiūtė, *The Power of Systems*.
133. It remains an issue for further research to reconstruct the involvement of the "second (socialist) world" with the Club of Rome and the debates triggered by *Limits to Growth* it published in 1972.

Bibliography

Andersson, Jenny. 2018. *The Future of the World: Futurology, Futurists, and the Struggle for the Post Cold War Imagination*. Oxford: Oxford University Press.

Baworowski, Jerzy Ludwik. 1974. "Charakterystyka i prognozowanie potencjału badań naukowych i prac rozwojowych" [Characterization and forecasting of the potential of scientific research and development works]. *Prace Naukoznawcze i Prognostyczne* III, no. 11: 29–43.

———. 1981. "Prognosticka studie svetoveho technickeho pokroku" [Prognostic study of world technical progress]. *Trend: prognostické informace* 13, no. 2: 1–5.

Baworowski, Jerzy Ludwik, and Leszek Kauf. 1979. "Metodyka wspólnego prognozowania rozwoju techniki w krajach członkowskich RWPG (opis i analiza realizacji eksperymentu)" [Methodology of joint forecasting of technological development in the COMECON member states (description and analysis of the experiment implementation)]. In *Z doświadczeń metodologicznych prognozowania społeczno-ekonomicznego i technicznego* [From the methodological experiences of socioeconomic and technical forecasting], edited by Andrzej Siciński and Maria Ptaszyńska, 179–85. Wrocław: Ossolineum.

Baworowski, Jerzy Ludwik, and Józef Kosiński. 1974. *Działalność Ośrodka Badań Prognostycznych Politechniki Wrocławskiej w 1973 roku* [Activities of the Centre of Prognostic Research at the Wrocław Technical University in 1973]. Warsaw: Ministerstwo Nauki, Szkolnictwa Wyższego i Techniki.

Becht, Lukas. 2017. "From Euphoria to Frustration: Institutionalizing Prognostic Research in the Polish People's Republic, 1969–76." *Acta Poloniae Historica* no. 116: 277–99.

Blažek, Petr, ed. 2017. *Akce "Sever": Státní bezpečnost a krize Polské lidové republiky 1980–1984* [Operation "North": State security and the crisis of the Polish People's Republic, 1980–1984]. Prague: Ústav pro studium totalitních režimů.

Brádlerová, Daniela, and Nataša Kmochová. 2001. "Nástin zahraničních styků ČSAV v letech 1962–1970" [An outline of international relations of ČSAV, 1962–1970]. In Česká věda a Pražské jaro (1963*–1970)* [Czech Science and the Prague Spring (1963–1970)], edited by Petr Svobodný and Blanka Zilynská, 107–18. Prague: Karolinum.

Bukowski, Maciej, and Wojciech Maciejewski. 2022. "Between Rationality and Reality: Economics and Central Planning in Poland (1945–1989)." In *Communist Planning versus Rationality: Mathematical Economics and the Central Plan in Eastern Europe and China*, edited by János Mátyás Kovács, 211–26. Lanham, MD: Lexington Books.

Catanus, Ana-Maria. 2015. "Official and Unofficial Futures of the Communist System: Romanian Future Studies between Control and Dissidence." In *The Struggle for the Long-Term in Transnational Science and Politics: Forging the Future*, edited by Jenny Andersson and Eglė Rindzevičiūtė, 170–94. New York: Routledge.

"Delfská prognóza vývoje železniční dopravy v PLR" [Delphi forecast for the development of rail transport in the People's Republic of Poland]. 1977. *Trend: prognostické informace* 9, no. 4: 17–28.

"Działalność PAN w 1984 r. LXII Sesja Zgromadzenia Ogólnego PAN (Warszawa, 31 V 1985 r.)" [Activities of the PAN in 1984. 62th General Assembly of the PAN (Warsaw, 31.05.1985)].1985. *Nauka Polska* 33, no. 5: 229–33.

Eberspächer, Achim. 2019. *Das Projekt Futurologie: Über Zukunft und Fortschritt in der Bundesrepublik 1952–1982* [The futurology project. On future and progress in the Federal Republic of Germany 1952–1982]. Paderborn: Ferdinand Schöningh.

Filipec, Jindřich. 1968. "Futurologie v Polsku" [Futurology in Poland]. *Sociologický Časopis* 4, no. 5: 617–18.

Friszke, Andrzej. 2015. *Między wojną a więżeniem 1945–1953: Młoda Inteligencja Katolicka* [Between war and prison 1945–1953: The young Catholic intelligentsia] Warsaw: Biblioteka "Więzi"/ISP PAN.

Gordon, Thomas I., and Olaf Helmer. 1965. *Prognozy wieloletnie (Próba przewidywania przyszłością metodą delficką).* Studia i materiały do perspektywicznego planu rozwoju nauki polskiej. Materiały zagraniczne 1 [Long-term forecasts (An attempt to forecast the future with the Delphi method). Studies and Materials for the Perspective Plan of Development of Polish Science. Foreign Materials 1]. Warsaw: Ośrodek Planowania i Koordynacji Badań Naukowych Polskiej Akademii Nauk.

Gotowski, Bogdan, Ryszard Kapuściński, Stefan Nowak, Jerzy Regulski, Andrzej Siciński, Witold Trzeciakowski, and Witold Gotowski. 1982. *Hipotetyczne Scenariusze Przemian Ekonomicznych, Społecznych i Politycznych w Polsce 1982/83* [Hypothetical scenarios of economic, social and political changes in Poland 1982/1983]. Warsaw: Warszawska Drukarnia Naukowa.

Gouarné, Isabelle. 2018. "Mandatory Planning versus Indicative Planning? The Eastern Itinerary of French Planners (1960s–1970s)." In *Planning in Cold War Europe: Competition, Cooperation, Circulations (1950s–1970s)*, edited by Ondrej Matejka, Sandrine Kott, and Michel Christian, 71–96. Berlin: De Gruyter.

Guth, Stefan. 2015. "One Future Only: The Soviet Union in the Age of the Scientific-Technical Revolution." *Journal of Modern European History* 13, no. 3: 355–76.

Henrich, Dieter. 2005. "Konstellationsforschung zur klassischen deutschen Philosophie: Motiv—Ergebnis—Probleme—Perspektiven—Begriffsbildung" [Constellation research on classical German philosophy: motive—result—problems—conceptualisation]. In *Konstellationsforschung* [Constellation Research], edited by Martin Mulsow and Marcelo Stamm, 15–30. Frankfurt a. M.: Suhrkamp.

Hermann, Tomáš, and Doubravka Ošáková, eds. 2013. *Plánování socialistické vědy: dokumenty z roku 1960 ke stavu a rozvoji přírodních a technických věd v Československu* [Planning of socialist science: Documents from 1960 on the state and development of the natural and technical sciences in Czechoslovakia]. Červený Kostelec: Pavel Mervart.

Hrzal, Ladislav. 1989. "O činnosti Problémové komise mnohostranné spolupráce akademií věd socialistických zemí 'Ideologický boj v současném světě'" [On the activities of the Problem Commission of Multilateral Cooperation of the Academies of Sciences of Socialist Countries: "Ideological Struggle in the Contemporary World"]. *Sociologický časopis* 25, no. 3: 324–27.

Jawłowska, Aldona. 1973. "Szkoła Prognozowania w Ryni" [School of prognostics in Rynia]. *Kultura i Społeczenstwo* 14: 295–303.

Jodkowska, Brygyda. 1972. "Antykomunistyczne Aspekty Futurologii" [Anti-Communist aspects of futurology]. In *Wybrane Problemy Współczesnej Futurologii Burżuazyjnej: Praca Zbiorowa [Tylko Do Użytku Wewnętrznego. Poufne]* [Selected Problems of Contemporary Bourgeois Futurology. Collective Work (For internal use. Confidential)], edited by Ośrodek Badań Stosunków Wschód-Zachód, 16–71. Warsaw: RSW Prasa.

Jungk, Robert, and John Galtung. 1969. *Mankind 2000*. Oslo: Universitetsforlaget/Allen & Unwin.

Kilias, Jarosław. 2016. "Socjologia polska i czeska: Wzajemne stosunki, ich charakter i konteksty" [Polish and Czech sociology. Mutual relationships, their character and contexts]. *Stan Rzeczy* 10: 283–315.

Kokowski, Michał. 2016. "The Science of Science (*Naukoznawstwo*) in Poland: Defending and Removing the Past in the Cold War." In *Science Studies during the Cold War and Beyond: Paradigms Defected*, edited by Elena Aronova and Simone Turchetti, 149–76. New York: Palgrave Macmillan.

Koryś, Piotr, and Maciej Tymiński. 2016. "Od socjalizmu do socjalizmu: Koncepcje reform gospodarczych w PRL po wybuchach społecznych w 1956 i 1980 r." [From socialism to socialism: Concepts of economic reform in the People's Republic of Poland after the social upheavals in 1956 and 1980]. *Dzieje Najnowsze* 48, no. 4: 125–40.

Kosiński, Józef Adam. 1981. "Dziesięciolecie działalności Ośrodka Badań Prognostycznych Politechniki Wrocławskiej 1971–1981" [Ten years of activity of OBP 1971–1981]. *Prace Naukoznawcze i Prognostyczne* 10, no. 1–2: 155–75.

Kosiński, Józef, Lech Ignaszewski, and Czesław Sztukowski. 1973. *Terminologicessij slovar' osnovnych ponjatij prognostiki* [Terminology: A glossary of basic concepts in forecasting] [Raport OBP, nr 4]. Wrocław: OBP Politechniki Wrocławskiej.

Kovács, Géza. 1975. "Naukowe przewidywanie przyszłości oraz długofalowe planowanie" [Scientific forecasting and long-term planning]. In *Problemy prognozowania w krajach socjalistycznych*, edited by Andrzej Siciński, Géza Kovács, and Władimir I. Starowierow, 49–68. Wrocław: Ossolineum.

Kozharov, Asen, ed. 1979. *Budushchee: Real'nye problemy i burzhuaznye spekuliatsii* [The future: Real problems and bourgeois speculations]. Sofia: Izdatel'stvo Bolgarskoj Akademii Nauk.

Kurczewski, Jacek, and Aleksander Lutyk. 1967. "Futurological Conference." *The Polish Sociological Bulletin* 16: 127–30.

Latour, Bruno. 1996. "On Actor-Network Theory: A Few Clarifications." *Soziale Welt* 47, no. 4: 369–81.

Lebow, Katherine. 2013. *Unfinished Utopia: Nowa Huta, Stalinism, and Polish Society, 1949–56*. Ithaca, NY: Cornell University Press.

Levčík, Bedřich, and Ota Klein. 1968 "Futurologický kongres v Oslu" [The futurological congress in Oslo]. *Sociologický časopis* 4: 230–32.

Machová, Vladimíra. 1974. "Projekt kierowania programami rozwoju nauki i techniki" [Project on conducting science and research development programs]. *Prace Naukoznawcze i Prognostyczne* 3, no. 11: 45–54.

Mazurek, Małgorzata. 2015. "Rewizyta etnograficzna: Jak się wytwarza wiedzę socjologiczną" [Ethnographical revisit: How to produce sociological knowledge]. *Kultura i Społeczenstwo* 59, no. 3: 31–62.

Mikiewicz, Jan. 1973. "Letnia Szkola Prognozowania w Ryni" [Summer School of Forecasting in Rynia]. *Prace Naukoznawcze i Prognostyczne* 2, no. 9: 81–84.

Mulsow, Martin. 2005. "Zum Methodenprofil der Konstellationsforschung" [The methodological profile of constellation research]. In *Konstellationsforschung*, edited by Martin Mulsow, and Marcelo Stamm, 74–97. Frankfurt a. M.: Suhrkamp.

Naumann, Katja. 2018. "International Research Planning across the Iron Curtain: East-Central European Social Scientists in the ISSC and Vienna Centre." In *Planning in Cold War Europe: Competition, Cooperation, Circulations (1950s–1970s)*, edited by Ondrej Matejka, Sandrine Kott, and Michel Christian, 98–122. Boston: De Gruyter.

"Oddělení společenských prognóz Ústavu filosofie a sociologie, Polská akademie věd, Varšava" [Department of Social Forecasting of the Institute of Philosophy and Sociology, Polish Academy of Sciences, Warsaw]. 1970. *Trend: prognostické informace* 2, no. 5: 34–35.

Ośrodek Badania Stosunków Wschód-Zachód, ed. 1972. *Wybrane problemy współczesnej futurologii burżuazyjnej: Praca zbiorowa [Tylko do użytku wewnętrznego.*

Poufne] [Selected problems of contemporary bourgeois futurology. Collective work (For internal use. Confidential)]. Warsaw: RSW Prasa.

Pajestka, Józef. 1963. "Rok 2000: Ku czemu zmierza świat?" [The Year 2000. Where is the world heading?]. *Przegląd Kulturalny*, 3 June.

Pelc, Karol. 1981. "Futures Research in a University: The Polish Experience." *Technological Forecasting and Social Change* 19, no. 3: 237–43.

Petrášek, František. 1974. *Systemowe ujęcie prognozowania z punktu widzenia wymogów kierowania rozwojem technicznym.* [A systemic take on forecasting from the point of view of technology development needs]. Warsaw: Ministerstwo Nauki, Szkolnictwa Wyższego i Techniki.

———. 1978. "On the Interpretation of the Forecasting Object as a Developmental Factor." *Prace Naukoznawcze i Prognostyczne* 8, no. 18: 41–54.

Petrášek, František, and Jan Šebelík. 1987. "Ze světa prognostiky" [From the world of prognostics]. *Prognostika* 2: 52–55.

Pietruska, Jamie. 2017. *Looking Forward: Prediction and Uncertainty in Modern America.* Chicago: The University of Chicago Press.

Pleskot, Patryk. 2010. "Jak wyjechać na Zachód? Procedury wyjazdów polskich uczonych do państw kapitalistycznych z ramienia uczelni wyższych i PAN w latach 1955–1989" [How to travel to the West? Procedures of travelling to capitalist countries for Polish scholars from universities and the PAN 1955–1989]. In *Naukowcy Władzy, Władza Naukowcom: Studia*, edited by Piotr Franaszek, 292–325. Wrocław: Instytut Pamięci Narodowej.

Pleskot, Patryk, and Damian Rutkowski. 2009. "Wstęp" [Introduction]. In *Spętana Akademia: Polska Akademia Nauk w dokumentach władz PRL, vol. 1: Materiały Służb Bezpieczeństwa (1967–1987)*, edited by Patryk Pleskot, and Damian Rutkowski, 15–33. Warsaw: Instytut Pamięci Narodowej.

Prognózování vědeckotechnického rozvoje v zemích RVHP: vybrané teze a referáty z 2. Mezin. sympozia zemí RVHP o problémech vědeckotechn. prognózování: Sofie, říjen 1980 [Forecasting of scientific-technological development in COMECON-countries: Selected positions and papers of the 2nd International COMECON-Symposium on Problems of Scientific-Technological Forecasting, Sofia, October 1980].1981. Prague: ÚVTR.

Richta, Radovan. 1969. *Civilization at the Crossroads: Social and Human Implications of the Scientific and Technological Revolution.* White Plains, NY: International Arts and Science Press.

Richta, Radovan, Ota Klein, Bedřich Levčík, and Irena Dubská. 1967. "The Perspectives of the Scientific and Technological Revolution." In *Mankind 2000*, edited by Robert Jungk and Johan Galtung, 198–204. Oslo: Universitetsforlaget/Allen&Unwin.

Rindzevičiūtė, Eglė. 2015. "The Future as an Intellectual Technique in the Soviet Union: From Centralized Planning to Reflexive Management." *Cahiers Du Monde Russe* 56, no. 1: 1–23.

———. 2015. "Toward a Joint Future beyond the Iron Curtain: East–West Politics of Global Modelling." In *The Struggle for the Long-Term in Transnational Science and Politics: Forging the Future*, edited by Jenny Andersson and Eglė Rindzevičiūtė, 115–43. New York: Routledge.

———. 2016. "A Struggle for the Soviet Future: The Birth of Scientific Forecasting in the Soviet Union." *Slavic Review* 75, no. 1: 52–76.

———. 2016. *The Power of Systems: How Policy Sciences Opened Up the Cold War World*. Ithaca, NY: Cornell University Press.

———. 2023. *The Will to Predict: Orchestrating the Future through Science*. Ithaca, NY: Cornell University Press.

Rocca, Gordon L. 1981. "'A Second Party in Our Midst': The History of the Soviet Scientific Forecasting Association." *Social Studies of Science* 11, no. 2: 199–247.

Rolbiecki, Waldemar. 1967. "Badania prognostyczne—prognozologia—futurologia (W związku z sympozjum prognostycznym w Pradze w dn. 21-24 II 1967 r.)" [Prognostic research—prognoseology—futurology (on the prognostic symposium in Prague, 21-24.02.1967]. *Nauka Polska* 15, no. 4: 95–99.

Schermer, Jindřich. 1963. *K problematice vědeckého předvídání*. [On the problem of scientific prediction]. Prague: Vysoká škola ekonomická.

Secomski, Kazimierz. 1970. "Prognostic Research in the Polish Academy of Sciences." *The Review* 3: 17–29.

———. 1971. *Prognostyka* [Prognostics]. Warsaw: Wiedza Powszechna.

Sedlak, Jaromir, and Ota Szulc [*sic!*]. 1972. "Powstanie futurologii burżuazyjnej i jej wpływy w Czechosłowacji" [The emergence of bourgeois futurology and its influences in Czechoslovakia]. In *Wybrane problemy współczesnej futurologii burżuazyjnej: Praca zbiorowa [Tylko do użytku wewnętrznego. Poufne]*, edited by Ośrodek Badania Stosunków Wschód-Zachód, 72–113. Warsaw: RSW Prasa.

Sedlák, Jaromír, and Ota Šulc. 1971. "Vznik buržoazní futurologie a její vliv v Československu" [The emergence of bourgeois futurology and its influences in Czechoslovakia]. *Sešity Ústavu pro filosofii a sociologii ČSAV: interní bulletin: příspěvky ke kritice buržoasních a revizionistických teorií* 7.

Seefried, Elke. 2017. "Der kurze Traum von der steuerbaren Zukunft: Zukunftsforschung in West und Ost in den 'langen' 1960er Jahren" [The short dream of the steerable future: Future research in the West and the East in the ‚long' 1960s]. In *Die Zukunft des 20. Jahrhunderts. Dimensionen einer historischen Zukunftsforschung*, edited by Lucian Hölscher, 179–220. Frankfurt: Campus.

———. 2024. *Shaping Tomorrow's World: A Twentieth-Century History of West German, Cold War, and Global Futures Studies*. New York: Berghahn Books.

Seefried, Elke, and Andreas Malycha. 2018. "Planen, um aufzuschließen: Forschungsplanung in der Bundesrepublik und der DDR" [Planning to catch up: Research planning in the Federal Republic and the Democratic Republic of Germany]. In *Plan und Planung: Deutsch-deutsche Vorgriffe auf die Zukunft* [Plan and Planning: German-German Anticipations], edited by Elke Seefried and Dierk Hoffmann, 35–67. Berlin: De Gruyter.

Siciński, Andrzej. 1969. *Prognozy a nauka* [Predictions and science]. Warsaw: Książka i Wiedza.

———. 1970. "Social Prognoses Division of the Institute of Philosophy and Sociology, Polish Academy of Sciences." *The Polish Sociological Bulletin* 10, no. 1: 166–68.

———. 1972. "Perspektivy přeměn životního stylu v Polsku" [Perspectives of changes in lifestyles in Poland]. *Sociologický časopis* 8: 282–90.

———. 1976. "Przedmowa" [Foreword]. In *Problemy metodologii prognozowania: Polska Akademia Nauk—Komitet Badań i Prognoz "Polska 2000"* [Problems of the Methodology of Forecasting: PAN Committee of Research and Prognostics "Poland 2000"], edited by Andrzej Siciński and Jerzy Gzula, 9–18. Warsaw: Książka i Wiedza.

———. 1981. "Two Sociologies: Of Times of Stability and of Times of Crisis." *The Polish Sociological Bulletin* 21, no. 55/56: 5–10.

———. 1990. "List otwarty do A. Gieysztora" [Open letter to A. Gieysztor]. *Gazeta Wyborcza* 25 June.

———. 2001. "Konwersatorium im. Bogdana Gotowskiego: Z dziejów opozycji intelektualnej sprzed upadku realnego socjalizmu w Polsce" [The seminar in honour of Bogdan Gotowski: From the history of intellectual opposition before the fall of real Socialism in Poland]. *Kultura i Społeczenstwo* 45, no. 3–4: 291–45.

Siciński, Andrzej, Géza Kovács, and Władimir I. Starowierow, ed. 1975. *Problemy prognozowania w krajach socjalistycznych* [Problems of forecasting in Socialist countries]. Wrocław: Ossolineum.

Šiklová, Jiřina. 1990. "The 'Gray Zone' and the Future of Dissent in Czechoslovakia." *Social Research: An International Quarterly* 57: 347–64.

Slíva, Jiří. 1973. "Práce prognostického výboru 'Polsko 2000'" [The work of the 'Poland 2000' forecasting committee]. *Trend: prognostické informace* 5, no. 5: 27–29.

———. 1975. "'Světová dynamika' a 'Meze růstu': Stručný výtah, ohlasy a některé souvislosti" ["World Dynamics" and "The Limits to Growth": A brief summary, reactions, and some context]. *Sešity Ústavu pro filosofii a sociologii ČSAV: interní bulletin: příspěvky ke kritice buržoasních a revizionistických teorií* 38: 5–43.

Sommer, Vítězslav. 2015. "Forecasting the Post-Socialist Future: *Prognostika* in Late Socialist Czechoslovakia, 1970–1989." In *The Struggle for the Long-Term in Transnational Science and Politics: Forging the Future*, edited by Jenny Andersson and Eglė Rindzevičiūtė, 144–68. New York: Routledge.

———. 2016. "Scientists of the World, Unite! Radovan Richta's Theory of a Scientific and Technological Revolution." In *Science Studies during the Cold War and beyond. Paradigms Defected*, edited by Elena Aronova and Simone Turchetti, 177–205. New York: Palgrave Macmillan.

———. 2017. "From Socialist Post-Industrialism to Market Economy: Futurology in Czechoslovakia (1960s–1980s)." *Bohemia* 57, no. 1: 55–81.

Sommer, Vítězslav, Matěj Spurný, and Jaromír Mrňka. 2019. *Řídit socialismus jako firmu: technokratické vládnutí v Československu, 1956–1989* [Running socialism like a business. Technocratic governance in Czechoslovakia, 1956–1989]. Prague: Ústav pro soudobé dějiny AV ČR, v.v.i.

Sowa, Ewa. 2012. "Jan Szewczyk: Teoretyczne postawy podmiotowości pracy" [Jan Szewczyk: Theoretical foundations of the autonomy of work]. *Kwartalnik Filozoficzny* 40, no. 4: 143–47.

Studený, Luboš. 2023. "Die Prognostik als Weg aus der Krise: Das Institut für Prognostik der Tschechoslowakischen Akademie der Wissenschaften als Instrument zur Krisenüberwindung" [Prognostics as a way out of crisis: The Forecasting Institute of the Czechoslovak Academy of Sciences as an instrument for crisis management]. In *Wissenskrisen—Krisenwissen: Zum Umgang mit Krisenzuständen in und durch Wissenschaft und Technik*, edited by Julie Engelschalt, Jason Lemberg, Arne Maibaum, Andie Rothenhäusler, and Meike Wiegand, 179–94. Bielefeld: transcript Verlag.

Šulc, Ota. 1967. "Sympozium o prognózách vědy a techniky v socialistických zemích" [Symposium on science and technology forecasts in Socialist countries]. *Politická ekonomie* 15: 592–93.

———. 1968. *Futurologie: přehled literatury a informací* [Futurology: Overview of literature and information]. Prague: Ústřední ekonomická knihovna.

———. 1970. "Budoucnost vědy o budoucnosti" [The future of future research]. *Tvorba: list pro kritiku a umění* 20: 7.
———. 1972. *Metody i zasady prognozowania rozwoju techniki (Komunikat nr. 29)* [Methods and rules of forecasting of technological development (Communication no. 29]. Wrocław: Ośrodek Badań Prognostycznych Politechniki Wrocławskiej.
———. 1971. "Marxisticko-leninská prognostika v boji proti buržoazním koncepcím výzkumu budoucnosti" [Marxist-Leninist Prognostics in the battle against bourgeois concepts of future research]. *Filosofický časopis* 19, no. 1: 102–8.
———. 1975. "Przyczynek do metodologii prognozowania stylu życia" [A contribution to the methodology of lifestyle forecasting]. In *Problemy prognozowania w krajach socjalistycznych*, edited by Andrzej Siciński, Géza Kovács, and Władimir I. Starowierow, 157–73. Wrocław: Ossolineum.
———. 1976. *Abeceda prognostiky* [ABC of Prognostics]. Prague: SNTL.
Sułek, Antoni. 2010. "'To America!' Polish Sociologists in the United States after 1956 and the Development of Empirical Sociology in Poland." *East European Politics and Societies* 2, no. 3: 327–52.
Surman, Jan. 2022. "Re-connecting Central European Science: An Introduction." In *Science Interconnected: German-Polish Scholarly Entanglements in Modern History*, edited by Jan Surman, 1–36. Erfurt: Verlag Herder Institut.
Szewczyk, Jan. 1971. "Jednotný ideologický postup" [A Unified Ideological Advance]. Interview by Milena Veselá. *Tvorba* 38 (22 September): 14.
Sztukowski, Czesław. 1974. "O metodyce prognozowania w skali międzynarodowej" [On the method of forecasting on an international scale]. *Prace Naukoznawcze i Prognostyczne* 3, no. 12: 99–102.
———. 1974. "Sofijskie spotkania prognostów krajów RWPG" [The Sofia meeting of COMECON forecasters]. *Prace Naukoznawcze i Prognostyczne* 3, no. 11: 65–70.
Tuchel, Klaus. 1967. "Die Zukunft der Zukunftsforschung" [The future of future research]. Die Zeit, 13 October. Retrieved 16 March 2025 from https://www.zeit.de/1967/41/die-zukunft-der-zukunftsforschung.
Zawadzki, Stanislaw. 1970. "Komise 'Polsko 2000'" [The Committee 'Poland 2000']. *Trend: prognostické informace* 2, no. 4: 31–32.
Zeman, Miloš. 1968. *Futurologický sborník* [Futurological anthology]. Prague: Futurologická společnost.
———. 1979. "Three Generations of Forecasting Methods." *Prace Naukoznawce i Prognostyczne* 9, no. 4: 31–38.
"Zpráva o konferenci 'Člověk a společnost ve vědeckotechnické revoluci'" [Conference report. Man and society in the scientific and technical revolution]. 1968. *Sociologický časopis* 4: 611–14.

Interviews

Kasprzak, Wacław. 2024. Interview by Lukas Becht. 11 November.
Petrášek, František. 2013. Interview by Vítězslav Sommer. 1 November. Audio.
Potůček, Martin. 2024. Interview by Luboš Studený. 5 June. Audio.
Slíva, Jiří. 2024. Interview by Luboš Studený. 10 October. Audio.
Zeman, Miloš. 2025. Interview by Luboš Studený. 19 February. Audio.

CHAPTER 4

Central Planning and Socialist Reciprocity

Maciej Górny

This chapter focuses on two interrelated topics: socialist planning and Polish–Czechoslovak academic cooperation in the period of post-Stalinist reformism.[1] Even a superficial overview of the literature reveals the apparent incongruity of these themes. Socialist planning (or, rather, its shortcomings and failures) was subject to numerous studies between the 1960s and the 1990s. Many offered a comparative view of the Eastern Bloc, which continues to make them interesting and useful despite the passage of time. The history of cooperation between socialist scholars, on the other hand, remains an understudied subject. Although the study of socialist internationalism has gained ground in recent years,[2] it has traditionally focused on the relations between the Soviet center and its satellites rather than horizontal transfers of ideas and expertise. Polish–Czechoslovak cooperation is no exception to this trend: studies devoted to bilateral contacts are few and far between and mostly focus on the pre-1945 period.[3] One of the reasons for this imbalance is the source base, which consists of two elements: the strictly formal documentation of bilateral agreements, academic events and cooperation on the one hand, and anecdotal and fragmented testimonies of individual scholars on the other. Moreover, with regard to the latter, the extent to which such personal reminiscences are influenced by the experience of 1989, still perceived (not unreasonably) as the result of the bankruptcy of the socialist economy, remains an open question. These two perspectives rarely come together in a single history of economic thought and social practice.[4] I believe that the international economic discussion of socialist planning offers such common ground.

This chapter focuses on the diachronic development of Marxist reformism in Poland and Czechoslovakia. It opens with a short overview of earlier studies devoted to reforms of the planned economy. This subdiscipline of economic history developed in concert with the reformist movements of the late 1950s and 1960s and, as such, had an impact not only on the subsequent historiography but also on the actors of history themselves, as this chapter will demonstrate. What follows is an overview of the answers of socialist economists to the crises of the mid-1950s that resulted in social tensions both in Poland and in Czechoslovakia. While there were many common elements, Poland's October economic liberalization was accompanied by a political thaw that resulted in a freer public sphere. Therefore, in the late 1950s, Polish economists were presenting a new model of academic debate, which liberal-minded Czechoslovak scholars strove to emulate. This was achieved with the aid of an exchange system developed by the Academies of Sciences as a form of practical "socialist reciprocity." The development and functioning of this exchange mechanism forms the second and largest section of this chapter. Finally, more light is shed on the Polish context of the Czechoslovak reformism of the 1960s. References to Polish Marxist economists played a role in legitimizing the ideas of Ota Šik's (1919–2004) team. Correspondingly, the latter's tenure as the head of the Economic Institute of the Czechoslovak Academy of Sciences resulted in an intensification of mutual relations.

Central Planning as an Object of Study

Cold War studies focusing on socialist reformist discourses offer a perfect starting point for the study of scholarly cooperation. Two of their principal strengths are their comparative approach and their very detailed analysis. The first of those virtues is partly a result of the topic of research. Although it had a local and international tradition, Socialist central planning of the 1950s and 1960s was in principle an adaptation (or, as many authors chose to claim, a dictate) of Stalinist modernization policy to the realities of the Eastern Bloc. Józef Wilczynski (or Wilczyński) (1922–1984), one of the most prolific academics writing in this vein, characterized the subject of his study as follows, which was roughly applicable to all of the socialist states:

> Central planning consists in the *ex-ante* setting of patterns of production, distribution and consumption. These plans cover the whole economy consisting of heterogeneous entities, and cover long periods . . . . Planners' decision may be erroneous at the time of plan construction or rendered obsolete by subsequent developments. The occurrence of bottle-necks, a familiar feature of centrally planned economies, provides evidence of the risk deriving from this source.[5]

Similar as they undoubtedly were, East Central Europe's socialist economies differed in many respects. Consequently, the timing of economic crises and reforms also differed from case to case. In the view of János Kornai (1928–2021), probably the most prolific critic of socialist economics, one of the key differences between these cases was the sincerity of socialist reformism. In some countries (the most "petrified" East Germany and Czechoslovakia are a case in point), the only objective of the repetitive attempts to adjust the system to the realities seemed to be to create a mirage of betterment; whenever the reforms were meant seriously—at least for a short time—they inevitably led to disappointment and deficits of legitimacy as demonstrated by the brutal suppression of the Prague Spring. Furthermore, the whole reform process remained fragile, as a return to Stalinist orthodoxy was always a viable option.[6] Other authors stressed such differentiating factors as social and income structure[7] or the level and range of liberalization (anticipating market relations between individual enterprises or between the associations thereof).[8]

Another characteristic feature of the Cold War history of socialist economy was the high level of specialization. More often than not, Western authors had a deep knowledge of at least one socialist country, having diligently read local economic publications and very resourcefully studied official statistics that, as most admitted, were not completely untrustworthy.[9] Biographies offer a simple answer to the question of why this professionalism was so widespread among Western specialists in socialist economies. Józef Wilczynski may again serve as an example: He was associate professor at the Military Academy in Duntroon, Australia, having served in the Polish Army on the Western Front during the war. Refusing to return to Poland, he defended his doctorate in London and immigrated to Australia. There he specialized in reform attempts in the socialist countries until his death in 1984. Besides his native Polish, he was able to read in other languages of East Central Europe. At times, such a biographical background seemed to distort the economic analysis, leading some authors to exaggerate the impact of their native country. Andrzej Brzeski (1925–2019), who emigrated from Poland in 1958 and continued his academic career at Berkeley and Davis, has been such a case. He offered a Poland-centered interpretation of the reformist attempts in claiming that "having in the mid-fifties spearheaded the movement for economic reform, Poland had provided a unique input into subsequent developments elsewhere." Subsequent reforms, be it in Czechoslovakia or Hungary, were seen by this scholar as mere re-readings of the Polish economists' revolutionary legacy.[10]

It is not difficult to see where most of the aforementioned Western economists were heading in this regard. Their respect, sometimes bordering on deference, was reserved for the most prominent socialist theoreticians of that time, such as Oskar Lange (1904–1965), Michał Kalecki (1899–1970), and Ota Šik. There was little respect for the real socialist economy and very little for socialism as a political philosophy. However, the socialist economists themselves did not subscribe to this crisis-centered perspective. Theirs was rather an optimistic view of a more humane economy, freed from the cruelties and deficiencies of capitalism. Planning represented the single most important factor in the process of humanizing market relations. In 1959 Lange wrote, "The essential difference in the operation of economic laws in a socialist society is that they do not operate there in an elemental way. Organised society shapes in a conscious, purposive way the circumstances which determine their operation."[11] Moreover, Lange and his peers exceled in criticizing the capitalist system and capitalist experts.[12] To understand the economic reformism of the 1950s to the 1970s it is necessary to see its protagonists as believers in the viability of socialist economy and central planning. In their eyes, both may have required certain adjustments here and there, but it did not mean that the system was not operational or based on false premises.

Crises and Crises-Slayers

It is probably the right moment to introduce the most prominent socialist reform projects in Poland and in Czechoslovakia. They had many features in common. Each was a response to the shortcomings of Stalinist industrialization based on massive investments in heavy industry. In the words of Bronisław Minc (1913–2004), longtime director of the Institute for Economic Science of the Polish Academy of Sciences (Zakład Nauk Ekonomicznych PAN), the dogma was that "the development of heavy industry makes it possible to achieve a much faster rate of development of both industry and the national economy as a whole than in the capitalist system."[13] However, this formula did not work half as well in reality as it did on paper. The shortcomings of the system had been analyzed by economists internally as early as 1954, but the impulse for open discussion came with Polish October.[14] In effect, the collectivization of agriculture was abandoned, light industry increased its share of state investment, and the standard of living became the concern of socialist economists.

One of the principal intellectual hubs in this process was the Planning Commission. There, a group of experts led by Kalecki designed a project for the thorough restructuring of Polish industry oriented toward modern branches of economy, such as semiconductors, automatized production, and computers. Although Lange supported this program during a parliamentary debate, the Party opted for a conservative policy and decided to return to a modified Stalinist model based on heavy industry and the exploitation of natural resources. Kalecki repeatedly (and critically) commented upon this decision, which finally led to his ousting from the Planning Commission in 1964.[15] The same year saw the publication of an important study by Włodzimierz Brus (1921–2007), which characterized the main weaknesses of the Stalinist model.[16] Brus's book would play a role in the economic debates in Czechoslovakia. Around the same time, a reformist model was developed by Aleksy Wakar (1898–1966), professor at the Warsaw Main School of Planning and Statistics, and his students. Their main focus was to reintroduce elements of market economy.[17]

The Polish crisis of 1956 was an alarm bell for the Czechoslovak Communist Party, which was facing similar problems in a slightly less drastic form. To confront the challenge the State Planning Authority (Státní úřad plánovací) initiated preparations for a reform. The essence of the resulting proposal by Kurt Rozsypal (1916–2013), echoing the ideas of Kalecki and Wakar, was to reintroduce market elements into industrial production, thereby replacing the previously dominant emphasis on the amount of goods produced and focusing on their value instead. Another similarity between the two countries was that the Czechoslovak reform also remained largely an intellectual exercise due to the implacable resistance of the socialist managerial elite and the Party leadership.[18]

Reform projects resurfaced again in early 1960, when a combination of various factors (including the split between USSR and China) led to disturbances throughout the Soviet sphere of influence. The Eastern Bloc as a whole experienced a decline in economic growth and Czechoslovakia was hit particularly hard: the country's GDP fell for the first time in more than a decade. This proved decisive; despite the Party elite's reluctance something had to be done.[19]

The Czechoslovak economic reform was preceded by the publication of a concept paper and extensive discussions between economic experts and Party functionaries. A new feature was that this time even radical ideas such as the liberalization of prices, profit-oriented production and fair competition for managerial posts had to be accepted.[20] A contemporary Western commentator was impressed:

> As an integral part of the economic reforms, enterprises have been granted a greater independence. Central planning is now concentrated on macroeconomic proportions and medium- and long-term developments. To achieve their objectives, the authorities now rely primarily on financial instruments (prices taxes, interest rates, capital charges, depreciation rates) designed to influence profits and thereby induce enterprises to act in desired directions.[21]

In effect, Czechoslovakia's economy underwent a process of decentralization that was soon accompanied by political liberalization. It would be destined to outlive the latter, as some elements of the reform survived the reintroduction of central planning in 1969.[22] To understand how profound (and, consequently, how divisive) the Czechoslovak volte-face had been, it is instructive to look at the language used by the main actors of the reform. Šik, the main author and key figure of the Prague Spring, enjoyed the status of a celebrity both at home and internationally. Whenever he traveled abroad, journalists and experts never missed the opportunity for an interview. Šik was very open, yet he carefully composed his words keeping in mind the particular audience. In an interview with the West German cultural journal *Osteuropa*, he spoke in the manner of a liberal, admitting that firms that did not make profits in Czechoslovakia would be closed down, because society could not afford to support ineffective businesses.[23] Such thoughts had never been so clearly articulated in the socialist bloc's economic press. Speaking with the editors of the very popular weekly Polish economic journal *Życie Gospodarcze*, Šik chose to formulate a basically identical message but in a different manner: "Thus, we attach great importance to the use of the market mechanism not to reconstruct the market economy, but to better align the interests of the individual, the enterprise and society."[24] The rules of discourse were adhered to or, at least, acknowledged.

Seen from the West, Czechoslovakia and Poland looked like two rather similar cases of a socialist economy struggling to reconcile planning with elements of market economy. From the mid-1950s onward, economic experts in both countries had been designing models that could potentially ease the systemic tensions and increase the standard of living of the population. While they differed as to the specific means to be used, they were not substantially different. Most probably almost all of them, at least in private, subscribed to the declaration Šik made in exile in 1971: "A socialist synthesis of plan and market could become the economic model of the near future. Combined with a scientifically and democratically organized political system, it is the social image of the future that can be called a humane socialist society."[25]

Close but Apart

Nevertheless, similar as they were, Polish and Czechoslovak economists belonged to Sovietized academic cultures, one of the primary traits of which was Party and state control over very limited international cooperation. Even so, beneath the surface the situation was dynamic, and this will first be approached through the mediation of economic journals with high circulations among economists and people interested in economic affairs. In this respect there was an imbalance between Poland and Czechoslovakia, which reflected the political developments in both socialist countries. As stated by Gil Eyal (1965–) and Larissa Buchholz, the public deployment of expertise should be the focus of a more inclusive intellectual history rather than the development of concepts and ideas.[26] Those interventions, however, require public space and a level of freedom of speech that allow experts to leave the narrow sphere of strictly academic circulation. In the case of Czechoslovakia, this precondition was not fulfilled until deep into the reformist period. *Ekonomická revue*, a widely read economic journal, was published between 1966 and 1968 and featured many of the Czechoslovak economists mentioned in this chapter.[27] In Poland, this role corresponds, for the most part, with the highly popular *Życie Gospodarcze*, which featured an interview with Šik in 1965. *Życie Gospodarcze* was established in 1945 and although it went through a phase of resolute Stalinism, it was also one of the first economic journals to embrace reform in 1956, from which time it remained interested in the Czechoslovak economy and economic thought.

In the early postwar period, *Życie Gospodarcze* strove to inspire mutual cooperation "in the spirit of cooperation between the Slavic states," as announced in one of the journal's first analyses of central planning.[28] The steady stream of correspondence and other texts ended abruptly in the Stalinist period in 1950. In subsequent years information from and about Czechoslovak affairs appeared sporadically, with the Soviet Union almost monopolizing the international pages of the journal. Scarce as they were, Czechoslovak contributions in the early 1950s were rather sterile in both an intellectual and political sense. The article by Jaromír Dolanský (1895–1973) (ex-minister of finance and chief planner) on Czechoslovakia's five-year plan conveyed nothing to the Polish reader other than the compulsory optimism.[29] In 1953, the thirteenth issue of the journal celebrated the memory of the deceased president Klement Gottwald (1896–1953) (who had "rescued his nation from Slanský's bunch of spies").[30] In the meantime the only real exchange of thoughts could be observed in a rather marginal discussion regarding a Polish–Russian economic dictionary, in which Antonín Radvanovský (1923–) (who would one day author a Polish–Czech and Czech–Polish economic dictionary) criticized the authors' approach for its lack of systematic categories.[31] The situation improved in 1954, when, on the occasion of the conclusion of the Czechoslovak five-year plan, several short texts devoted to Poland's neighbor appeared in *Życie Gospodarcze.* Although most consisted of general slogans, an unusually long article by Czech agricultural economist Jan Krblich (1911–1965) hinted at one of the structural problems of the Czechoslovak socialist economy: the relative underdevelopment of agricultural production, which lagged behind industry. Despite all the accompanying words of praise for the accomplishments of the five-year plan, Krblich's voice was finally adding some substance to the dull formulas of late Stalinism.[32]

The paucity of open debate in this period can be largely attributed to the specificity of Czechoslovak Stalinization. Although the Polish communist authorities were prone to brutality, they rarely selected their victims from among the professoriate. Wakar, whose Russian ethnic origin and personal contacts to Russian "white" emigration attracted the attention of the Soviets, was among the few academics actually persecuted. In contrast, the communist takeover of Czechoslovakia demolished the cadres and institutions, leading to what Catherine Albrecht (1968–) calls "de-professionalisation of economics."[33] The main economic journal of the country, *Politická ekonomie*, enthusiastically participated in all of the witch-hunts, specifically targeting economic experts such as Josef Goldmann (1912–1984). As Albrecht noted, the resulting intellectual misery could not be compensated for by the specialist *Plánované hospodářství: Politickoekonomický časopis SPK*, which was published by the Planning Commission. Quite contrary to the Party's official internationalism, practical isolation from international (including socialist) science was an important factor in the Sovietization of Czechoslovakia until around 1953.[34]

Reciprocal Years

In the early postwar period in Poland and Czechoslovakia, interest in and cooperation between the neighbors was widespread and supported by the new authorities.[35] This resulted in flourishing cooperation between artists and scholars based on the organizational and financial support of several institutions, even before the signing of the agreement on friendship and mutual assistance in 1947.[36] Academic events and collaborations were part of the activities of the Slavic Committee, a body designed by Stalin but populated with representatives of the intellectual elites in both countries.[37] The Polish–Czechoslovak Society of Friendship was another institution that actively promoted academic exchange, until its dissolution in 1951.[38] However, the advent of Stalinism in the late 1940s interfered with these activities and had reduced them to a minimum by the turn of the decade.

The main factor in the incipient change was the establishment of centralized Soviet-style Academies of Sciences in the first half of the 1950s. The newly established institutes, committees, and research centers reinvigorated the international exchange that had fallen victim to Stalinism. The first sign of the new era was the exchange of information and publications between the academies. The initial indications of mutual interest were the greeting letters that were sent out on the occasion of the New Year and

Labor Day.[39] The first invitations to conferences and for research stays followed soon after. These were based on annual cultural agreements between the two states, which stipulated the (limited) number of researchers that would be hosted by each of the academies.[40]

The beginnings were very modest. Prior to 1953, academics took part in official exchanges under the interstate agreement, which resulted in a relatively small number of travels to conferences.[41] In 1953, in the framework of academic exchange, only fifteen Czechs and Slovaks visited Poland and six Poles visited Czechoslovakia, and these negligible figures were only marginally higher the following year. According to a report by the ČSAV, "Against the backdrop of other countries, such as the Hungarian People's Republic or the German Democratic Republic, these contacts failed to achieve the level and dimension corresponding to the plans and needs of academic research in both countries."[42]

In November 1955, the leadership of ČSAV hosted the PAN for a long and lively discussion on future cooperation. Both sides agreed that the current state of affairs was far below their expectations, although they prioritized different elements. The Czechoslovaks focused on personal exchanges of young scholars, whereas the Polish delegation (whose most active speaker was the historian and Party functionary Henryk Jabłoński [1909–2003]) stressed the importance of decentralization. Nevertheless both sides agreed that for actual exchange to flourish, the institutes needed to take the initiative.[43] The numbers were still modest, at best, but they were growing: in 1954, twenty-four Czechs and Slovaks visited the institutes of PAN, while in the following year the number had risen to forty-one (the respective Polish numbers for the two years were twenty-eight and fifty-five).[44]

Economists were among the first to act. In March 1955, Vladimír Kaigl (1918–), the first director of the EÚ ČSAV and his counterparts from Poland and East Germany met in Berlin, half a year before the academy leadership, to discuss ways out of the isolation. All three economic institutes agreed to exchange their publications and journals. Additionally, the EÚ ČSAV decided to form a separate research unit documenting the development of brotherly economies.[45]

In parallel to these international developments, the position of both economic institutes within the domestic science noticeably strengthened. The academies of which the ZNE PAN and EÚ ČSAV were part gradually became stronger and replaced traditional institutions. In addition, these economic institutes began to play the role of coordinating bodies, notably with the 1957 law in Czechoslovakia that identified the EÚ ČSAV as the think tank for the country's economic policy.[46]

Nevertheless, neither the Czechoslovak *Politická ekonomie*, nor the Polish *Życie Gospodarcze* and *Handel Wewnętrzny* (established in 1955 and devoted to the problems of internal trade) visibly increased their interest in the neighboring economy after 1955. Instead, more subtle changes brought about by destalinization resulted in certain structural shifts within those journals. Beginning in 1955 *Politická ekonomie* included a section entitled "Science in the Socialist Countries," which featured some texts by authors from Poland and the GDR. Reviews of Polish publications were largely absent, with the notable exception of a short description of the specialist journal *Zeszyty Naukowe SGPiS*, in 1958.[47] *Życie Gospodarcze* maintained a limited interest in the practical questions of Czechoslovak economy and planning and published roughly one text every two issues. However, it placed more importance on the overall political climate of Polish October. *Życie Gospodarcze* would become the stage for reformist voices such as Edward Lipiński (1888–1986), who harshly criticized the shortcomings of Stalinist modernization. He stated, "The 'plan' became the economy's main goal instead of the much needed products, bonuses [were prioritized over] a real increase in productivity and wastefulness [prevailed over] maximal thriftiness."[48] This degree of criticism was probably far beyond that which would have been tolerated from Lipiński's Czechoslovak colleagues. Aside from his frank criticism of Stalinist industrialization, Lipiński voiced his belief in the close association between the liberalization of the economy and the democratization of the country and culture. "Socialism is freedom," he asserted in 1956.[49]

There were also direct references to bilateral relations, some of which may have raised more than an eyebrow. In a 1956 article by Juliusz Mikołajski (1901–1987), the blame for the difficulties in bilateral contacts in the Cieszyn/Český Těšín area in the Stalinist period was placed squarely on the shoulders of the ruling Party elite.[50]

The paucity of Czechoslovak topics after 1956 would not have been surprising, were it not for the fact that it contrasted with considerable interest in Yugoslav economy. Indeed, in the period of Polish October *Życie Gospodarcze* abounded in analyses, reports, and technical information. There was nothing similar to be found in the Czechoslovak economic journals. Uninterested in the Yugoslav experiments in workers' self-government, they were also very cautious in their reactions to the events in Poland. In a review of *Zeszyty Naukowe SGPiS*, Vladislav Pavlát (1929–2021), then a young economist who in the distant capitalist future would become one of the driving personalities behind the Czech stock exchange, also commented on ongoing debates concerning Poland's Marxist economy around 1956. His language was carefully measured in such a way that his reverence for his more internationally established Polish colleagues did not overshadow his socialist orthodoxy. In the first installment of his report, Pavlát criticised the virtual non-existence of Czechoslovak theoretical debate, which stood in sharp contrast to the vigorous discussion among Polish economists in the years 1956 and 1957.[51] In another article on the same topic, published less than a year later, he again emphasized the vitality of the discussion, but also the predilection of certain "rightist" Marxists to veer in the direction of bourgeois econometrics. Revisionism among Polish economists was a real threat, which clearly outweighed any benefits of free discussion. This was manifested in the popular belief (held by Lipiński and Jan Drewnowski (1908–2000), one year before the latter emigrated, in 1958) that any viable economic theory should be explainable in the language of another viable economic theory. Pavlát claimed that this infected Marxism with capitalist ideas, as it was a pipedream to believe that it was possible to use bourgeois mathematic methods without committing bourgeois errors.[52]

Pavlát's balancing act between ideological caution and his fascination with the open scholarly debate in Poland was typical of the Czechoslovak economic milieu as a whole. Sometimes Polish economists were solicited with the explicit intention of breathing life into domestic discussions. One example was the invitation extended to Minc for a conference at the Prague University of Economics (Vysoká škola ekonomická) in December 1956. As director of the ZNE PAN, Minc could not be described as a revisionist troublemaker. Nevertheless, in the invitation letter from Vladimír Sedlák (1913–1986), president of Prague University of Economics (notably written in Polish, which was not standard practice between the academies; Kaigl's letters were mostly in Russian), Minc was encouraged to "actively" participate in an "exchange of views" with leading Czechoslovak specialists in productivity.[53] Having taken this request at face value, he advocated wage increases in proportion to increases in productivity, which—contrary to Stalinist principles—he did not consider a threat to the fulfillment of the plan. In the lively discussion that ensued most Czechoslovak economists opposed this view, with the exception of Šik, whose future impact on international relations in science is difficult to overestimate.[54]

The gradual opening up of Polish–Czechoslovak contacts remained formalized on many levels. Economics did not substantially differ from other fields of activity in this respect. The numbers of travelling specialists were regulated by agreements between the academies, institutes, and ministries. The documents suggest that, despite the travel formalities and the related costs, it was the host institutions that faced the greatest practical difficulties in accommodating their guests. This was the reason that exchange plans were sometimes not fulfilled. For instance, in 1961 a number of employees of PAN had to cancel their planned visit because they had not received a timely response from the ČSAV bureau for international relations, while others were rejected due to the lack of accommodation.

Both academies did their best to improve the system and the numbers of specialists involved in exchanges grew.[55] Departments for international cooperation divided the sum total of what were known as "reciprocal weeks" between the institutes, offering economists in the early 1960s typically around eight weeks a year.[56] Thus, within this particular segment of academic exchange, Poland was the second most visited socialist country after the Soviet Union (thirteen weeks) and ahead of the GDR and Hungary (both three weeks).[57] According to the summary from 1969, Poland was the only socialist country to exceed the plan, which forced the Czechoslovak institute to apply for additional travel funds.[58] Typically, the departments for international cooperation agreed on limited numbers of persons, but they continued to receive applications from institutions and companies eager to send their employees abroad. The Polish State Planning Commission (Państwowa Komisja Planowania Gospodarczego), which until its liquidation in November 1956 was responsible for the six-year plan, issued sample applications and formulated a set of fundamental rules for all members of Polish delegations. Members were required to be able to communicate in the receiving country, to possess specialist knowledge and to be trusted to keep state secrets.[59] Similar sets of rules were also gradually introduced by academic institutions. One of the most sophisticated was the Czechoslovak Ministry of Finance's eight-page-long instruction concerning the rules for visiting foreign institutions and hosting foreign delegates.[60] In practice, the employees of most institutes of the academy followed very similar rules to those specified by the secretary of the biological and medical section of ČSAV, Otto Jírovec (1907–1972), in September 1958. Upon returning from a research visit:

> The researcher submits . . . a written report to his/her section, department, and foreign department, in which he/she is obliged to state the main contribution of his/her trip. He/she shall present his/her experience both to the Departmental Council and to the staff of his/her laboratory. Each trip is evaluated by the department in terms of the professional, political, and cultural benefits to the staff member and the research tasks on which the laboratory is working.[61]

Both sides filled in application forms of varying types and forms, which gradually changed over time. Typically, they contained basic biographical information, languages spoken, estimated travel costs (local costs were covered by the hosting institution and in some cases the delegated employees financially contributed to their trips), a short description of the research topic, and a list of the institutions and individuals to be consulted during the trip. Due to travel restrictions, it was necessary to be accepted and placed on the travel list to obtain a passport for foreign travel. The reports that were delivered upon the employee's return were typically rather short and reproduced the information from applications, with alterations wherever necessary. In general, Czechoslovak bureaucracy was quicker and more determined to formalize the whole procedure, while PAN tended to leave traveling economists and other experts more freedom in composing their applications and reports. In effect, Czechoslovak applications tended to be longer than many of the concluding reports, while in Poland the opposite tendency prevailed. The Polish model could be seen as the more logical and desired approach, as much more can be written about a visit after it has been concluded. Even so, when in the mid-1960s new printed forms were introduced at PAN, they included an A6 sheet of paper for a shortened version of the concluding report, thereby allowing academy bureaucrats to avoid reading the longer texts that accompanied it.

Timelines

Much of the content of those documents concerning economic experts is simple enumerations and predominantly pertains to the two central economic institutes EÚ ČSAV and ZNE PAN. However, beneath the surface certain developments can be identified and located in time. In the early period of growing bilateral exchange, Polish economic libraries featured prominently among the most popular destinations of Czech and Slovak travelers. References to the post-1956 economic discussions in Poland were occasionally made, though with the necessary caution: "These issues [price policy] have received considerable attention in Poland in recent years; a considerable number of economists have been researching them and have achieved certain results. It is necessary to become acquainted with Polish knowledge."[62] Later, certain Czech and Slovak applicants (notably Šik) openly cited better access to Western literature as their main incentive for research travel to Poland.[63]

In the 1960s, programs of Czechoslovak visits became more sophisticated and included not only academic collaboration but also visits to selected enterprises to analyze wages, bonuses, and other management strategies: problems that plagued socialist economies in the 1950s. Šik's appointment to the post of director at EÚ ČSAV in 1961 was seemingly the decisive factor in this. Under his leadership, a new group of younger scholars entered exchange programs with socialist countries. His impact was also discernible in the selection of research topics, which were predominantly related to basic problems of economic reform. One such issue was on the agenda of Čestmír Kožušník (1928–2010) during his visit to Warsaw in the spring of 1963. He attended discussions at ZNE PAN and at

SGPiS, met with Minc and Brus, and visited selected factories. Kožušník's conclusion was that Polish experiments with binding wage increases, not related to production increases but in relation to the value of that production, should be replicated in Czechoslovakia. He was also impressed by the Poland's decision to centrally increase the price of meat and reduce the price of industrial products in 1958. Even so, Brus assured the Czech guest that the decentralized model of socialist economy was not tantamount to eliminating centralized price policy.[64]

Though critical in the transfer of reformist ideas, bilateral relations were not the only channel for international contacts. There were two other notable forms of cooperation. First, both Polish and Czechoslovak economists took part in joint projects coordinated by one or more of the COMECON members. In July 1962, during an international conference in Warsaw, socialist economists debated the new economic policy based on the principles of the 22nd Congress of the Communist Party of the Soviet Union. The organizing committee was packed with prominent politicians and scientists. Eugeniusz Szyr (1915–2000), deputy prime minister responsible for economy, and Tadeusz Kotarbiński (1886–1981), president of PAN, served as co-chairs, and the committee included several ministers and high officials. Kalecki and Lange were also present. One of their main concerns was the wasteful allocation of research capacity in the economy, that is, the unnecessary duplication of specialized research in more than one socialist country. Division of labor was the order of the day. Poland and Czechoslovakia were made responsible for the conceptual development of cooperation between various branches of industry. This required not only participation in joint conferences but also regular meetings in smaller working groups. As demonstrated by the travel documentation of Jaroslav Habr (1911–1998), whose earlier career had been disrupted in the 1950s and who had returned to an academic position in 1962, such obligations offered rich opportunities for the pursuit of individual research. In one of his first research trips to Warsaw he acquainted himself with new publications on structuralist models and their optimalisation, both of which were subjects of his long-term interest.[65] Similar opportunities emerged with joint publication projects such as collectively edited volumes that required reciprocal visits, although such travels were among the first to be cut back on.[66] At the other end of the scale, large gatherings including all or most of the socialist countries could always count on sufficient funding. This was clearly the case with the prestigious conference on balancing national economies in October 1959, which was attended by economists from China and North Korea.[67]

The second channel that opened up in the late 1950s, and to a much greater extent in the 1960s, was contact with Western scholars. In Czechoslovakia, the numbers of experts involved in this exchange grew rapidly during the Prague Spring and remained surprisingly high after the military intervention and occupation of the country. In 1969 the Economic Institute of ČSAV organized more than fifty research trips abroad, including destinations such as Santiago de Chile, Tokyo, Istanbul, and most countries of Western Europe. The costs of these visits were mostly covered by the hosting institutions, while Czechoslovak visits to West Germany were organized within the framework of an agreement between ČSAV and German Academic Exchange Service. Ten foreign professors and four foreign students from the "capitalist" countries spent several weeks each in Prague at the invitation of the Institute.[68] Perhaps even more telling was the shift in academic savoir vivre. When, in May 1966, the president of the ČSAV, František Šorm (1913–1980), addressed Minc with an invitation to an international panel of economic experts on the management and planning of science, he used English rather than Russian, which had formerly played the role of the inter-academic lingua franca: "I think that the use of a single language would simplify the discussions very much—read the invitation—and propose the use of English language for this purpose."[69]

The Reform Talk

A new phase coincided with the general crises of the socialist economies in the early 1960s. As previously mentioned, this state of affairs prompted reformist ideas. In terms of personalities, the downturn of the Czechoslovak economy elevated Šik to the highest echelons of power. His expert voice became heard beyond the borders of the country, as evidenced by his 1965 interview with *Życie Gospodarcze*, a genre that was extremely rare in this journal. The interview was preceded by a 1964 article by Marek Misiak (1933–2013), a prolific economic journalist. The author analyzed the new Czechoslovak economic planning with a focus on the policy of liberalizing wages. The idea of omitting some elements of central planning and allowing enterprises and branch unions to decide for themselves was greeted with enthusiasm.[70] At the same time, the English language *Czechoslovak Economic Papers* (launched in 1959) published an article by Jiří Kosta (1921–2015) announcing a groundbreaking manifesto by Šik to be published in the next issue of the journal.[71] Indeed, Šik used radical language. He spoke against the fetish of central planning, advocating instead decentralization as the only cure in the "difficult process of regenerating our whole economic way of thinking."[72] In describing the previously dominant management philosophy, he did not hesitate to use both criticism and irony: "Economic theory of the Stalin era, with its purely apologetic and propagandistic character could not gain the respect and recognition of practical economic workers and so no one even used it when there were practical economic problems to be solved."[73]

The Czechoslovak reform enjoyed the support of significant Polish economists. This support was communicated on more than just the pages of specialist journals, such as *Życie Gospodarcze*. In the preface to the Czech translation of his *Political Economy* (published in 1966), Oskar Lange supported the Czechoslovak reform generally, and specifically the flexible approach to central planning.[74] Kalecki also spoke warmly of Šik and enjoyed reciprocal respect from Czechoslovak economists. Both he and Lange were among the most cited economists on the pages of Czechoslovak journals and the frequency of their appearance continued to increase until the period of "normalization."[75] A fascinating document of this informal association is the article by Josef Goldmann (1921–1984) and Josef Flek (1913–1975) in *Życie Gospodarcze* (in 1966), in which they applied Kalecki's growth theory to Czechoslovak statistical data. In the final paragraph of the text, they aptly noted: "However, if the successful introduction of a new planning and management system is to occur as soon as possible, far-reaching changes are required in the thinking of the people involved in both the research and the practice of planning and management."[76]

The Polish reactions to the Czechoslovak reform in many ways mirrored the Czechoslovak reactions to Polish October. Travel reports delivered to PAN upon the return of scholars in late 1960s mentioned extraordinarily lively discussions with and among Czechoslovak colleagues. And it was not only Czechoslovaks. Józef Zagórski (1906–1978) summarized a joint Polish–Slovak seminar in September 1965 as an extremely interesting experience partly due to "Yugoslav colleagues who made for a nice, friendly atmosphere."[77] As with the situation around 1956, part of the milieu viewed Šik's activity with hope, believing that he was about to accomplish what Poland's economic reformers had planned for their own country in 1956. But there was also a significant group that opposed the main elements of his reform. They formed the core of the Polish delegation to the international symposium in Gösing (Austria) at the end of September 1965. Minc, Zygmunt Wyrozembski (1907–1979), and Józef Pajestka (1924–1994) opposed Šik's idea of decentralized planning. In their view, leaving this task to socialist enterprises and branch associations was a mistake.[78]

This opposition was partly the result of the rhetorical radicalism of the Czechoslovak reform. In May of the same year, Šik had visited Warsaw at the invitation of the Polish Economic Association (Polskie Towarzystwo Ekonomiczne). His aim was to promote his ideas, which he duly did by giving speeches at various gatherings and the aforementioned interview with *Życie Gospodarcze*. On this occasion he spoke with Brus, Kalecki, and Minc, among others. In summarizing his impressions, he chose to concentrate on the first two, who openly supported Czechoslovak reforms, while ignoring the skeptical opinion of Minc. In effect he convinced himself that only radical and coordinated reform could succeed and—more importantly—that he enjoyed the support of the majority of reasonable Marxist economists:

> Meetings with Polish economists confirmed that the envisaged changes in the management system in many socialist countries are moving in the same direction. In Poland, a number of experiments are being prepared to strengthen the power of associations and to increase the role of certain economic instruments. However, the measures in the management of Czechoslovak industry differ substantially in that they represent a comprehensive solution to the problems of the central plan and the set of economic instruments. They therefore represent an internally combined system, the practical application of which significantly changes the economic conditions for reconciling the interests of enterprise and society. Discussions with Polish economists from the field of theory and practice confirmed that only a synthetic approach to the problem of management makes it possible to affect the now widely known shortcomings of the administrative-directive management system.[79]

However, the reality was less sanguine. There was neither universal support for the Czechoslovak reform program nor universal acclaim for Šik. Moreover, some Polish economists felt increasingly concerned about the scope of the economic and political changes in their neighboring country. This feeling permeated their reports from research travels to Czechoslovakia and casual contacts with Czechs and Slovaks abroad. Wyrozembski took part in an international conference in Smolenice Castle (the usual meeting place of the Slovak Academy of Sciences) in 1966. He noted, in a slightly vague but sufficiently understandable manner, that "I found it characteristic that the representatives of capitalist countries proved much more skeptical toward the new structural tendencies of capitalism than some of the representatives of the socialist countries."[80] As if to reinforce the point, he characterized his own contribution to the conference as "warning mostly young Czech economists not to overestimate the output of Western economists in the area of theory of growth."[81] In the autumn of that year a major conference of the International Economic Association took place in Nice with four Czechs and Slovaks and three Poles among the speakers. The theme of the conference was planning and the market. The joint report written by Minc, Maksymilian Pohorille (1915–2003), and Bohdan Gliński (1926–2009) praised the convivial atmosphere and character of the discussions with colleagues from the West. However,

> In the course of the discussion, significant differences in views on the role of the plan and the market emerged among economists of the socialist coun-

> tries. In the speeches of the Yugoslav and Czechoslovak economists there were clear tendencies to diminish the role of the central plan (treating the plan as a picture of the future market). Czech comrades tended to overemphasize the mistakes of the earlier period and idealize the role of the market, competition, and far-reaching automation of enterprises.[82]

Given that there was but one representative of Yugoslavia present in Nice, most of the criticism was directed toward the Czechs and Slovaks.

Beyond the Reform Period

The Prague Spring ended in failure and the same can be said about socialist planning. Neither were Šik's concepts of a "third way" successful in the long run. Therefore, it is tempting to bring the economic chapter of the history of "Slavic reciprocity" to a close with the intervention of the Warsaw Pact armies in August 1968. However, the history of science does not strictly follow the timeline of political events. In fact, the legacy of the intensive Polish–Czechoslovak exchange lasted well into the 1970s. It was true that ZNE PAN recorded no international visits during the whole of 1968, the year that was memorable not only for the intervention, but also for the antisemitic campaign in Poland that resulted in the forced emigration of many scholars including Brus. Nevertheless, in subsequent years the number of "reciprocal weeks" grew significantly in comparison with the modest numbers of the mid-1950s. According to the agreement between the Czechoslovak and Polish academies, there were 300 reciprocal weeks every year just for short visits, in addition to thirty reciprocal months for longer stays in the neighboring country. There was agreement on the organizing of joint summer schools in various disciplines and the exchange of publications through institute libraries lasted well into the post-1989 era.[83] Both economic institutes continued their cooperation in the form of an agreement for 1971–75. It stipulated joint research into two main spheres of economic theory: planning and socialist management and contemporary capitalism. The agreement also referred to a commitment to increase academic mobility between the two countries. It was not Šik but Josef Nikl (1916–1985), the Institute's new director, who signed the joint Russian-language version of the agreement.[84]

As suggested by the wealth of literature devoted to the crises of socialist economy, there were both analogous circumstances and relations between the economies and economists of the Eastern Bloc. The similar structural problems of the Stalinist model generated a set of compatible responses. Therefore, it was customary in the earlier literature (and often still is)[85] to see them as a continuity of reform attempts that repeatedly failed either because they were too limited (and could not work) or because they were too radical (and generated a political backlash). Interestingly, this was the perspective adopted by Leszek Balcerowicz (1947–) in his contribution to the Festschrift for Šik, which was published in 1990. Perhaps unknowingly repeating Šik's arguments from his discussion with Polish economists in the mid-1960s, Balcerowicz considered the crucial question of the scope of the reform:

> The thesis about the high reform threshold suggests that in assessing the changes of the proposed reforms in socialist countries, one should be wary of using the hitherto popular formulation that a given proposal is a "step in the right direction." For this misses the essential issue of the magnitude of the step. It may be that a proposed reform, although more radical than its predecessors, is still not enough to propel it beyond the minimum threshold.[86]

More recent research takes a much more distanced approach. A fundamental study by Jiří Havel (1957–2012), Jan Klacek (1942–), Jiří Kosta (1921–2015), and Zdislav Šulc (1926–2015) focuses on the impact of non-Marxist economic theories that trickled into Czechoslovakia in the 1960s, while acknowledging the influence of the Soviet School of Optimal Planning in the subsequent period of normalization.[87] Similar reference is made in a remarkable study on socialist management by Vítězslav Sommer (1981–), Matěj Spurný (1979–), and Jaromír Mrňka (1988–).[88] Václav Rameš (1958–), who analyzes post-1989 publications on and by socialist Czechoslovak economists, identifies a similar tendency. Zuzana Richterková's (1981–) biographical study on Kožušník is an illustrative example of the psychological mechanism at play. Based on interviews with Kožušník, she depicts his 1952–55 stay in the Soviet Union as a formative experience. Faced with the realities of the preeminent socialist economy, the Czech economist began to develop a critical attitude toward the system's shortcomings. This intellectual development concluded in the 1970s when, sidelined after the intervention of the Warsaw Pact armies, Kožušník rejected both the Optimal Planning and Lange's model of a socialist economy with centrally controlled prices. Writing on his own initiative and with no hope of publishing his work, Kožušník arrived at the arguments initially put forward by Ludwig von Mises (1881–1973) and Friedrich Hayek (1899–1992) in their polemics with Lange in the interwar period.[89]

What most of such biography-oriented texts gloss over is the temporal and ideological middle ground between Stalinist orthodoxy and open revisionism (or outright rejection of socialism). In Kožušník's case there was clearly a phase in which he and his colleagues (Goldmann and Flek among others) were attempting to operationalize Lange's model for the sake of the Czechoslovak economic reforms.[90] Perhaps even more interestingly, it was in Brus's *Modely socialistické ekonomiky* (published in Czechoslovakia in 1964) that Kožušník and his colleagues first came across Mises's and Hayek's polemics with the Marxists.[91] While it would probably be an overstatement to compare the significance of bilateral contacts between Czechoslovak and Polish economists to the flourishing cooperation between philosophers—with Adam Schaff (1913–2006), Leszek Kołakowski (1927–2009), and Zygmunt Bauman (1925–2017) all visiting Prague more than once in early 1960s—it is still worthy of note.[92]

Conclusion

Within the strict rules of Stalinist internationalism there was very little room for any kind of intellectual transfer between the officially brotherly socialist nations. The initial impulses for the new chapter of socialist "Slavic reciprocity" came from the leadership of the Polish and Czechoslovak Academies of Sciences. Simultaneously, in the mid-1950s, the economic institutes of PAN and ČSAV established closer links and developed a mechanism for personnel exchange. Surprisingly, the first item to travel from Poland to Czechoslovakia was not an economic theory or a mathematical method but the spirit of open academic debate. In the late 1950s, Polish visitors, even those from the conservative wing of socialist economics, carried it with them. This was the case of Minc who, though far from susceptible to revisionism, contributed to reviving the Czechoslovak scientific community. At the same time, the animated post-1956 discussions among Polish economists were cautiously followed and commented on in Czechoslovakia. From the outset planning was the central issue of debate among socialist scholars.

In the late 1950s and early 1960s, the exchange system developed under the umbrella of both academies played a role in internationalizing Czechoslovak economists. At that time, the People's Republic of Poland's libraries were better supplied with international literature, but opportunities to talk to Polish colleagues were also highly valued by open-minded Czech and Slovak economists. One such individual, Šik, presided over the Economics Institute at its most international moment, at a time when dozens of guests from all over the world were visiting Prague and many Czechoslovaks were conducting their research abroad. Nevertheless, even at that time the Polish connection was held in high esteem and the bilateral exchange grew substantially. In the search for unquestionably Marxist points of reference for their reformist ideas, in the 1960s Šik and other Czechoslovak economists often referred to Kalecki and Lange.

However, neither Kalecki nor Lange played a major role in shaping Polish economic policy. Moreover, during the Prague Spring it was Polish economists, including, perhaps unsurprisingly, Minc, who expressed concerns about Czechoslovak revisionism. Ironically, the cause for concern was not primarily differences in their views on socialist planning. Although Šik's decentralization efforts were controversial, the more significant issue was the receptiveness with which young Czechoslovak economists embraced Western scholarship in their discipline. The experience of post-1956 discussions among Polish economists had demonstrated that dialogue between the systems without strict Party control was too dangerous to be permitted.

Author Biography

Maciej Górny is a historian specializing in the intellectual history and historiography of Central and Eastern Europe. He holds a PhD (2006) and habilitation (2015) from the Institute of History, Polish Academy of Sciences (IH PAN), where he now serves as a full professor and vice-director. His research explores national stereotypes, Polish–German relations, and Marxist historiography. Górny has led and contributed to several major research projects and is the author of influential works on World War I, historical geography, and the history of historical thought.

List of Abbreviations

AAN	Archiwum akt nowych (Central Archives of Modern Records)
APAN	Archiwum Polskiej Akademii Nauk (Archive of the Polish Academy of Sciences)
COMECON	Council for Mutual Economic Assistance
ČSAV	Československá akademie věd (Czechoslovak Academy of Sciences)
EÚ ČSAV	Economic Institute of the Czechoslovak Academy of Sciences (Ekonomický ústav Československé akademie věd)
MÚA AV ČR	Masarykův ústav a Archiv AV ČR (Masaryk Institute and Archives of the Czech Academy of Sciences)
SGPiS	Szkoła Główna Planowania i Statystyki (Main School of Planning and Statistics)
ZNE PAN	Economic Institute of the Polish Academy of Sciences (Instytut Ekonomii Polskiej Akademii Nauk)

Notes

1. Kolář, "Post-Stalinist Reformism"; Kolář, *Der Poststalinismus.*
2. Babiracki, "Two Stairways to Socialism"; Kunakhovich, "Ties That Bind"; Rieber, "Afterword"; Solovey and Dayé, "Introduction," 1–41; Olšáková, "Expert Cul-

tures"; Thoms, "Approaches to the History," 11–27; Mazurek, "Polish Economists," 588–610.

3. Baron and Madecki et al., *Česká polonistická studia*; Borák and Gładkiewicz, *Česká polonistika a polská bohemistyka*; Janko and Stasiewicz-Jasiukowa, *K dějinám československo-polských vědeckých styků.*
4. Dommann, Chassé, and Suter, "Einleitung," 109.
5. Wilczynski, *Profit, Risk and Incentives*, 54.
6. Kornai, *Das sozialistische System*, 457.
7. Ellman, *Socialist Planning*, 276–78.
8. Dobb, *Socialist Planning*, 50.
9. Wilczynski, *Socialist Economic Development*, 13.
10. Brzeski, "Poland as a Catalyst," 6.
11. Lange, "Political Economy of Socialism," 5.
12. See Lange's pre-World War II polemics with Mises and Hayek in Lange and Taylor, *On the Economic Theory of Socialism.*
13. "szybki rozwój przemysłu ciężkiego umożliwia osiągnięcie o wiele szybszego niż w ustroju kapitalistycznym tempa rozwoju zarówno przemysłu, jak i gospodarki narodowej jako całości." Brus, Minc, and Pohorille, *Materiały szkolenia partyjnego*, 95.
14. Kaliński, *Gospodarka Polski*, 80.
15. Ellman, *Socialist Planning*, p. 42.
16. Brus, *Ogólne problemy funkcjonowania gospodarki socjalistycznej.*
17. Kaliński, "Model gospodarki centralnie kierowanej," 53.
18. Jančík and Kubů, "Der erste Versuch," 3–61.
19. Wilczyński, *Profit, Risk and Incentives*, p. 17.
20. Jančík and Kubů, "Zwischen Planbefehl," 63–123. See also Šulc, *Dokumenty k hospodářské politice.*
21. Wilczyński, *Profit, Risk and Incentives*, 19.
22. Jančík and Kubů, "Zwischen Planbefehl und Markt," 108.
23. Shaffer, "Das neue ökonomische Modell," 211.
24. "Przywiązujemy więc wielką wagę do wykorzystania mechanizmu rynkowego nie w celu rekonstrukcji gospodarki rynkowej, lecz lepszego skojarzenia interesów jednostki, przedsiębiorstwa i społeczeństwa." "Czechosłowackie koncepcje reform," 1–2.
25. "Eine sozialistische Synthese von Plan und Markt könnte das wirtschaftliche Modell der nächsten Zukunft werden. Vereint mit einem wissenschaftlich und demokratisch organisierten politischen System, ist es das gesellschaftliche Zukunftsbild welches als humane sozialistische Gesellschaft bezeichnet werden kann." Šik, *Demokratische und sozialistische Plan- und Marktwirtschaft*, 9.
26. Eyal and Buchholz, "From the Sociology of Intellectuals," 117–37.
27. Sommer, "The Economics of Everyday Life."
28. Nowotny, "Organizacja planowania," 907.
29. Dolanský, "Podwyższenie wskaźników."
30. "Klement Gottwald wiecznie."
31. Radvanovský and Kučera, "Czechosłowacki głos," 1332–3.
32. Krblich, "Rozwój socjalistycznego rolnictwa."
33. Albrecht, "Professionalism in the Economic Institute," 266.

34. Míšková and Kunštát, "Mezinárodní vědecká spolupráce."
35. Kořalková, "K československo-polským vztahům."
36. "Smlouva č. 154/1947 Sb., Smlouva o přátelství a vzájemné pomoci mezi Československou republikou a republikou Polskou." *Zákony pro lidi.* Retrieved 7 February 2024 from https://www.zakonyprolidi.cz/cs/1947-154.
37. Pręcikowski, "Polska w 'nowym ruchu słowiańskim.'"
38. Vaculík, "Společnost polsko-československého přátelství."
39. APAN, ZNE PAN, sign. W. 419 Współpraca z Czechosłowacją 1955–1964, Vladimír Kaigl to Bronisław Minc, 30.12.1955.
40. APAN, ZNE PAN, sign. W. 419 Współpraca z Czechosłowacją 1955–1964, Plan realizacji umowy o współpracy kulturalnej między rządem PRL a rządem Republiki Czechosłowackiej za okres 1 stycznia do dnia 31 grudnia 1956 r.
41. Masarykův ústav a Archiv Akademie věd České republiky (thereafter MÚA AV ČR), Zahraniční odbor Úřadu presidia ČSAV, 11, Hodnocení zahraničních styků 1954–1957, Přehled o vědeckých stycích s Polskou lidovou republikou.
42. Ibid.
43. MÚA AV ČR, Zahraniční odbor Úřadu presidia ČSAV, 17, Vědecké spolupráce se zahraničními akademiemi 1953–1959, Společná schůze delegace Polské akademie věd se zástupci Československé akademie věd 12. listopadu 1955.
44. MÚA AV ČR, Zahraniční odbor Úřadu presidia ČSAV, 11, Hodnocení zahraničních styků 1954–1957, Přehledná tabulka o osobních stycích ČSAV a SAV za rok 1955.
45. MÚA AV ČR, Zahraniční odbor Úřadu presidia ČSAV, 17, Vědecké spolupráce se zahraničními akademiemi 1953–1959, Návrh na spolupráci mezi ekonomickými ústavy Akademií věd zemí socialistického tábora.
46. Albrecht, "Professionalism in the Economic Institute," 276.
47. Pavlát, "Zeszyty naukowe SGPiS."
48. Lipiński, "Ekonomia i zadania ekonomistów."
49. Lipiński, "Rewizje."
50. Mikołajski, "Po obu stronach Olzy."
51. Pavlát, "Na okraji diskuse."
52. Pavlát, "Z polské diskuse."
53. APAN, ZNE PAN, sign. W. 419, 59 Współpraca z Czechosłowacją 1955–1964; Vladimír Sedlák to Bronisław Minc, 13 November 1956.
54. Petrovský, "Ekonomická konference."
55. MÚA AV ČR, Zahraniční odbor ČSAV k. 75, sign. 436; K hodnocení realizace plánů vědecké spolupráce mezi ČSAV a PAV 1961 a k jednání o realizaci plánů v roce 1962.
56. MÚA AV ČR, Zahraniční odbor 376 Polsko; Plán zahraničních styků na rok 1963—Polsko.
57. MÚA AV ČR, Zahraniční odbor Úřadu presidia ČSAV, k. 47, sign. 399; Plán zahraničních styků na rok 1963.
58. MÚA AV ČR, Zahraniční odbor Úřadu presidia ČSAV 1969, k. 75, sign. 537 Hodnocení vědeckých styků ústavů ČSAV za rok 1969; EÚ ČSAV, Spolupráce s vědeckými institucemi socialistických zemí v r. 1969.
59. AAN, Państwowa Komisja Planowania Gospodarczego, sign. 2/274/0/5/1456 Współpraca gospodarcza i naukowo-techniczna z zagranicą 1954–1956.

60. MÚA AV ČR, Zahraniční odbor Úřadu presidia ČSAV 1969, Směrnici o maximální výši jednotlivých druhů výdajů spojených s návštěvami zahraničních osob a výdajů na representaci.
61. "Vědecký pracovník podává . . . písemnou zprávu své sekci, pracovišti I (OR i—JS) zahraničnímu odboru, ve které je povinnen uvést hlavní přínos ze své cesty. O svých zkušenostech přednáší jak na ústavní radě, tak v kolektivu pracovníků své laboratoře. Každá cesta je hodnocena na pracovišti s hlediska odborného, politického i kulturního pro pracovníka i s hlediska přínosu pro výzkumné úkoly, na kterých laboratoř pracuje." MÚA AV ČR, Zahraniční odbor ÚP ČSAV 1953–1959, k. 12 Hodnocení zahraničních styků 1958–1959; Zpráva o organisaci a zajišťování využití zahraničních zkušeností, 5 September 1958.
62. "Týmto otázkam bola v Poľsku v posledných rokoch venovaná značná pozornosť; ich výskumom sa zaoberá značný počet ekonómov, ktorí dosiahli určité výsledky. S poľskými poznatkami sa potrebuje nutne zoznámiť." MÚA AV ČR, Zahraniční odbor 1953–1959, k. 40 Studijní pobyt v zahraničí a jiné zahraniční styky. Polsko, 16. II. 1959 Návrhový list "A" pre vyslanie vedeckého pracovníka do Polska, Jaroslav Němec.
63. Šik and Kouba, *Poznatky z Varšavy*.
64. MÚA AV ČR, Zahraniční odbor ÚP ČSAV 1960–1964, 169 Cestovní zprávy: Polsko, Kožušník Čestmír, Zpráva o studijním pobytu v Polsku ve dnech 31 May–13 June 1963.
65. MÚA AV ČR, Zahraniční odbor ÚP ČSAV 376 Polsko; Návrhový list "A" Jaroslav Habr.
66. Such was the case of Miroslav Rosický whose travel to Poland was, first, canceled and later realized in a reduced form paid with travel funds from another international conference in GDR that had been canceled. MÚA AV ČR, Zahraniční odbor ÚP ČSAV 376 Polsko; Jiří Karlík to Zahraniční odbor presidia ČSAV, 2 April 1963.
67. APAN, ZNE PAN w. 419, sign. 22 Materiały z Konferencji nad zagadnieniami bilansu gospodarki narodowej, 22 October 1959.
68. MÚA AV ČR, Zahraniční odbor Úřadu presidia ČSAV 1969, k. 75, sign. 537 Hodnocení vědeckých styků ústavů ČSAV za rok 1969; EÚ ČSAV, Spolupráce s vědeckými institucemi socialistických zemí v r. 1969.
69. APAN, ZNE PAN, w. 419, sign. 60 Sprawozdania zagraniczne Czechosłowacja od I 1965 do końca 1967; František Šorm to Bronisław Minc, 27 May 1966.
70. Misiak, "Gospodarka CSRS," 5.
71. Kosta, "Czechoslovak Economists," 139–48.
72. Šik, "Problems of the New System," 33.
73. Ibid., 8.
74. Lange, "Předmluva k českému vydání," 5–6.
75. See, for example, Tous, "[Review of] Zdeněk Vergner and Miroslav Souček."
76. "Jeśli jednak udane wprowadzenie nowego systemu planowania i zarządzania ma nastąpić możliwie najszybciej, wymagane są daleko idące zmiany w sposobie myślenia ludzi zaangażowanych zarówno w badania naukowe, jak i w praktykę planowania i zarządzania."
77. APAN, ZNE PAN, w. 419, sign. 81 Sprawozdania z wyjazdów zagranicznych 1962–1966; Józef Zagórski, Sprawozdanie z udziału w łącznym seminarium Zakładu nauk

Ekonomicznych PAN i Instytutu Ekonomicznego Słowackiej Akademii Nauk w sprawie postępu technicznego w Bratysławie we wrześniu 1965 r.

78. Šik, Levčík, and Suchan, *Mezinárodní symposium*, 10.
79. "Setkání s polskými ekonomy potvrdila, že předpokládané změny v soustavě řízení v řadě socialistických zemí se ubírají shodným směrem. V Polsku se připravuje řada experimentů, které mají posílit pravomoc sdružení a zvýšit úlohu některých ekonomických nástrojů. Opatření v řízení československého průmyslu se však podstatně liší v tom, že představují komplexní řešení problémů centrálního plánu a souboru ekonomických nástrojů. Představují tedy vnitřně skloubený systém, jehož praktické uplatnění podstatně mění ekonomické podmínky pro sladění zájmů podniku a společnosti. Diskuse s polskými ekonomy z oblasti teorie i praxe potvrdila, že jedině syntetický přístup k problematice řízení umožňuje postihnout dnes již všeobecně známé nedostatky administrativně-direktivní soustavy řízení." Šik and Kouba, *Poznatky z Varšavy*, 5–6.
80. APAN, ZNE PAN, w. 419, sign. 81 Sprawozdania z wyjazdów zagranicznych 1962–1966; W. J. Wyrozembski, Sprawozdanie z uczestnictwa w Międzynarodowej Konferencji w Bratysławie (Smolenice) 10-16.10.1966 r., poświęconej zmianom strukturalnym we współczesnym kapitalizmie i nowym tendencjom w ekonomicznej teorii na Zachodzie.
81. Ibid.
82. "W toku dyskusji ujawniły się znaczne różnice poglądów na rolę planu i rynku wśród ekonomistów krajów socjalistycznych. W wystąpieniach ekonomistów jugosłowiańskich i czechosłowackich występowały wyraźne tendencje pomniejszania roli planu centralnego (traktowanie planu jako obrazu przyszłego rynku). Towarzysze czescy mieli tendencję do przesadnego akcentowania błędów poprzedniego okresu i idealizowania roli rynku, konkurencji i daleko posuniętej automatyzacji przedsiębiorstw." APAN, ZNE PAN, w. 419, sign. 81 Sprawozdania z wyjazdów zagranicznych 1962–1966; B. Minc, M. Pohorille, B. Gliński, Notatka z konferencji ekonomistów krajów socjalistycznych i kapitalistycznych (zorganizowanej pod auspicjami Międzynarodowego Towarzystwa Ekonomicznego) na temat: "Planowanie w rynek—współczesne tendencje w różnych systemach ekonomicznych."
83. APAN, ZNE PAN, w. 419, sign. 61 Współpraca naukowa z Czechosłowacją 1968–1972; Protokół o współpracy naukowej między Polską Akademią Nauk i Czechosłowacką Akademią Nauk na lata 1968–1970, 19.01.1968.
84. Ibid., Projekt porozumienia między instytutami, 5 August 1971.
85. See Kovács, *Communist Planning*, with separate chapters devoted to individual socialist states.
86. Balcerowicz, "On the Reformability," 199.
87. Havel et al., "Economics and Systemic Change"; Rameš, "(Post)socialističtí ekonomové."
88. Sommer et al., *Řídit socialismus jako firmu*.
89. Richterková, *Přínos Čestmíra Kožušníka*, 29–37.
90. See ibid., 70.
91. Ibid., 65.
92. Kopeček, *Hledání ztraceného smyslu*, 298–300.

Bibliography

Albrecht, Catherine. 2000. "Professionalism in the Economic Institute of the Czechoslovak Academy of Sciences, 1953–1963." In *Věda v Československu v letech 1953–1963: Sborník z konference (Praha, 23.-24. listopadu 1999)* [Science in Czechoslovakia in the years 1953–1963: Conference Proceedings (Prague, 23–24 November 1999)], edited by Hana Barvíková, Marek Ďurčanský, and Pavel Kodera, 265–77. Prague: Academia.

Babiracki, Patryk. 2016. "Two Stairways to Socialism: Soviet Youth Activists in Polish Spaces, 1957–1964." In *Socialist Internationalism in the Cold War: Exploring the Second World*, edited by Patryk Babiracki and Austin Jersild, 79–105. Cham: Palgrave Macmillan.

Balcerowicz, Leszek. 1990. "On the Reformability of the Soviet-Type Economic Systems." In *The Evolution of Economic Systems: Essays in Honor of Ota Sik*, edited by Kurt Dopfer and Karl-F. Raible, 193–201. New York: Palgrave Macmillan.

Baron, Roman, and Roman Madecki. 2014. *Česká polonistická studia: tradice a současnost (filologie, historie, politologie, právo)* [Czech Polish studies: Tradition and present (Philology, history, political science, law)]. Prague: Historický ústav.

Borák, Mečislav, and Ryszard Gładkiewicz. 2009. *Česká polonistika a polská bohemistika na přelomu století: Czeska polonistyka i polska bohemistyka na przełomie stuleci* [Czech Polish Studies and Polish Bohemian Studies at the turn of the century]. Prague: Univerzita Karlova v Praze.

Brus, Włodzimierz. 1964. *Ogólne problemy funkcjonowania gospodarki socjalistycznej* [General problems with the functioning of the socialist economy]. Warsaw: Państwowe Wydawnictwo Naukowe.

Brus, Włodzimierz, Bolesław Minc, and Maksymilian Pohorille, eds. 1954. *Materiały szkolenia partyjnego: Podręcznik dla szkół politycznych. II rok nauki* [Party training materials: A handbook for political schools. second year of study.]. Warsaw: Książka i Wiedza.

Brzeski, Andrzej. 1971. "Poland as a Catalyst of Change in the Communist Economic System." *The Polish Review* 16, no. 2: 3–34.

"Czechosłowackie koncepcje reform (wywiad z Otą Szikiem i docentem Karolem Koubą)" [Czechoslovak Reform Concepts (Interview with Ota Šik and Associate Professor Karol Kouba]. 1965. *Życie Gospodarcze* 15, no. 25: 1–2.

Dobb, Maurice. 1970. *Socialist Planning: Some Problems*. London: Lawrence & Wishart.

Dolansky, Jaromir. 1951. "Podwyższenie wskaźników 5-letniego planu gospodarczego CSR w przemyśle" [Increasing the targets of the 5-year economic plan of CSR in industry.]. *Życie Gospodarcze* 6, no. 11: 609–10.

Dommann, Monika, Daniel Speich Chassé, and Mischa Suter. 2014. "Einleitung: Wissensgeschichte ökonomischer Praktiken" [Introduction: History of knowledge of economical practices]. *Beiträge zur Wissenschaftsgeschichte* 37: 107–11.

Ellman, Michael. 2014. *Socialist Planning*. Cambridge: Cambridge University Press.

Eyal, Gil, and Larissa Buchholz. 2010. "From the Sociology of Intellectuals to the Sociology of Interventions." *Annual Review of Sociology* 36: 117–37.

Havel, Jiří, Jan Klacek, Jiří Kosta, and Zdislav Šulc. 1998. "Economics and Systemic Change in Czechoslovakia, 1945–1992." In *Economic Thought in Communist and Post-Communist Europe*, edited by Hans-Jürgen Wagener, 213–63. London: Routledge.

Jančík, Drahomír, and Eduard Kubů. 2006. "Der erste Versuch einer Reform der zentralen Planwirtschaft in der Tschechoslowakei" [The first attempt at reforming the central planned economy in Czechoslovakia.]. In *Sozialistische Wirtschaftsreformen: Tschechoslowakei und DDR im Vergleich,* edited by Christoph Boyer, 3–61. Frankfurt am Main: Vittorio Klostermann.

———. 2006. "Zwischen Planbefehl und Markt: Der Diskurs der zweiten tschechoslowakischen Wirtschaftsreform" [Between planning directives and the market: the discourse of the second Czechoslovak economic reform]. In *Sozialistische Wirtschaftsreformen,* edited by Christoph Boyer, 63–123. Frankfurt am Main: Vittorio Klostermann.

Janko, Jan, and Irena Stasiewicz-Jasiukowa, eds. 1989. *K dějinám československo-polských vědeckých styků* [Toward a history of Czechoslovak-Polish scientific relations]. Prague: Ústav československých a světových dějin ČSAV.

Kaliński, Janusz. 1996. *Gospodarka Polski w latach 1944–1989. Przemiany strukturalne* [The Polish economy in the years 1944–1989: Structural transformations]. Warsaw: Państwowe Wydawnictwo Ekonomiczne.

———. 2013. "Model gospodarki centralnie kierowanej w Polsce w ujęciu 'szkoły wakarowskiej'" [Model of the centrally planned economy in Poland from the perspective of the 'Wakar School']. In *W poszukiwaniu modelu gospodarki centralnie kierowanej,* edited by Piotr Jackowicz, 51–62. Warsaw: Oficyna Wydawnicza Szkoła Główna Handlowa.

"Klement Gottwald wiecznie żyć będzie w sercach narodów CSR" [Klement Gottwald will live forever in the hearts of the peoples of the CSR]. 1953. *Życie Gospodarcze* 8, no. 13: 386.

Kolář, Pavel. 2016. *Der Poststalinismus: Ideologie und Utopie einer Epoche.* Cologne: Böhlau Verlag.

———. 2017. "Post-Stalinist Reformism and the Prague Spring." In *The Cambridge History of Communism,* vol. 2: *The Socialist Camp and World Power 1941–1960s,* edited by Norman Naimark, Silvio Pons, and Sophie Quinn-Judge, 170–95. Cambridge: Cambridge University Press.

Kopeček, Michal. 2009. *Hledání ztraceného smyslu revoluce: Zrod a počátky marxistického revizionismu ve střední Evropě 1953–1960* [In search of the lost meaning of the revolution: the birth and early years of marxist revisionism in Central Europe, 1953–1960]. Prague: Argo.

Kořalková, Květa. 1963. "K československo-polským vztahům v letech 1945–1948" [On Czechoslovak-Polish relations in 1945–1948]. *Slovanské historické studie*: 309–36.

Kornai, János. 1995. *Das sozialistische System: Die politische Ökonomie des Kommunismus* [The socialist system: The political economy of Communism]. Baden-Baden: Nomos Verlag.

Kosta, Jiří. 1964. "Czechoslovak Economists Discuss Ways of Improving the System of Planned Management." *Czechoslovak Economic Papers* 4: 139–48.

Kovács, János Mátyás, ed. 2022. *Communist Planning versus Rationality: Mathematical Economics and the Central Plan in Eastern Europe and China.* Lanham, MD: Rowman & Littlefield.

Krblich, Jan. 1954. "Rozwój socjalistycznego rolnictwa w pierwszym pięcioletnim planie rozwoju czechosłowackiej gospodarki narodowej" [The development of socialist agriculture in the first five-year plan for the development of the Czechoslovak national economy]. *Życie Gospodarcze* 9, no. 17: 673–76.

Kunakhovich, Kyrill. 2016. "Ties That Bind, Ties That Divide: Second World Cultural Exchange at the Grassroots." In *Socialist Internationalism in the Cold War: Exploring the Second World*, edited by Patryk Babiracki and Austin Jersild, 135–59. Cham: Palgrave Macmillan.

Lange, Oskar. 1966. "Předmluva k českému vydání" [Introduction to the Czech edition]. In *Politická ekonomie: Obecné otázky*, by Oskar Lange, trans. Ivan Figura, 5–6. Prague: Academia.

———. 1959. "The Political Economy of Socialism." *Science & Society* 23, no. 1: 1–15.

Lange, Oskar, and Fred M. Taylor. 1964. *On the Economic Theory of Socialism*, edited by Benjamin E. Lippincott. Minneapolis: University of Minnesota Press.

Lipiński, Edward. 1956. "Ekonomia i zadania ekonomistów" [The economy and tasks of economists]. *Życie Gospodarcze* 14, no. 3: 8.

———. 1956. "Rewizje" [Revisions]. *Nowa Kultura*, 9 September, 1.

Mazurek, Małgorzata. 2018. "Polish Economists in Nehru's India: Making Science for the Third World in an Era of De-Stalinization and Decolonization." *Slavic Review* 77, no. 3: 588–610.

Mikołajski, Juliusz. 1956. "Po obu stronach Olzy" [On both sides of Olza]. *Życie Gospodarcze* 11, no. 15: 12.

Misiak, Marek. 1964. "Gospodarka CSRS poszukuje nowych rozwiązań" [The economy of CSR looks for new solutions]. *Życie Gospodarcze* 19, no. 46: 5.

Míšková, Alena, and Miroslav Kunštát. 2019. "Mezinárodní vědecká spolupráce ČSAV v letech 1952–1956" [International cooperation of the Czechoslovak Academy of Sciences 1952-1956]. In *Dějiny Československé akademie věd, vol. I, 1952–1962*, edited by Martin Franc and Věra Dvořáčková, 292–306. Prague: Academia.

Nowotny, S. 1945–46. "Organizacja planowania w ministerstwie przemysłu" [The organisation of planning in the ministra of industry]. *Życie Gospodarcze* 1, no. 22: 905–7.

Olšáková, Doubravka. 2018. "Expert Cultures and Infrastructural Globalism: Socialist Experts under Restricted Internationalism." In *Experts and Expertise in Science and Technology in Europe since the 1960s: Organized Civil Society, Democracy and Political Decision-Making*, edited by Christine Bouneau and David Burigana, 53–68. Brussels: Peter Lang Verlag.

Pavlát, V[ladislav]. 1958. "Zeszyty naukowe SGPiS" [Scholarly notes of the SGPiS]. *Politická ekonomie* 6, no. 5: 459.

Pavlát, Vladislav. 1957. "Na okraji diskuse polskich ekonomů o předmětu politické ekonomie" [On the margins of the discussion among Polish economists regarding the subject matter of political economy]. *Politická ekonomie* 5, no. 7: 533–43.

———. 1958. "Z polské diskuse o marxistické a soudobé buržoasní politické ekonomii" [From the Polish discussion on Marxist and contemporary bourgeois political economy.]. *Politická ekonomie* 6, no. 8: 706–20.

Petrovský, V[lastimil]. 1957. "Ekonomická konference o produktivitě prace" [Economic conference on labor productivity]. *Politická ekonomie* 6, no. 3: 214–20.

Pręcikowski, Leszek Sławomir. 2000. "Polska w 'nowym ruchu słowiańskim': Z działalności Komitetu Słowiańskiego w Polsce w latach 1945–1947" [Poland in the 'New

Slavic Movement': On the activities of the Slavic Committee in Poland in 1945–1947]. *Acta Universitatis Lodziensis, Folia Historica* 69: 173–90.

Radvanovský, Antonín, and Antonín Kučera. 1952. "Czechosłowacki głos w dyskusji o Słowniku" [The Czechoslovak voice in the discussion on the Dictionary]. *Życie Gospodarcze* 7, no. 23: 1332–33.

Rameš, Václav. 2017. "(Post)socialističtí ekonomové jako objekt historického zkoumání. Současný stav bádání a perspektivy" [(Post)Socialist economists as the subject of historical inquiry. The current state of research and perspectives]. *Soudobé dějiny* 24, no. 3: 389–402.

Richterková, Zuzanna. 2013. *Přínos Čestmíra Kožušníka českému ekonomickému myšlení (v kontextu diskuze o plánu a trhu)* [The contribution of Čestmír Kožušník to Czech economic thought (in the context of the debate on planning and the market)]. Prague: Národohospodářský ústav Josefa Hlávky.

Rieber, Alfred. 2016. "Afterword: Promises and Paradoxes of Socialist Internationalism (Personal and Historical Reflections)." In *Socialist Internationalism in the Cold War: Exploring the Second World,* edited by Patryk Babiracki and Austin Jersild, 327–43. Cham: Palgrave Macmillan.

Shaffer, Harry G. 1968. "Das neue ökonomische Modell: Probleme und Aussichten. Ein Interview mit Ota Šik" [The new economic model: Problems and prospects. an interview with Ota Šik]. *Osteuropa* 18: 208–19.

Šik, Ota. 1971. *Demokratische und sozialistische Plan- und Marktwirtschaft* [Democratic and socialist planned and market economy]. Zürich: Arche.

———. 1965. "Problems of the New System of Planned Management." *Czechoslovak Economic Papers* 5: 7–33.

Šik, Ota, and Karel Kouba. 1965. *Poznatky z Varšavy (Zpráva o návštěvě v Polsku ve dnech 17.—21.5.1965), Inform. Publ. Č. 23* [Findings from Warsaw (report on a visit to Poland, May 17–21, 1965), Inform. Publ. No. 23]. Prague: Ekonomický ústav ČSAV.

Šik, Ota, Bedřich Levčík, and Karel Suchan. 1965. *Mezinárodní symposium na téma "Hospodářské plánování a hospodářský růst" (Gösing, Rakousko, 28. září—1. října 1965)* [International symposium on "Economic planning and economic growth" (Gösing, Austria, September 28–October 1, 1965)]. Prague: Ekonomický ústav ČSAV.

"Smlouva č. 154/1947 Sb., Smlouva o přátelství a vzájemné pomoci mezi Československou republikou a republikou Polskou" [Treaty No. 154/1947 Coll., Treaty of Friendship and Mutual Assistance between the Czechoslovak Republic and the Republic of Poland]. Retrieved 7 February 2024 from https://www.zakonyprolidi.cz/cs/1947-154.

Solovey, Mark, and Christian Dayé. 2021. "Introduction: Cold War Social Science: Transnational Entanglements." In *Cold War Social Science: Transnational Entanglements,* edited by Mark Solovey and Christian Dayé, 1–41. Cham: Palgrave Macmillan.

Sommer, Vítězslav. 2019. "The Economics of Everyday Life in 'New' Socialism: Czechoslovak Public Economics and Economic Reform in the Prague Spring Era." In *Economic Knowledge in Socialism, 1945–89,* edited by Till Düppe and Ivan Boldyrev, 52–72. Durham, NC: Duke University Press.

Sommer, Vítězslav, Matěj Spurný, and Jaromír Mrňka. 2019. *Řídit socialismus jako firmu: Technokratické vládnutí v Československu, 1956–1989* [Running socialism like a business. technocratic governance in Czechoslovakia, 1956–1989]. Prague: Nakladatelství Lidové noviny.

Šulc, Zdislav, ed. 1998. *Dokumenty k hospodářské politice v Československu z let 1963–1969* [Documents on economic policy in Czechoslovakia, 1963–1969]. Prague: Vysoká škola ekonomická v Praze.

Thoms, Ulrike. 2018. "Approaches to the History of International Scientific Congresses: Opportunities and Problems." In *Science Overcoming Borders*, edited by Věra Dvořáčková and Martin Franc, 11–27. Prague: Masarykův ústav a Archiv AV ČR.

Tous, Miroslav. 1968. "[Review of] Theoretical Problems of Economic Growth in Czechoslovakia, by Zdeněk Vergner and Miroslav Souček." *Czechoslovak Economic Papers* 8: 155–60.

Vaculík, Jaroslav. 2012. "Společnost polsko-československého přátelství v letech 1946–1951" [The Polish-Czechoslovak Friendship Society, 1946–1951]. *Sborník prací Pedagogické fakulty Masarykovy univerzity (Řada společenských věd)* 26, no. 1: 154–59.

Wilczynski, J. 1973. *Profit, Risk and Incentives under Socialist Economic Planning*. London: Palgrave Macmillan.

———. 1972. *Socialist Economic Development and Reforms: From Extensive to Intensive Growth under Central Planning in the USSR, Eastern Europe and Yugoslavia*. London: Palgrave Macmillan.

Archival Sources

Archiwum Akt Nowych (AAN), Państwowa Komisja Planowania Gospodarczego, sign. 2/274/0/5/1456 Współpraca gospodarcza i naukowo-techniczna z zagranicą 1954–1956.

Archiwum Polskiej Akademii Nauk, Warszawa (APAN), Zakład Nauk Ekonomicznych, sign. W. 419 Współpraca z Czechosłowacją 1955–1964, Vladimír Kaigl to Bronisław Minc, 30.12.1955.

APAN, Zakład Nauk Ekonomicznych, sign. W. 419 Współpraca z Czechosłowacją 1955–1964, Plan realizacji umowy o współpracy kulturalnej między rządem PRL a rządem Republiki Czechosłowackiej za okres 1 stycznia do dnia 31 grudnia 1956 r.

APAN, Zakład Nauk Ekonomicznych PAN, w. 419, sign. 81 Sprawozdania z wyjazdów zagranicznych 1962–1966; Józef Zagórski, Sprawozdanie z udziału w łącznym seminarium Zakładu nauk Ekonomicznych PAN i Instytutu Ekonomicznego Słowackiej Akademii Nauk w sprawie postępu technicznego w Bratysławie we wrześniu 1965 r.

APAN ZNE w. 419, sign. 22 Materiały z Konferencji nad zagadnieniami bilansu gospodarki narodowej, 22 X 1959.

APAN ZNE, w. 419, sign. 61 Współpraca naukowa z Czechosłowacją 1968–1972; Protokół o współpracy naukowej między Polską Akademią Nauk i Czechosłowacką Akademią Nauk na lata 1968–1970, 19.01.1968.

APAN ZNE PAN, w. 419, sign. 60 Sprawozdania zagraniczne Czechosłowacja od I 1965 do końca 1967; František Šorm to Bronisław Minc, 27 May 1966.

APAN ZNE w. 419, sign. 81 Sprawozdania z wyjazdów zagranicznych 1962–1966; W. J. Wyrozembski, Sprawozdanie z uczestnictwa w Międzynarodowej Konferencji w Bratysławie (Smolenice) 10–16.10.1966 r., poświęconej zmianom strukturalnym we współczesnym kapitalizmie i nowym tendencjom w ekonomicznej teorii na Zachodzie.

MÚA AV ČR, Zahraniční odbor Úřadu presidia ČSAV, 11, Hodnocení zahraničních styku 1954–1957, Přehled o vědeckých stycích s Polskou lidovou republikou.

MÚA AV ČR, Zahraniční odbor Úřadu presidia ČSAV, 17, Vědecké spolupráce se zahraničními akademiemi 1953–1959, Společná schůze delegace Polské akademie věd se zástupci Československé akademie věd 12. listopadu 1955.

MÚA AV ČR, Zahraniční odbor Úřadu presidia ČSAV, 11, Hodnocení zahraničních styků 1954–1957, Přehledná tabulka o osobních stycích ČSAV a SAV za rok 1955.

MÚA AV ČR, Zahraniční odbor Úřadu presidia ČSAV, 17, Vědecké spolupráce se zahraničními akademiemi 1953–1959, Návrh na spolupráci mezi ekonomickými ústavy Akademií věd zemí socialistického tábora.

MÚA AV ČR, Zahraniční odbor ČSAV k. 75, sign. 436; K hodnocení realizace plánu vědecké spolupráce mezi ČSAV a PAV 1961 a k jednání o realizaci plánu v roce 1962.

MÚA AV ČR, Zahraniční odbor 376 Polsko; Plán zahraničních styků na rok 1963—Polsko.

MÚA AV ČR, Zahraniční odbor Úřadu presidia ČSAV, k. 47, sign. 399; Plán zahraničních styků na rok 1963.

MÚA AV ČR, Zahraniční odbor Úřadu presidia ČSAV 1969, k. 75, sign. 537 Hodnocení vědeckých styků ústavů ČSAV za rok 1969; EÚ ČSAV, Spolupráce s vědeckými institucemi socialistických zemí v r. 1969.

MÚA AV ČR, Zahraniční odbor Úřadu presidia ČSAV 1969, Směrnice o maximální výši jednotlivých druhů výdajů spojených s návštěvami zahraničních osob a výdajů na representaci.

MÚA AV ČR, Zahraniční odbor ÚP ČSAV 1953–1959, k. 12 Hodnocení zahraničních styků 1958–1959; Zpráva o organisaci a zajišťování využití zahraničních zkušeností, 5.9.1958.

MÚA AV ČR, Zahraniční odbor ÚP ČSAV 1953–1959, k. 40 Studijní pobyt v zahraničí a jiné zahraniční styky. Polsko, 16. II. 1959 Návrhový list "A" pre vyslanie vedeckého pracovníka do Polska, Jaroslav Němec.

MÚA AV ČR, Zahraniční odbor ÚP ČSAV 1960–1964, 169 Cestovní zprávy: Polsko, Kožušník Čestmír, Zpráva o studijním pobytu v Polsku ve dnech 31.5.–13.6.1963.

MÚA AV ČR, Zahraniční odbor ÚP ČSAV 376 Polsko; Návrhový list "A" Jaroslav Habr.

MÚA AV ČR, Zahraniční odbor ÚP ČSAV 376 Polsko; Jiří Karlík to Zahraniční odbor presidia ČSAV, 2. April 1963.

CHAPTER 5
Asymmetries in Mathematical Knowledge

Tomáš W. Pavlíček

Knowledge Crossing Borders

In 1959, shortly after defending his doctoral thesis in Poznań, Zbigniew Semadeni (1934–) traveled to Czechoslovakia for a short study visit. Over sixty years later, during our oral history session, Professor Semadeni recalled names of at least seven Prague mathematicians with whom he had discussed his research at the time. Although they shared a deep interest in mathematical analysis, some accounts indicate that Semadeni was in tourist mode. The scholar Ilja Černý (1929–) took him to the Bohemian castle Karlštejn and on a tour of Prague's restaurants. However, the real value of Semadeni's visit materialized on the return rail journey from Prague to Wrocław, which traversed mountainous terrain. In my interview with him, he recalled the moment:

> ZS: Can I tell you about the best theorem of my life? Someone even put it on Wikipedia. I came up with it at a border crossing, on a train from Prague to Wrocław, at midnight.
>
> TWP: How did that happen?
>
> ZS: They woke me up for the border check, and there was nothing to do. I was sitting alone in the dark. So I started thinking about mathematics and this theorem came to me. It was the best in my lifetime. This also shows how strange it is. Obviously, that study visit must have affected me in some way. A breakthrough. That's how mathematics works—it had been there in my head. I had to write it all down afterward—that's another story—but I already knew, more or less, what it was.[1]

Inspired by his time with colleagues in Prague, Semadeni derived a novel theorem in functional analysis.[2] The circumstances of its origin were not known publicly, partly because Semadeni did not continue with this research, changing his focus to other disciplines (topology and category theory). Finally, he became involved in the didactics of mathematics and, beginning in the 1970s, he was regularly invited to participate in the International Congress of Mathematicians (hereafter ICM; he participated in the organization of the ICM in Warsaw in 1983, and he was an invited speaker at the ICM in Berkeley in 1986) and was sent by the Ministry of Science, Higher Education and Technology to represent the People's Poland on the Executive Committee of the International Commission on Mathematical Instruction (member 1979–1982, vice president 1983–1986).[3]

This brief outline of Zbigniew Semadeni's biography illustrates reciprocal international cooperation beyond the Iron Curtain and reciprocal practices between scientists in Central and Eastern Europe. It shows how cooperation beyond the Iron Curtain was associated with entanglements within the Eastern Bloc, which is a recurring theme of this chapter.

After World War II, students and future mathematicians in both Poland and Czechoslovakia were trained through close collaboration between universities and the state mathematical institutes, which were soon incorporated into their respective academies of sciences (1952–53). The framework of their education was defined as part of the communist construction of socialism, serving as a dynamic component of internationalization, reciprocity and comradely cooperation.[4] However, when examining personal encounters and the circulation of practices among socialist countries, I prefer to use the concept of Olympic internationalism,[5] as it was clearly shaped by the idea of competition and scholars representing their nations, and it served the intellectual elites in training new cadres, despite many shortcomings. In the Eastern Bloc, scholars had little access to original mathematical monographs from the West, and they were only available in Russian translation. However, Russian was not a lingua franca for many scholars, certainly not for Hungarians and Germans. Sovietization progressed only gradually, and language played a crucial role in communication among mathematicians.[6]

The formation of the first postwar youth generation was shaped by new collective norms, including the democratization of education, cadre approval for admission to studies and employment assignments for graduates. The mental legacy of the war, combined with demands for socially justified reforms and the need to catch up with the Western world led to shared youth experiences across the Eastern Bloc. Juliane Fürst argues that fear of this "Stalin's last generation" is one factor that contributed to the Cold War atmosphere.[7] The generation that followed established their professional credentials through uniformly structured careers, as is often evident from archival sources. Yet the question remains: to what extent did they prioritize sharing older experiences over pursuing deeper forms of generational emancipation?[8]

The aim of this chapter is to analyze the circumstances and spaces in which mathematical research was practiced at the intersection between Poland and Czechoslovakia and the factors that facilitated the positioning of a new generation of mathematicians in new mathematical disciplines. As Christopher Hollings has argued, traditional historiography often assumes that the natural sciences transcended the boundaries of languages and states, while the humanities were politically restricted.[9] Yet politics also played a significant role in shaping the internationalization of the natural sciences[10] and, as Hollings notes, the mathematical community was also defined by the processes of connection and isolation.[11] However, binary thinking is not always productive, particularly in the context of the Cold War, as argued by Deleuze and Guattari in *A Thousand Plateaus* (1980).[12] Following their call to move beyond rigid dichotomies, this study will demonstrate how shifts in language, national community, or disciplinary focus in Central and Eastern Europe could at times benefit scholars.

The chapter follows developments in the mathematical community from the end of World War II, focusing on key aspects of the emergence of the new mathematical generation that gradually took over the leading roles in the discipline from those socialized during the interwar period. It begins by discussing the origins of mathematical institutes in Poland and Czechoslovakia, the transformation of the educational systems for training new cadres, and the individual practices within mathematical seminars (specifically in the three interrelated fields of functional analysis, topology, and logic).[13] It then goes on to show the asymmetries between seminars. Drawing on oral history research, the chapter demonstrates the importance of personal encounters for generating new knowledge, as well as the significant role of local conferences in facilitating cross-border collaboration in the Eastern Bloc and also across the Iron Curtain. The final part of the chapter addresses the perceived gender imbalance in academia from the perspective of female scholars.[14]

Entangled History "From Below"

In researching this chapter, I examined the training of new generations, the organization of research, and collaboration with foreign partners in syntheses of the history of mathematics and in the archival collections of the Institute of Mathematics of the Polish Academy of Sciences (hereafter IM PAN) and the Institute of Mathematics of the Czechoslovak Academy of Sciences (hereafter MÚ ČSAV). I also included correspondence in personal collections, and biographical texts containing personal reminiscences. However, it should be noted that both official and informal sources are selective and tend to emphasize achievements (not failures).

To address the inherent selectivity of institutional archives, I combined the prosopography of institutional archives and discourse analysis of personal collections with an anthropological approach rooted in oral history research.[15] The purpose of conducting interviews with experts was not to produce prosopographical documentation for the history of science,[16] but rather to gain a deeper understanding of the processes and events experienced by the individuals. It also allows a shift of focus from a linear narrative to a reconstruction of past experiences enriched by the accompanying emotions.[17]

Combining spoken and written sources allowed for a "history from below" perspective, focusing on communication in seminars and informal international contacts. The research involved interviews with a dozen scientists in Poland and Czechia, two-thirds of whom were male and one-third female.[18] The research also encompassed the Slovak context of the former Czechoslovak federation, but it was not possible to conduct interviews with Slovak mathematicians.

During the interviews, the conversations typically included a biographical narrative (childhood, studies, academic career), discussion of knowledge production (organization of research, publications), and experiences of everyday life in research institutes, mobility, and transnational encounters during the Cold War. In order to achieve this, it was necessary to reinforce semantic networks with episodic elements.[19] Interviewees were identified through the recommendations of previous participants and through an exploration of archival materials. This approach made it possible to engage scientists who were constrained, marginalized, or excluded.[20] When addressing the issue of gender, I focused on postwar academic networks, particularly male dominance in the newly emerging fields.[21]

Given that critics of oral history often point to the difficulty of verifying the narratives, it is necessary to adopt a model of autobiographical memory with both semantic and episodic elements. While semantic elements provide verifiable details such as dates and formal milestones (for example, year of doctoral defense), episodic elements express atmosphere and subjective experience (for example, where others were sitting and how a debate unfolded). These elements are embedded in historical time and carry emotional significance, making them more memorable. Narrators can switch between the two as they recall their experiences.[22] This approach to entangled knowledge allows for the formulation of research questions and nuanced interpretations that enrich our understanding of the trajectories of cooperation between scholars behind the Iron Curtain and their strategies (or lack of) in the new, postwar academic landscape. The research focuses on both prominent and minor scholars, and it recognizes that where marginalization occurred, it may have been caused by gender, social, or geographic factors. However, it was also shaped by forms of entanglement that were recorded in formal archival collections by those who restricted the circulation of new knowledge during the Cold War and by the ways the Academies of Sciences themselves regulated the emerging disciplines.

Educational Reforms and the "History from Below" Perspective

After World War II, scholars in Poland and Czechoslovakia called for academic reforms, including democratizing access to education and addressing knowledge gaps caused by the war. There was a widespread belief that scientific planning exerted a significant influence on both the national economy and global peace. However, the transition to a system of planned science proved difficult to implement in both Eastern and Western blocs, prompting demands for a reform of the system of education.[23] My research seeks to explore the defining features of the postwar educational reform, the revival of mathematical communities in Poland and Czechoslovakia, their first postwar congress, the transfer of new forms of education, and reciprocal exchanges in the training of young scholars.

Following the resumption of studies at Czech universities in Prague and Brno during the shortened summer semester of 1945 (Slovak universities had remained open during the war and German-language universities were dissolved in May 1945), a process of nationalist purges of professors and students was initiated in relation to their wartime activities, which included the expulsion of Germans and members of other minority groups (including Jews). This fostered a climate of distrust that prevailed in Czechoslovak (and Polish) academia throughout the Stalinist period.[24] This climate of suspicion was sustained by continued ideological vetting in the decades that followed.[25]

Some literature claims that the totalitarian era began with the communist coup d'état in Czechoslovakia in 1948 (and the 1946 referendum in Poland), and the distrust among scholars with regard to the implementation of Marxist reforms in higher education in 1950 (in both states). However, as Christiane Brenner has argued, totalitarian thinking, which excluded alternative perspectives, was already present in the media discourse immediately after the war.[26] My aim is to critically examine the perspective of "captive universities" and "chained academies" in Central and Eastern Europe in the late 1940s.[27] It should be noted that educational reform and planned research were promoted by both Marxist and other scientists. Agata Zysiak demonstrates that many researchers at Poland's first socialist "red university" in Łódź considered themselves to be participating in an extensive process of social transformation and technological revolution.[28]

In the context of socialism, both the economy and scientific research were subject to central planning. The potential risks associated with this approach stemmed from the possibility that state institutions might fail to ensure the reliability of the scientific results or the timely development of new methods. The official plan was to be aligned with the education of new scholars, and the internationalization of scientific research was to be facilitated by new academies of sciences. Mathematical research had been typically carried out in institutes without the need for substantial laboratories, while the findings were disseminated remotely, primarily through journals, correspondence, and participation in international congresses.

The reform of the curriculum was not a necessary requirement for socialist nation-building, but rather a consequence of the recognition that students were not acquiring adequate knowledge.[29] Postwar Poland experienced an urgent demand for young specialists, because many scientists had died in the Holocaust or as a result of the Nazi and Soviet occupation (approximately a thousand professors plus other scholars).[30] Therefore, as a first step, the diplomas of all young people who attended underground universities were officially recognized through a process of nostrification. The postwar narrative of youth was particularly shaped by the consequences of war, emphasizing disrupted education and delayed academic development.[31]

In People's Poland, a debate emerged on the future direction of mathematical research. Mathematicians noticed how the new regime valued intellectuals, and in this context Stanisław Mazur emphasized the development of new disciplines current in the world, especially in the Soviet Union. Kazimierz Kuratowski, on the other hand, defended the continuity of the interwar Polish School of Mathematics but officially supported adopting reforms at all levels: "The curricula for higher as well as secondary education have an essential significance for the growth of properly educated scientific cadres. In the postwar period (unlike before the war) the improvement of the mathematics curriculum became one of the most important aspects of the Polish Mathematical Society's activities."[32]

Most professors insisted on the continuity of the core disciplines, such as functional analysis, topology, mathematical logic, and set theory, particularly Władysław Orlicz, Karol Borsuk, Andrzej Mostowski. As director (1948–1968), Kuratowski oversaw the incorporation of the State Mathematical Institute (established in 1948) into the Polish Academy of Sciences in 1952, where it became the Institute of Mathematics of the Polish Academy of Sciences (IM PAN).[33]

In parallel, some mathematicians emphasized the application of mathematics for the economy, industry and medicine, while also calling for greater support for the education of young scholars. The leading figure among this group was Hugo Steinhaus (1887–1972), who emphasized probability theory, statistics, and computing.[34] In Wrocław, he established a branch of IM PAN with a focus on applied mathematics, where he organized conferences for experts from various fields. He was firmly convinced that without modern mathematics, the country would remain technologically backward. Steinhaus's vision of dynamic modernity was shaped by the perspective of (communist) transformation but also incorporated capitalist elements for the advancement of society. He envisioned promising academic careers for professors.[35] While even US factories were employing Polish mathematicians, the directors of domestic factories were failing to utilize the potential of mathematicians. At his inauguration as rector of the recently established University of Wrocław he stated: "Our state educates them and acclaims them but does not find a proper role for them." He invoked Tadeusz Kotarbiński's essay *A Treatise on Good Work*,[36] which stated that "a good organizer does nothing: he just keeps an eye on things."[37] In his view, this was the role of a mathematician in factory production. This change was not only advocated by intellectuals but also enacted at the level of factories. As Nečasová noted for the 1950s, there was a notable shift in evaluation of shock workers (*udarnicy*) in the Eastern Bloc. The focus began to shift from celebrating physical labor to honoring intellectuals who achieved outstanding results with their minds.[38]

In Czechoslovakia, the Central Institute of Mathematics[39] was established in 1950 at the initiative of the topologist Eduard Čech (1893–1960). It devised an innovative training program for doctoral students. In 1950, there was a strong group of twenty doctoral students, supported by only eleven faculty members. Professors conducted seminars on advanced topics to help students keep pace with the latest developments in mathematics and start international cooperation. This is why the first postwar cohort held Čech in high regard and praised the integration of the Institute into the Czechoslovak Academy of Sciences in 1953 (MÚ ČSAV). Young scholars became heads of the new departments and launched new journals focused on applied mathematics, which benefited the national economy. Still, as a young editor of one of the new journals wrote: "We lack mathematicians who understand the technical sciences enough to understand the problems of an engineer."[40] The Czechoslovak economic plan to rapidly construct new dams in the 1950s required young mathematicians capable of calculating the mechanical stresses involved in the solidification of concrete. Young mathematicians were directly engaged in solving such problems, and for them, Hugo Steinhaus was an iconic figure of progressive expertise.

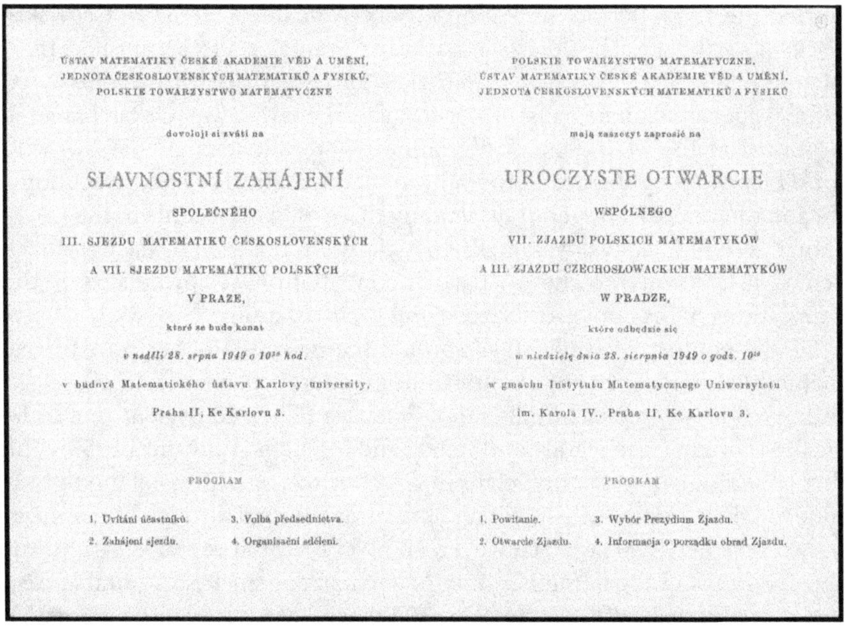
ÚSTAV MATEMATIKY ČESKÉ AKADEMIE VĚD A UMĚNÍ,
JEDNOTA ČESKOSLOVENSKÝCH MATEMATIKŮ A FYSIKŮ,
POLSKIE TOWARZYSTWO MATEMATYCZNE

dovolují si zváti na

SLAVNOSTNÍ ZAHÁJENÍ

SPOLEČNÉHO

III. SJEZDU MATEMATIKŮ ČESKOSLOVENSKÝCH

A VII. SJEZDU MATEMATIKŮ POLSKÝCH

V PRAZE,

které se bude konat

v neděli 28. srpna 1949 o 10³⁰ hod.

v budově Matematického ústavu Karlovy university,

Praha II, Ke Karlovu 3.

PROGRAM

1. Uvítání účastníků.
2. Zahájení sjezdu.
3. Volba předsednictva.
4. Organisační sdělení.

POLSKIE TOWARZYSTWO MATEMATYCZNE,
ÚSTAV MATEMATIKY ČESKÉ AKADEMIE VĚD A UMĚNÍ,
JEDNOTA ČESKOSLOVENSKÝCH MATEMATIKŮ A FYSIKŮ

mają zaszczyt zaprosić na

UROCZYSTE OTWARCIE

WSPÓLNEGO

VII. ZJAZDU POLSKICH MATEMATYKÓW

A III. ZJAZDU CZECHOSŁOWACKICH MATEMATYKÓW

W PRADZE,

które odbędzie się

w niedzielę dnia 28. sierpnia 1949 o godz. 10³⁰

w gmachu Instytutu Matematycznego Uniwersytetu

im. Karola IV., Praha II, Ke Karlovu 3.

PROGRAM

1. Powitanie.
2. Otwarcie Zjazdu.
3. Wybór Prezydium Zjazdu.
4. Informacja o porządku obrad Zjazdu.

Figure 5.1. Bilingual invitation to the Joint Congress of Czechoslovak and Polish Mathematicians in Prague 1949. Courtesy of MÚA AV ČR, collection Jednota československých matematiků a fyziků, inv. nr 365.

Mutual cooperation between Czech and Polish mathematicians began earlier, in the interwar period, when two congresses of Slavic mathematicians were held in Prague and Warsaw. As a result of these relationships, Czechoslovak and Polish topologists (Čech, Kuratowski and Knaster) decided that the first postwar congress of mathematicians would be organized jointly in Prague in 1949. A dictionary of mathematical terms in Polish and Czech for congress participants was printed, and a protocol on cultural cooperation was signed on 7 September 1949.[41] In the spirit of reciprocity, state officials and the directors of the mathematical institutes in Prague and Warsaw signed the agreement while visiting Bratislava. The cooperation primarily consisted of joint conferences and reciprocal exchange of scholars for study stays and was supported by scholarships and intergovernmental cultural cooperation agreements.[42]

In June 1950, Steinhaus organized a large workshop of Polish–Czechoslovak statisticians on the application of nomograms in economics. The conference received generous financial support from the Polish Ministry of Higher Education and Science and was held in Wrocław. The participation of twenty-one Czechoslovak mathematicians had been envisaged, but almost all of them canceled,[43] citing research and academic responsibilities during the exam period. In response, Vice-Minister Eugenia Krassowska-Jodłowska and Czechoslovak ambassador František Píšek canceled their planned flight from Warsaw to Wrocław and Steinhaus was able to avoid official speeches on socialist fraternity. Ultimately, it was Oskar Lange, a professor at the Main School of Planning and Statistics in Warsaw, who most clearly recognized both the value of mathematical science in economics and the value of cooperation, sending three of his assistants to the workshop.[44] While Czechoslovak mathematicians did not attend, Czechoslovak economists seemed to cherish Lange's contributions, as evidenced by the translations of his books into Czech and their reception.[45]

In the context of study stays abroad, the principle of reciprocity was maintained, with exchange opportunities offered primarily to those aspiring to the degree of Candidate of Sciences (CSc., equivalent to a PhD; awarded in the Czech lands until 1998, and in Poland only until 1958). The first generation of mathematicians in Czechoslovakia expressed their gratitude to Eduard Čech, who facilitated their stays abroad through his contacts in Moscow and Warsaw.[46] However, archival research revealed significant shortcomings in the administration of such arrangements. A notable case was that of Jaroslav Kurzweil (1926–2022), who graduated in mathematics and physics (compulsory combination) in Prague in 1949 and continued as an aspirant in the MÚ ČSAV. In 1953, Čech arranged a study visit to Poland for him. However, Čech announced Kurzweil's arrival in Warsaw by telephone at the last minute. The lack of prior notice left the coordinator at IM PAN unprepared and therefore they temporarily accommodated him in the expensive Bristol Hotel, which the Ministry of Culture later refused to pay for. Kurzweil reflected on the dynamic shift from the initial disarray to more purposeful action, when the coordinator redirected him to Professor Władysław Orlicz (1903–1990) in Poznań, remarking: "Here in Warsaw we only hold meetings, conferences, and talk—but nothing really happens, nobody does anything."[47] Under Orlicz's supervision, Kurzweil turned his attention to functional analysis and focused on analytical operations in Banach spaces. Orlicz arranged for him to give lectures on his findings in various institutes across Poland. Kurzweil recalled: "I talked about this theorem everywhere, but I only had a few brief notes on one little piece of paper."[48] Compared to the early phase of Kurzweil's scholarship in Warsaw, his time in the more peripheral Poznań was better organized. The only question that remains is how much this different academic environment influenced the development of his scholarly work. Upon his return to Prague, Kurzweil was distressed to find that he was not able to reconstruct the mathematical proof of his theorem from his notes. However, during a sleepless night, he made a breakthrough and reconstructed the proof. This led him to publish two articles and obtain his Candidate of Sciences

Figure 5.2. Czech mathematician Jaroslav Kurzweil on a reciprocal exchange visit to Poland, consulting with Polish mathematician Andrzej Alexiewicz. Poznań 1953. Jaroslav Kurzweil Family Collection, published with permission.

degree in 1955. Such experiences of an overnight break though, on a train at midnight (in Zbigniew Semadeni's case), highlight how, in this period of centrally planned research, the production of mathematical knowledge followed an informal and spontaneous procedural course.[49] A few years later, Semadeni was sent from Poznań to Prague in a reciprocal exchange.

Perspective of Youth: Mathematical Olympiad and Student Circles

During the joint congress held in Prague in 1949, the Polish delegates shared with Czech and Slovak counterparts their plans for organizing a Mathematical Olympiad (MO). Inspired by the ideal of peaceful contests between "athletes" from various cities, in the spirit of the ancient tradition of the Olympic Games, the idea was first realized in Leningrad (1934) and Moscow (1935), but there it was limited to city-level competitions. The Polish mathematicians also drew on the tradition of modern Hungarian correspondence competitions that began in 1894.[50]

However, in People's Poland, the first nationwide MO was organized in 1949–50, in response to the call for equal access to education and the search for talented pupils beyond elites and ethnical or regional minorities. It involved not only scholars but also students and teachers. As Kazimierz Kuratowski noted in *Half Century of Polish Mathematics,* the competitions "make it possible to select from hundreds of participants, those who are most talented and devoted to mathematics. This is a very important and—as it has turned out—very useful way of adding to our corps of young mathematicians."[51]

The idea of the Polish MO was shared in a spirit of solidarity with other states within the Eastern Bloc (and took on an international dimension in 1959). It was eventually extended to other disciplines such as physics and history.[52] The Olympiad model was not directly adopted from the Soviet approach, as one might assume from the Cold War context: the first All-Russian Mathematical Olympiad was not held until 1961, and the first all-Soviet competition was not held until 1967.[53] Instead, it was the Polish model that met with widespread acceptance among mathematicians in neighboring countries, most notably in Czechoslovakia, where the first MO was held in 1951.[54] This initiative was warmly welcomed by domestic authorities and teachers and scholars at all levels and from 1959 national communities reciprocally changed in the organization of the International MO.

The reformers of the school system in both Poland and Czechoslovakia also drew on the Soviet experience of mathematics clubs for gifted pupils led by well-known academicians.[55] In Poland, one exceptional student and later famous mathematician, Wiesław Żelazko (1933–2025), praised this practice at his former secondary school, where he became an honorary member of its club due to his outstanding grades in mathematics. He recalled how the club was initially organized exclusively by his schoolmates: "I received a letter from more senior schoolmates who had a maths club . . . I felt obliged, so I started reading various books. I learnt differential and integral calculus."[56] After the war, he moved with his parents from Łódź to Warsaw and was grateful that schoolmates had recognized his talent. He also mentioned that his interest in mathematics had been partly stimulated by a change of teacher: "it made me curious and I started reading various mathematical books. In particular, Sierpinski's book, *Introduction to Set Theory and Topology*."

Żelazko recalls the high demand for new engineers at that time, but thanks to his success in the final stage of the MO in 1951, he was allowed to choose mathematics instead of a technical university and was admitted to the Warsaw University. After two years of study, he was selected to join an elite group of the best students to receive training in mathematical sciences. Competition between students was deliberately fostered by professors. Although many students of applied mathematics and probabilistics—fields with proclaimed benefits for the economy—had to study for five years, a few selected talents such as Żelazko studied a year less, in order to accelerate their scientific careers.

After defending his PhD under the supervision of Stanisław Mazur in 1960, Wiesław Żelazko traveled widely and became an internationally recognized expert in functional analysis. By contrast, most of his peers from university were not invited to participate in international research networks. Hanna Stande (1933–2025), Żelazko's schoolmate and later his wife, did not complete a doctoral degree and became a teacher. Nevertheless she gained recognition in the mathematical community because she organized a math club and five of her students became professors of mathematics. Considering the new system of co-educational schools and reformed universities in People's Poland, Stande had to conform to the male model of a scientist. Actually, a female academic competing with her male colleagues fit the socialist atmosphere of late Stalinism, as daily magazines advised parents to encourage in their daughters "the noble ambition to compete in science and social work."[57]

Shortly thereafter, the model of traditional gender roles returned. One of Stande's students, Zofia Adamowicz (1950–), had a worse experience. As a professor of mathematical logic, she reflected on her position at the Polish Academy of Sciences in the 1970s, 1980s, and even 1990s: "You know, women didn't have it easy at all. I needed to be three times better than a man to be justly appreciated."[58] Then she also recalled being humiliated by her teacher. Nevertheless, Adamowicz's later accomplishments were attributable to her own perseverance and to a close circle of collaborators around Żelazko at the IM PAN.[59]

Meanwhile, in Czechoslovakia, the public discourse discussed the role of mathematics in the socialist state. One media commentary observed:

> How many of us even realize that a Mathematical Olympiad is held in our country every year for thousands of young people, from whom future outstanding mathematicians, physicists, inventors, engineers, directors, economists will emerge. Wouldn't it be worthwhile for factories to sponsor the district and regional rounds? And wouldn't it be nice if all of us started valuing mathematics in accordance with its importance in our society?[60]

Finalists of the MO were expected to their express gratitude to the working class for giving them the opportunity to pursue studies and subsequently apply their knowledge to the production process with minimal delay. The Olympiad introduced a different principle of student selection than that of working-class origin, and consequently academically gifted high school students, as potential future communists, were often invited to study in the Soviet Union, which was advantageous in the field of mathematics. They received generous scholarships, whereas international travel for scholars during late Stalinism was rare.[61]

In Czechoslovakia and Poland, university students in general were encouraged to participate in student scientific circles (import from Russian *kruzhok*). In Poznań, Zbigniew Semadeni was a keen reader of books in foreign languages and was able to explain Kuratowski's German monograph *Topologie I* to other students.[62] His preparation paid off when, during an exam he was able to respond convincingly when Professor Orlicz attempted to challenge him. Subsequently, Semadeni was admitted to a doctoral studies program.[63] After defending his dissertation and visiting Prague, Semadeni moved away from functional analysis to explore new disciplines.

The new arrangements in socialist education and science had its own rules and peculiarities, but simultaneously the reciprocal system allowed for non-standard and informal forms of activity, such as elite student circles, informal contacts, and organized study stays, which turned out to be surprisingly fruitful. Researching this environment through a lens of generational consciousness helps to clarify how young scholars developed a sense of belonging among their peers and how modern knowledge was transmitted among them.

Asymmetrical Decentralization and a Mathematical Seminar in a New City

Mathematicians in Poland and Czechoslovakia recognized that in an era defined by reciprocity the selection of the most able students was important for two reasons: providing new scholars for academia and supplying new cadres for the economy and industry. The results of the Olympiads could be used for the initial selection of promising scholars. The next stage of selection began after students had entered the university. Professors leveraged students' motivation by creating excellence courses and determining which students should be assigned to a particular seminar or supervisor. The goal was to educate new cadres quickly and responsibly to prepare them for their further careers. While it was necessary to maintain a certain ratio between academicians and people going into production and economic sectors, there were still questions about what factors influenced the selection of further scientists and how students could navigate the process.[64]

Students could participate in the Socialist Youth Unions. Moreover, the youth unions, in collaboration with the faculty party cells, played a significant role in university admissions, viewing family origin as a key factor. This included the rigorous screening of students from resettled families, which was a particular issue in Poland, where the forced resettlement of families from the eastern parts of the Polish lands (those annexed by the Soviet Union) after 1945 had caused significant displacement.[65]

While conducting interviews about childhood with the first postwar student generation it became clear that most families had little or no choice in the site of resettlement. One experience of deterritorialization was provided by Roman Duda, born in 1935 in Brody near Lviv, who was resettled in Kluczbork, which was part of the German Province of Upper Silesia prior to World War II. In 1952 he was refused admission to the University of Wrocław due to a personal grudge held by a person from Kluczbork. Only after inquiries from his parents was he finally accepted. With this peculiar experience, he became a member of the Socialist Youth Union. He arrived to Wrocław to find it in ruins. Duda witnessed how professors from Lviv[66] were establishing a new mathematical center, which was partly a rival to Warsaw, partly pursuing new directions of collaboration, most notably with Prague and Hugo Steinhaus.

Duda recalls how the students appreciated the committed approach of the founders of the mathematical center in Wrocław. They provided extensive insights into the achievements of the Lwów [Lviv] School of Mathematics, but remained silent about the interwar academic climate, the Polish–Ukrainian conflicts, and the German and Soviet occupations.[67] Their silence was prescient, as was confirmed at the unveiling of the monument to the twenty-five professors of Lviv universities executed by the Nazis in July 1941. In 1964, the idea of erecting a monument in front of Wrocław Polytechnic was conceived, given the fact that such a memorial was forbidden in Lviv itself. However, because Polish state officials prohibited the inclusion of names on the plaque, there were two expressive ceramic figures remaining "mute" and it deliberately avoided any reference to Lviv.[68] It was only with the advent of Solidarność, a movement in which Duda participated in Wrocław, that names were added to the plaque and the title changed. Thus, it was not the resettled professors, but the resettled students who restored the remembrance of the Lviv academic community in Wrocław.

While in Poland progressive mathematical knowledge circulated among various cities, particularly through transfers from the East (in Lviv and Vilnius) to newly established academic centers (in Wrocław or the "red universities" in Lublin and Łódź), in Czechoslovakia the movement was more unidirectional. Professors from Masaryk University in Brno, such as Eduard Čech, took up vacant positions at Charles University in Prague thereby reinforcing the asymmetry between the capital and the smaller cities but increasing the mobility of scientists in Brno, Olomouc, and in the Slovak part of the state. In 1945, Slovakia had only two universities, Comenius University and the Technical University, both of which were in Bratislava. The head of the Department of Mathematics, Jur Hronec (1881–1959), had to rely on his colleague from Brno, Otakar Borůvka (1899–1995), to travel regularly to Bratislava to supervise doctoral students.[69] The West–East academic asymmetry in Czechoslovakia was further reinforced by the habit of gifted students leaving Slovakia to study mathematics in Prague, after the establishment of a separate Faculty of Mathematics and Physics there in 1953.[70]

However, in Slovakia, Hronec and other seminar leaders had to resist pressures to centralize mathematical research solely in the MÚ ČSAV in Prague. Consequently, they advocated for the establishment of the Mathematical Institute of the Slovak Academy of Sciences in Bratislava (1959) and concentrated on new disciplines. The advent of graph theory represented a novel intellectual frontier, one that had not yet been explored in Prague or Poland. However, Otakar Borůvka had previously engaged with graphs in 1924 and was able to train Slovak students in the 1950s and 1960s. Those who understood Hungarian (due to the minority in southern Slovakia) could benefit from the exceptional expertise in graph theory in neighboring Hungary. Because Budapest was closer than Prague, the world-famous Slovak School of Graph Theory was established in Bratislava.[71]

The mathematics seminars in Wrocław emerged, to some extent, as a rival to the central institutes in Warsaw. Roman Duda attended the topology seminar of Professor Bronisław Knaster (1893–1980). For health reasons, Knaster was the only mathematician in Poland who was officially permitted to hold a seminar at home. In this domestic setting, he conducted the seminar more informally, openly sharing his own correspondence with mathematicians, which he read to his students. Duda recalled:

> They were very interesting seminars, because Knaster usually began by discussing what was interesting in the world of mathematics, in the mathematical community, and in mathematics itself. He talked about his correspondence, publishing intentions, and the books that were about to be published. Sometimes he interwove his own personal memories, also very interesting, from his time in Warsaw—though never from Lviv.[72]

Knaster's home seminars allowed him to exert more influence on his students. In contrast, most other seminars were held in the buildings of institutes, where there was little space for individual attention. Therefore, apart from the scheduled access to the large computing laboratory, which was available once a week, mathematicians were forced to conduct their research at home. The housing conditions for scientists under mature socialism evoke emotional memories. For example, Roman Duda had to conduct his research in the small apartment in which his four children grew up. The same was true of Zofia Adamowicz in Warsaw, who raised two sons and did much of her academic work between the kitchen and the children's bedroom, and recalled that many of her the best ideas came to her while cycling.[73] In another recollection, Semadeni begged the IM PAN Director Kazimierz Kuratowski to approve his request for a larger apartment. Even in the 1980s, the MÚ ČSAV in Prague also had a shortage of working space for its employees.

In contrast to such difficulties, Knaster's seminar was comfortable, and he made an enormous effort to work individually with each student in correcting drafts of their articles. Nevertheless, his interests had fallen out of step with the global development of general topology since the 1950s. As Duda observed: "Unfortunately we didn't see it at the time, it had already become a fringe area of mathematics. Knaster's contact with world mathematics had ended in 1939. And I remember that in his seminar we studied Hausdorff's book, which was great, but it was from 1914. Still, we learned topology from it."[74] In a more detailed reflection, Duda described his contrasting experiences with Knaster: his insistence on rigorous precision in the writing of articles (and in French), but at the same time, clinging to outdated analyses. Nevertheless, students were reluctant to challenge the professor in his seminar.[75]

Like Zbigniew Semadeni, Roman Duda benefited from a visit to Prague in 1961, where the first International Symposium on General Topology and its Relations to Modern Analysis and Algebra was held. Knaster was not able to attend himself but was obviously curious about the event as the American topologist R. H. Bing (1914–1986) attended. In earlier articles (1948, 1951) Bing had provided proof of Knaster's continuum, a construct known as a pseudo-arc that he had introduced in 1922.[76] This rare coincidence in science offered a unique opportunity to bring the mathematicians together. Knaster arranged for Bing to travel from Prague to Poland with Duda in the latter's car. Nevertheless, the meeting between the two topologists remained formal, restrained, and academically unproductive. Bing employed the progressive method of chain structures, while Knaster used older geometric methods. Moreover, Knaster did not speak English, while Bing spoke neither French nor German.[77] Although Knaster enjoyed sharing news and correspondence with his students in Wrocław, he had lost contact with global developments in topology, while linguistic divisions further deepened the asymmetry.

Prague Topological Symposia since 1961: A Transfer of Knowledge

The idea of organizing the Symposium was conceived by Eduard Čech. Before the war, he was the only Czech topologist to achieve worldwide recognition, because he independently derived homology theory at the same time as Russian topologist Pavel Alexandrov. This theorem generated significant enthusiasm at the First International Topological Conference held in Moscow in 1935.[78] After Čech attended this event, he went on a scholarship to the United States, which resulted in broader international cooperation. However, the next topological conference was not held until twenty-six years later in Prague. Following Čech's sudden death in 1960, the conference committee was led by his younger colleague Miroslav Katětov. Although the First Topological Symposium in Prague took place after the fall of Stalinism, Roman Duda recalls being appalled by the presence of the socialist–realist monument to Stalin overlooking the Vltava River (it was removed in 1962).[79] His first conference abroad allowed him to delve into new areas of study in topology.

Zbigniew Semadeni, who attended the symposium with his mentor Orlicz, also took advantage of new contacts. As a result of his lecture in Prague, he was invited to the summer school at Stanford University, and immediately upon his return he shifted away from functional analysis and completed his habilitation in category theory in 1963. He noted that "the habilitation made a good impression, partly because nobody knew about [category theory]."[80] This was not a new discipline, but rather a new general theory of natural equivalences introduced in 1945 by two US scholars, Saunders Mac Lane and Samuel Eilenberg, the latter of whom was a mathematician of Polish–Jewish origin who had managed to leave Poland in April 1939.[81]

The Symposium had another beneficial impact on Duda,[82] as it marked his first trip abroad. Two years later, he secured a study visit to Cambridge, which inspired him to pursue his own direction. He recalled:

> I learned a lot there. There were already new directions in topology, some of which were of great interest to me. However, I was still completing various things under the influence of Knaster, which took up a great deal of my time but brought me no recognition.[83]

Duda referred the rigidity of Knaster's topology in a paper,[84] but only in the interview did he explicitly highlight the asymmetry between Knaster's seminar and the knowledge he gained from his trips to Prague and Cambridge.[85] Such a critical reflection is not to be found in the archive materials, as in his correspondence with Karol Borsuk the leader of Warsaw topology, Duda refrained from criticizing the aging Knaster, and instead relayed his messages to Borsuk. Of course, the recollection in the interview is a later reflection, but Duda's disillusionment is most clearly reflected in his decision to be involved in the Committee for the Protection of Workers in the 1970s and also his less productive career in Wrocław topology.

Nevertheless, there were some Polish scholars of topology who were productive in the late 1960s. For example, Borsuk's work on shape theory received considerable interest among topologists in Moscow, who translated his English monograph into Russian.[86] Moscow, which was regarded as the leading center of topology in the world, absorbed the work of mathematicians from Central and Eastern Europe but did not allow them to write for Soviet magazines. Only mathematicians from the Soviet republics were allowed to publish there.[87] The persistence of internal borders of the Eastern Bloc was clearly apparent from Duda's experiences. Before attending conferences in Moscow or Kyiv, he always asked to be allowed to visit Lviv to foster collaboration with his Ukrainian colleagues. His requests were routinely denied on the grounds that Lviv was a marginal university and that a Polish university professor should engage with higher-ranking institutions. Apart from a single study visit to Cambridge in 1963, he was not allowed to travel outside of the Eastern Bloc until 1988 (due to his involvement in Solidarność). In 1990, after the collapse of the communist regime, he finally visited his birthplace of Lviv and resumed cooperation with the

Figure 5.3. Collage from the 80th anniversary of Bronisław Knaster, Wrocław, 25 May 1973. Left: Roman Duda stands in front of the blackboard. Right: Bronisław Knaster is monitoring the ongoing development of Wrocław topology by his students. Roman Duda's private archive, published with permission.

university, which culminated in his authorship of a monograph on the Lviv Mathematical School.[88]

Contribution of the Prague Topological Symposia to Other Disciplines

The list of participants to Prague Topological Symposia shows the strong interest in the contributions of scholars of topology from both Central and Eastern Europe and the West.[89] Notably, during the ICM in Moscow in 1966, one of the fathers of category theory, Saunders Mac Lane, proposed the establishment of an international center of category theory in Prague. This envisioned a special seminar including the young Prague mathematicians Zdeněk Hedrlín (1933–2018), Věra Trnková (1934–2018), and Aleš Pultr (1938–). As a non-Soviet location, Prague promised the possibility of access by researchers from both sides of the Iron Curtain. However, the occupation of Czechoslovakia in 1968 put an end to the plan. In the wake of the invasion, there were political audits in the name of "normalization." As a consequence, Trnková, who in 1967 became the first habilitated female professor of mathematics in Prague, was demoted to a basic research position and was not permitted to teach.[90] The Prague Topological Symposium continued to be held, but the center was never established. In Poland, such a progressive specialization would probably not have gained traction, as the seminars were controlled by the Lviv school professors. Even so, in 1967, a new summer school of algebraic topology was held near the Baltic coast.[91]

The significant Polish presence at the Symposium in Prague in 1961 was appreciated by domestic scholars because it introduced new directions in research. Although Aleš Pultr attended Čech's lecture, he was in Katětov's seminar.[92] Pultr's interest diverged from that of his teacher, but he was given the freedom to pursue his own path.[93]

After graduating in 1961, he was able to attend the first symposium. He then defended his doctoral thesis in 1964 and presented a paper at the second symposium.[94] As his focus shifted from topology to category theory and graphs, he appreciated the foreign contacts he was able to make at the symposium.[95] It was not through Katětov's seminar that Pultr acquired new knowledge but through interactions with the symposium's participants. In interviews, mathematicians frequently reference colleagues outside their immediate area of expertise. Correspondingly, Pultr expressed his appreciation for scholars from the USSR and Poland. Among others, he mentioned: "Definitely Semadeni—his great hobby was category theory and that was rather interesting for us, this collaboration."[96]

Zbigniew Semadeni clearly earned respect in Prague. His shift to a new discipline helped him to distance himself from the tensions he had experienced in Poznań around Władysław Orlicz.[97]

Semadeni was assigned to Second Department led by Aleksander Alexiewicz, a former student of Orlicz in Lviv.[98] Orlicz hierarchized the students, what led to personal animosities with Alexiewicz.[99] When Alexiewicz became the leader of both departments, an internal conflict between him and Orlicz appeared. A younger student of functional analysis, Andrzej Sołtysiak (1950–) observed that "the gentlemen did not like each other." Therefore he began to seek guidance for his thesis in functional analysis in Warsaw from young Professor Żelazko.[100] The tight control of academic positions meant that talented scholars had to make additional efforts to circulate their knowledge through new intellectual networks.[101] Of course, tensions also existed between professors in Prague, but Eduard Čech, known affectionately as "Big Eddie," set a different tone. He allowed younger professors at the faculty to have their choice of doctoral students, while he supervised those that remained (such as Trnková in 1957), thereby setting an example for younger generations as the respected leader of the community.[102]

Language Reorientation in Mathematics in the Late 1960s

Considering the asymmetrical circulation of knowledge in Central and Eastern Europe, it is worthwhile to discuss the use of languages in mathematicians' communication and the travel opportunities for scientists during the Cold War. First, within the mathematical community the scheduled day and time of professors' seminars at the major institutes were widely known, and therefore anyone—even a foreign scholar—could be invited to attend. Archive sources confirm that frequent exchanges between Poland and Czechoslovakia were facilitated by geographical and linguistic proximity.[103] The role of language is illustrated in the correspondence of Orlicz and Borůvka. Several Czech and Slovak mathematicians wrote to Orlicz in Polish with the aim of fostering connections, most likely in preparation for personal encounters.[104] Conversely, the correspondence with Borůvka indicates, Polish mathematicians tended to write him in Polish or French, in a more elaborate style.[105] Unlike other professors of his generation, Borůvka was able to partially transition from French to English, aided by his wife Milada, who was an English philologist. In the 1970s, English became the lingua franca of mathematics worldwide, and therefore the use of minor Slavic languages in correspondence declined. The formulation of mathematical problems in Russian, German, French, and Italian also soon disappeared from articles and world congresses. An analysis of Borůvka's correspondence reveals that, in retirement, he seldom replied in writing, preferring to use the telephone. Although informal communication is not archived, it has had a strong influence on the dynamics of scholarly exchange.

The second important factor was the motivation of younger scholars to acquire new knowledge through participation in international congresses. Following the political upheavals of 1968,[106] many academics left Czechoslovakia and Poland, which diminished also the effectiveness of domestic seminars. The plans for an international center for category theory in Prague were also abandoned. The restricted opportunities for older mathematicians to travel and share new perspectives from foreign conferences in their seminars were even more limiting. In contrast, Polish mathematicians succeeded in establishing the Banach Center for International Mathematical Cooperation of the Socialist Countries in Warsaw in 1972. This institution provided excellent conditions for scholars from Central and Eastern Europe and the Soviet Union to meet their peers from the West.[107]

Local Conferences in the Mountains and Other Alternative Centers of Entangled Knowledge

The desire of mathematicians to create alternative spaces for interaction most likely stemmed from a variety of reasons: the pursuit of new knowledge in specific disciplines, the state's interest in the development of applied research, the limited mobility of seminar leaders and students seeking an escape from personal conflicts. The perspective of entangled knowledge allows us to observe both the sites of interaction and the personal encounters. Local conferences were a stimulating environment for inviting guests from abroad and were characterized by a more informal atmosphere than congresses organized with the official support of state institutions.[108]

The informal setting of such gatherings created opportunities for becoming familiar with current trends. It would be inaccurate to suggest that talented individuals only left mathematical analysis or general topology in favor of new disciplines. I would rather argue that local conferences were beneficial, even for disciplines with a long tradition of the Polish School of Mathematics—namely, functional analysis and mathematical logic. Two prominent representatives of functional analysis have already been mentioned, namely, Władysław Orlicz and Wiesław Żelazko. In Czechoslovakia, Vlastimil Pták (1925–1999), Jaroslav Kurzweil's colleague in the IM PAN held the leading position in functional analysis. He spent a study visit with Orlicz (Poznań) and Eilenberg (Stanford) and absorbed their practices in leading a seminar and developed his own way of informal interaction to promote new trends in the key discipline intended for talented students.

Beginning in 1970, Pták regularly organized the Functional Analysis Seminar for the training of his students, it was always a stay in the mountains in May. From 1975 Polish mathematicians from various institutes participated in it. It was there that Sołtysiak met Żelazko and chose him as his mentor. The bilateral event in Central and Eastern Europe quickly reached an international level thanks to Pták's reciprocal visits abroad and inviting of individual participants to Czechoslovakia. After sessions in English, informal discussions over a beer in Polish, Czech, and Slovak played an important role. Sołtysiak recalled at least five Czech and Slovak mathematicians who spoke Polish fluently.[109]

Because Pták loved to hike, he was always looking for a location in the mountains. His goal was for students to learn a challenging discipline in an undisturbed environment, away from central institutional influence, but under his guidance. Pták had established an international reputation as a specialist in functional analysis and also inherited an "old style" professorial attitude from his doctoral stay with Orlicz in 1954. In order to achieve the same level of prestige as the matador of the Lwów School of Mathematics, Pták expected his students to work hard and be completely obedient:

> Do you know what it was like? In the evening, after the lectures—because the lectures took place in the morning with a break for lunch, and then again in the afternoon—we would go for a beer in the *hospoda*. Then the Czechs would return to Professor Pták to go over the key points in the lectures. Professor Pták wanted them to revise certain things again, so that they would remember them, and so that they could explain to each other what they hadn't understood in the lectures. What's more, it wasn't easy, at least, not for our colleagues from Czechoslovakia. Only some of the older ones were allowed to skip [the revision sessions], but the young ones weren't, as they didn't have that option. And at the first conference I attended, it was even worse, because we started at nine in the morning, but all our colleagues from Czechoslovakia had to report to the lecture hall by eight for their classes in Czech or Slovak.[110]

Pták demanded discipline but realized that the key was for students to understand the theorems in their native language. As excessive discipline tends to lead to tensions, tensions arose between professors and students. One such case was Jaroslav Zemánek (1946–2017), an outstanding contestant at the Czechoslovak MO, who Pták persuaded to enter his seminar, where he received supervision on his doctoral topic.[111] However, Zemánek's dynamic personality clashed with his professor's strict approach. Therefore, in 1976, Zemánek took advantage of a reciprocal academic visit to Warsaw, where he remained and defended his dissertation under the supervision of Wiesław Żelazko. The essence of this circulation of knowledge was not the clash of two different characters, but the fact that the traveling entities (concepts and researchers) were not limited to a particular location.[112] The industrious Zemánek established himself by moving from Czechoslovakia to Poland. Whether Zemánek was a Czech or a Polish mathematician is of no consequence. In Warsaw, as secretary of the editorial board of *Studia Mathematica*, he developed a strong international network.[113]

It is clear that Żelazko played a key role as a mediator between Poland and Czechoslovakia. The case of Jaroslav Zemánek demonstrates that there was a certain asymmetry between the countries. In Czechoslovakia, doctoral theses and habilitations were typically evaluated by domestic scholars, while Polish institutions invoked the principle of reciprocity, and, from the 1970s, invited foreign experts as reviewers (at least for habilitations). In Zemánek's case, Żelazko invited the prominent Cambridge specialist Barry Johnson to serve on the committee.[114]

Wiesław Żelazko developed extensive contacts in the USSR, Western Europe, and the United States, and from 1963 he was regularly invited to give guest lectures abroad. He also hosted scholars from socialist countries in Warsaw. During our interview he mentioned having visited fifty-six countries and showed me two notebooks as documentation of his broad contacts: one a guestbook of scholars who had visited him and the other a travel itinerary, which also included the venues for Pták's seminars in Czechoslovak mountains.[115]

It is worth noting that the mathematicians' visits to his home were for playing bridge as well as academic collaboration. Outside of Poland, informality may have been motivated by particular security concerns. In the Soviet Union, certain dissidents and Jewish scholars were not allowed to participate in seminars. In Moscow, seminars of the Academy of Sciences were less closely supervised than those at the universities, but even some seminars from the Steklov Mathematical Institute of Russian Academy of Sciences were moved to the mathematicians' kitchens or their dachas. Work conducted at the kitchen table either had to be completed before the meal or the meal would be delayed.[116]

The memory of informal phenomena in the USSR left a lasting impression on visitors from Poland and Czechoslovakia. The circulation of new knowledge was shaped in several informal settings: seminars in the mountains with Pták (as recalled by Sołtysiak), conversations with Alexandrov after his river swimming sessions (mentioned by Duda) and the summer schools on algebraic topology held on the shores of the Baltic Sea in Gdańsk.[117]

Geographies of knowledge, which examine the relationship between science and power, and place and space, show how political ruptures in Central and Eastern Europe changed the connections between formal academic environments and informal interactions beyond academia.[118]

Exchanges and local conferences generated collaboration and the circulation of new knowledge. The mathematicians' closest disciplinary colleagues, astronomers, participated in observing days (or nights) at foreign observatories, where open discussions on non-scientific topics would sometimes take place. Physicists were similarly tied to the instruments and laboratories of their specialization. Having spent long periods at major research centers such as Dubna in the Soviet Union and CERN in Switzerland, they became mediators of knowledge between East and West.[119]

By contrast, mathematicians enjoyed more freedom in the choice of venues for meetings. Although they could theoretically meet in spa towns or in the conference centers of the Czechoslovak Academy of Sciences (for example, Liblice near Prague or Smolenice in Slovakia), they usually preferred to go to affordable mountain resorts. Pták and his seminar moved between resorts on the long Czechoslovak–Polish border. This was not only a Carpathian phenomenon. As Żelazko's itinerary shows, they also met in places in the Bohemian Forest and the Ore Mountains on the borders with West and East Germany. Thus, in the mountains, new knowledge related to metaphorical borders was created in close proximity to political borders, and from there it circulated into the hierarchical spaces of universities, academic institutes, and international congresses to be validated or contested.[120]

Agents of New Knowledge and Power and Gender Tensions

Local conferences in the mountains contributed to the decentralization of knowledge and facilitated personal encounters, which also opened opportunities for female scholars, although women's participation in mathematical seminars under mature socialism has not been researched.

The following section, focusing on the student generation of the 1970s, will therefore highlight the personal mobility of female scholars of mathematical logic who managed to forge alternative patterns of interaction.[121]

Poland had a rich tradition of mathematical logic, which led to the development of branches of theoretical and applied computer science, probability, graph theory, artificial intelligence, and the computational complexity of mathematical operations.[122] Warsaw was home to the world renowned Professor Andrzej Mostowski (born in Lviv in 1913) who played a fundamental role in advancing mathematical logic, set theory, and the foundations of mathematics.[123]

At Warsaw University and in the IM PAN Mostowski supervised many doctoral theses.[124] Zofia Adamowicz, his final doctoral student before his sudden death in 1975, recalls his mentoring style in the seminar sessions but also the moments, when male students were sometimes aggressive toward their female colleagues during discussions. It is noteworthy that the seminars were held in the Palace of Culture and Science, which served as the headquarters for the Polish Academy of Sciences.[125] However, Mostowski made efforts to curb this behavior, urging the participants to maintain a respectful discourse.[126] He also drew on his international connections to organize a semester-long transnational lecture series in Warsaw in 1973 (similar to those held in the United States). It brought together logicians from different countries, who presented papers and engaged in discussions together in Warsaw.

Due to the lack of specialization in mathematical logic in Czechoslovakia, the experience of the Logic Seminar in 1973 in Warsaw proved crucial for logician Petr Hájek (1940–2016).[127] Upon returning to Prague, he applied the knowledge he had acquired to his own work. He also drew on the academic experience of his wife, Marie Krčálová, who had studied mathematics at the prestigious Moscow State University. In the MÚ ČSAV, Hájek developed his concept of syntactic models (interpretations), which became the flagship of the Institute's research. His contributions to applications of logic (for computer science) and theoretical arithmetic are well documented, and therefore I will focus more on his role as a facilitator of entangled knowledge. At MÚ ČSAV, Petr Hájek managed to establish and lead two parallel seminars with two different teams, actively involving master's students and inviting them to pursue doctoral studies.[128] One team, focused on the application of logic, developed the GUHA (General Unary Hypotheses Automaton) method, which automatically generated hypotheses from empirical data mining concerning patients and medical diagnoses. This had practical applications for socialist healthcare in Czechoslovakia.[129] The second team worked on models of Peano arithmetic and included Kamila Bendová, Vítězslav Švejdar, Hájek's wife Marie, and most notably Pavel Pudlák, who was later joined by his student Jan Krajíček. Initially, this seminar was so marginalized within the Institute that it was difficult to find a room to hold it in.[130] Nevertheless, thanks to Hájek's meticulous working methods, it produced significant results.[131] However, the informal nature of this research runs the risk of overlooking the genesis of the team, particularly Hájek's role in organizing the seminar in such a way that the women scholars were not forced to delay their research (for example, the doctoral assistant Kamila Bendová was raising several children on her own).[132] Moreover, to enable his ailing wife Marie to attend, he even hosted the seminar in his own home for a short period.

Petr Hájek's main collaborator, Pavel Pudlák (1952–), recalls how their academic environment was shaped by power dynamics. Certain professors tended to select the most accomplished male students, while overlooking the achievements of female students.[133] Therefore, it is noteworthy that the leader of the logic seminars at MÚ ČSAV succeeded in creating a gender-balanced team. The atmosphere was warmer and more supportive than in Warsaw at the time, as the pregnant Zofia Adamowicz observed during her study visit with Hájek in 1981 in Prague.[134]

Students of logic within Pudlák's circle were aware of the philosophical associations within mathematical science, but they sought more stimulating material than the mandatory course in Marxism–Leninism. This led them to invite philosophy graduate and participant in the theoretical cybernetics course, Václav Benda, to give lectures on Martin Heidegger. Benda, the husband of logician Kamila Bendová, spent the years 1979–1983 in jail because of his dissident activities (which included his signing of Charter 77, the prominent protest against human rights violations in Czechoslovakia). This had certain implications for Hájek's seminar. Although he did what he could to protect Bendová, he was not able to secure approval for her doctoral defense, so she remained in an assistant position until 1990. However, as a mother of six children, the State Security service did not dismiss her from the Academy.[135]

Political tensions in the 1980s also found their way into Petr Hájek's second seminar, although police surveillance was only imposed later, because the motivation for scientific interaction was not primarily political in its origin.[136] In addition to learning about the forcing method,[137] logicians wanted to understand the proof of the independence of certain mathematical statements from the axioms of Peano arithmetic developed by Alex Wilkie and Jeff Paris (1981). This led Prague logicians to organize winter seminars at the chalet of the Academy of Sciences in Alšovice in the Jizera Mountains, which also featured cross-country skiing. These gatherings were attended by Polish logicians, and in turn, they invited Czechs the next year to their seminar in Karpacz on the other side of the border. This reciprocal exchange of knowledge created an opportunity to invite guests from Western countries. Naturally, the State Security service took a keen interest and sent a spy to the conference. However, the agent was a mathematician from a different field, who was quickly identified by the logicians.[138]

It should be noted that these logic seminars did not always have "a warm, informal atmosphere."[139] Zofia Adamowicz recalled moments of gender tension during discussions in Karpacz when some male scholars harshly criticized their female colleagues. However, the presence of foreign guests helped bring an appropriate tone to the proceedings. While some of the logicians in attendance were already married couples, the mountain seminars formed the backdrop for new partnerships and even marriages. Most notably, Alistair Lachlan formed a relationship with Małgorzata Dubiel from Poland, and Jeff Paris became involved with Alena Vencovská from

Czechoslovakia. Other Prague logicians took advantage of Paris's regular romantic visits to Prague and attempted to prolong his stay. They even officially petitioned the Institute's director to arrange for Paris to deliver an extra lecture on Peano arithmetic at the Institute.[140] From such exchanges the circulation of new knowledge in logic gave rise to several progressive directions in the field.[141]

Among the various initiatives, it was probably the Mathematical Foundations of Computer Science Polish–Czechoslovak conference, held regularly since 1972 with international participation, that had the most significant international impact.[142] The organization was based on the principle of reciprocity between institutes of the Polish, Slovak, and Czechoslovak academies of sciences (which reflected the federal arrangement emphasizing the parity of Czechs and Slovaks).[143]

From the 1980s onward, the networks of mathematicians naturally became increasingly international, thereby attracting the attention of the State Security services. In 1980, the authorities forced Hájek to cancel the planned International Logic Colloquium in Prague, citing the "provocation" of Kamila Bendová's participation in the organizing committee. Political tensions disrupted events elsewhere. The 1982 ICM scheduled to be held in Warsaw was also canceled for fear that Western participants would publicly express support for Solidarność following the imposition of martial law in Poland. This complicated the considerable organizational efforts of both Zbigniew Semadeni and Wiesław Żelazko. Both later recalled how they eventually managed to negotiate with the Polish state and the International Mathematical Union an exceptional agreement to postpone the congress to 16–24 August 1983. When the event was finally held, the French delegation used it to gather information on who among the Polish intelligentsia had been interned for dissident activities (such as mathematician Roman Duda).[144]

The imposition of martial law in Poland prompted many scholars to emigrate. The mathematicians who left often experienced a derailment of their academic careers and uncertainty regarding a future position abroad. By contrast, after the invasion of Czechoslovakia in 1968, many departing scholars left for specific academic positions they had been offered. Even so, the interviews revealed that male scholars were favored, while women who emigrated had to settle for positions in teaching or librarianship.[145]

Conclusion

This overview of the circumstances and spaces in which new mathematical knowledge emerged reveals how displacement, asymmetry, and internal constraints within the Eastern Bloc shaped the particular mathematical communities in the socialist states. A certain asymmetry between Polish and Czech or Slovak communities stemmed from wartime losses and waves of emigration. The careers of Polish mathematicians were more deeply affected by the experience of deterritorialization. There were significant differences in the structure of academic studies and seminars, professorial practices, and foreign study visits, which meant that socialist reciprocity did not function as intended. In Warsaw, the Banach Center for International Mathematical Cooperation of the Socialist Countries was founded (1972) and the successful organization of the ICM in 1983 (despite its near cancellation due to martial law) positioned Poland as a hub for international mathematical exchange. The center provided access to foreign literature and facilitated prestigious foreign internships for Polish scholars both in the Eastern Bloc and in the West. No similar center was established in Prague.

Through the center, the researcher received a mentor who was working in the same way. However, prior to this, there had been unnecessary academic tensions during reciprocal exchanges, such as when Professor Pták was reportedly displeased that the Prague student Jaroslav Zemánek defended his dissertation in Warsaw before an international commission. Biographies of scholars such as Zbigniew Semadeni indicate that international recognition could be attained by shifting one's focus to a new discipline, as his work was not primarily associated with his former institute or mentor in Poznań.

Despite the official rhetoric that socialist reciprocity would benefit all scholars equally, the institutes of the academies of sciences in Prague (MÚ ČSAV) and Warsaw (IM PAN) held the central positions, and therefore only certain regional teams managed to establish international networks (such as the Slovak School of Graph Theory). For individual scholars outside these centers, excellence often depended on personal encounters. As such, conferences held in the mountains (or summer schools in the Baltic States) were particularly valuable, as they facilitated contact with foreign guests and reciprocal exchanges of knowledge. Most mathematicians in the Soviet Union experienced greater isolation and their opportunities to participate in conferences abroad were more restricted. When Central and Eastern European mathematicians were permitted to travel for reciprocal exchanges, having the opportunity to discuss and exchange new literature and share experiences with new practices, they were able to achieve important advances in knowledge. Initiatives such as the Mathematical Olympiad, regular conferences in mountains and informal spaces of cooperation can be considered platforms for knowledge distribution between Central and Eastern Europe and the West.

Thanks to personal contacts and local conferences, Czechoslovak mathematicians were able to invite their Western colleagues to the Czech or Slovak mountains. This made the development of new theories and methods, especially in the fields of logic and theoretical cybernetics, all the more remarkable. Moreover, the informal atmosphere at conferences in mountains fostered greater participation of female scholars.

Returning to socialist-era innovations, such as Mathematical Olympiads and student science circles, a range of informal pedagogical practices emerged. Notable examples were camps for outstanding talents in mathematics and physics, student-led publications, and the early identification of promising students for postgraduate study. In contrast to formal seminars in the cities, local conferences held in the mountains offered a more effective form of training. Such gatherings were often attended by scientists from neighboring countries and sometimes from the West, providing a kind of mentoring that often surpassed that of the authoritarian professor. The town–mountain asymmetry is largely metaphorical, as informal education was manifested in other forms and places. Therefore, I have highlighted the entangled forms of knowledge, forms that fostered the dynamic development of new concepts, notions, and theories. The forms of education and seminar formats that evolved reveal the significant asymmetry between the scientific practices of the first postwar generation of mathematicians and the experiences of their students.

In addition to the aforementioned contexts and spaces, there were four important factors that shaped the careers of the new generation of scholars in the emerging mathematical disciplines.

First, the ruptures of the Cold War affected not only the humanities but also the natural sciences. Mathematicians recalibrated their positions as they observed, on the one hand, the involvement of physicists and astronomers in international peace diplomacy and, on the other, developments in philosophy and logic. Within this landscape, they found a wider scope for applying their work in cybernetics, graph theory, and didactics. Second, these new disciplines contributed not only to applied mathematics but also to the general development of the mathematical sciences. By contrast, the traditional form of seminars delivered by interwar professors lagged behind the global development in knowledge. Young scientists in Central and Eastern Europe had to rely on mutual contacts to defend their positions in their academic field. Third, interviews with female scientists highlight the challenges they faced in advancing their careers, because research institutes were rigidly hierarchical and male colleagues could foster their careers without interruption for maternity leave. Fourth, the 1960s and 1970s witnessed fundamental changes in the language of science. European mathematicians could no longer use French or German to disseminate their results internationally, as US scholars communicated exclusively in

English. Meanwhile, mathematicians in the Soviet Union were constrained by the prescribed use of Russian. The language of mathematics and the national languages of the scholars were therefore determining factors in the opportunities for mutual cooperation.

Although the individual fates of scholars in Central and Eastern Europe were significantly influenced and often constrained by Communist Party politics, the ruptures of the Cold War did not halt efforts to disseminate knowledge or diminish the perceived value of what scholars could gain from personal encounters.

Author Biography

Tomáš W. Pavlíček specializes in the social and cultural history of Central Europe, oral history, and the history of science. He earned his PhD through a Czech–German program (LMU Munich–Charles University). As a historian at the Institute of History of the Polish Academy of Sciences he leads international research on Czech–Polish scholarly networks and has published extensively on science, religion, and memory in nineteenth- and twentieth-century Europe.

List of Abbreviations

APAN	Archiwum Polskiej Akademii Nauk (Archive of the Polish Academy of Sciences)
CERN	Conseil européen pour la Recherche Nucléaire (European Organization for Nuclear Research)
CSc.	candidatus scientiarum (Candidate of Sciences)
ČSAV	Československá akademie věd (Czechoslovak Academy of Sciences)
GUHA	General Unary Hypotheses Automaton
ICM	International Congress of Mathematicians
IM PAN	Instytut Matematyczny Polskiej Akademii Nauk (Mathematical Institute of the Polish Academy of Sciences)
MO	Mathematical Olympiad
MUNI	Masarykova univerzita v Brně (Masaryk University Brno)
MÚ ČSAV	Matematický ústav Československé akademie věd (Mathematical Institute of the Czechoslovak Academy of Sciences)
MÚA AV ČR	Masarykův ústav a Archiv AV ČR (Masaryk Institute and Archives of the Czech Academy of Sciences)

Notes

1. Tomáš W. Pavlíček, interview with Zbigniew Semadeni, 12.10.2023. Unless otherwise stated, all translations are by the author. See Wikipedia, "Zbigniew Semadeni."
2. The proposition for the case of non-isomorphic Banach spaces was published in 1960, and it is still considered a breakthrough in functional analysis. Semadeni, "Banach Spaces Non-Isomorphic." See Candido, "On the Semadeni Derivative." While Google Scholar is not the principal repository for mathematicians, Semadeni's paper has attracted ninety citations, which is a notable achievement.
3. On ICM in the 1960s, see Furinghetti and Giacardi, *The International Commission.*
4. Nečasová, *Nový socialistický člověk*, 110.
5. Kreuder-Sonnen, "From Transnationalism to Olympic Internationalism."
6. Russian mathematicians were obliged to publish in Russian (after the Stalinist purge involving the academician Luzin). However, the door to foreign study visits was open to linguistically talented scholars from minor states. Duda, *Historia matematyki*, 342–45; Hollings, Siegmund-Schultze, and Sørensen, *Meeting under the Integral Sign?*, 89–102.
7. Fürst, *Stalin's Last Generation*, 13.
8. Klich-Kluczewska, "Kobieta wobec rodziny," 291–99.
9. Hollings, *Scientific Communication*, 24–28, 101–2. The debate concerning the natural sciences and the humanities has traditionally drawn on two general approaches: while the historical approach values complexity and particularity, the structural approach attributes higher value to simplicity and generality. McAllister and Kornet, "Historical and Structural Approaches."
10. Krige, "Introduction: Writing the Transnational," 1–31.
11. Hollings et al., *Meeting under the Integral Sign?*, 5.
12. Deleuze and Guattari, *A Thousand Plateaus.*
13. Duda, *Historia matematyki*, 357–65.
14. Surman, "Discussion with."
15. Although I used a chronological list of questions, the goal was not to produce a questionnaire, but to conduct semi-structured in-depth interviews. In the second part and in the additional session, I focused my research questions on mathematics education, the relationship with mentors, contacts with classmates, and the circulation of knowledge realized through scholarships, conferences, and publications. Thomson, "Four Paradigm."
16. Hamilton and Shoeps, "Building Partnerships."
17. Thompson, "The Voice of the Past."
18. The table of interviewees presents two scholarly generations and various mathematical disciplines, see Entanglements, "Asymmetries in Mathematical Knowledge." Some interviewees recommended that I contact their colleagues, others I contacted independently. The short time of the project (2021–24) and also COVID conditions limited contacts with some people. I spoke to one mathematician, Jaroslav Kurzweil, in 2017 and remained in regular phone contact with him until 2021.
19. The concept of active semantic networks was outlined in 1975 by Norman and Rumelhart, *Exploration in Cognition*. See Maruszewski, *Pamięć autobiograficzna*, 16.
20. Kurkowska-Budzan, Soroko, and Stasiak, "Wywiad historyczny."

21. Traditionally, from the nineteenth century, women had been employed in mathematical institutes or astronomical observatories in "calculator" (*rachmistrz* in Polish, *počtářka* in Czech) positions, in the stereotypical belief that calculations were done more precisely by women and that such work could be paid less, almost as little as those women who washed chemical glass apparatus in "dishwasher" positions in laboratories. Men were not asked to perform such tasks, although there were "masculine" positions such as handling telescopes, operating generators, engines, and so forth. Zrull, "Women in Glass." Literature based on written documents frequently highlight the male-orientated understanding of successful science, often based on institutional cooperation and interactions between academies or institutes. On Lynn Bloom (1977), see Kałwa, "Herstoria mówiona w Polsce."
22. Maruszewski, *Pamięć autobiograficzna*, 18–20.
23. For example, James B. Conant as US High Commissioner observing the transformation in Germany criticized the education system in the United States and posited the Eastern Bloc's experience as a model for consideration. Olšáková, "Pugwash in Eastern Europe."
24. Hudek and Kováč, *Dejiny Slovenskej Akadémie vied*; Jareš and Franc, *Mezi konkurencí a spoluprací*.
25. Pavlíček, "Memento obětí totalitních."
26. Brenner, *Mezi Východem a Západem*; similarly Zysiak, "Modernizing Science"; Lochmatow, *Public Knowledge in Cold War*, 7.
27. Connelly, *Captive University*; Herczyński, *Spętana nauka*.
28. Zysiak, *Punkty za pochodzenie*, 29, 42. Social scientists were more inclined toward progressive modeling of forecasts regarding the development of socialist society. See the chapter in this volume by Jan Surman.
29. This was the experience of one of the interviewees. Pavlíček, "Rozhovor s prof. Jaroslavem Kurzweilem."
30. Steinhaus, *Wspomnienia i zapiski*, 321.
31. Havránek, "Kontakty s vědou Východního bloku," 46.
32. Kuratowski, *A Half Century of Polish Mathematics*, 91. The same was true for Czechoslovakia.
33. The leading mathematicians contributed articles on the role of mathematics in the construction of socialism. Professor Mazur (1946–48 in Łódź, thereafter in Warsaw) wrote several articles on the implementation of Marxism in mathematics. (See Mazur, "Walka o postępową naukę.") In Poland, Mazur, Kuratowski, and Borsuk joined the Polish United Workers' Party. For Czechoslovakia, see similar articles by E. Čech, V. Jarník, M. Katětov, and O. Borůvka (they joined the Communist Party of Czechoslovakia). However, only a few were politically active. Interviewees from both academy institutes of mathematics recalled that their workplace did not have its own party cell and that they shared the cadre officer with another institute. See Tomáš W. Pavlíček, interview with Zofia Adamowicz, 14 July 2022; Tomáš W. Pavlíček, interview with Kamila Bendová, 9 September 2024.
34. On Steinhaus see Duda, *Słownik biograficzny*, 395–97.
35. See Arnason, "Communism and Modernity," 71.
36. *Traktat o dobrej robocie* (1955); English translation: Kotarbiński, Tadeusz. 1965. *Praxiology: An Introduction to the Science of Efficient Action*. New York: Pergamon Press.

37. On 6 October 1958. Steinhaus, "Matematyka wczoraj i dziś," 16, 18. Compared with Steinhaus's memories, which are important because of his criticism of the academic policy in Poland. Steinhaus, *Wspomnienia i zapiski.*
38. Nečasová, *Nový socialistický člověk,* 112.
39. The former Mathematical Institute of Czech Academy of Sciences and Arts (1947) was transformed into the state institute in 1950 (Ústřední ústav matematický).
40. Babuška, "Úvodem," 1. The editor was the mathematician Ivo Babuška.
41. Archiv MUNI, Bohuslav Hostinský collection, box 8, nr 12, Sjezd matematiků čsl. a polských 1949, congress dictionary.
42. Pavlíček, "Czech and Polish Mathematicians."
43. Apart from František Vyčichlo, Antonín Špaček, and Václav Pleskot. See the correspondence on the workshop held 14–15 June 1950: Archiwum PAN, collection IM PAN, wykazy 275, akta nr 2, nr 28–36 współpraca bilateralna, Czechosłowacja, sygn. Nr 2.16 konferencja robocza/robotnicza polsko-czechosłowackich statystyków w Wrocławiu 14–15.06.1950.
44. Tadeusz Czechowski, Jędrusz [maybe Jędrzej/Jerzy Lewandowski], Kazimierz Stachiewicz.
45. Górny, "The Decade of Reciprocity," 61–86.
46. MÚA, A AV ČR, I sekce matematicko-fyzikální ČSAV collection, inv. nr 73, contacts abroad, correspondence.
47. Kurzweil quoted the coordinator Marcelli Stark. In Warsaw he participated in a Polish language course. Kulawiaková and Pavlíček, "The Training of the Czech," 195.
48. Pavlíček, "Rozhovor s prof. Jaroslavem Kurzweilem," 114. See Kurzweil, "A Characterization of Analytic Operations."
49. Dupré and Somsen, "The History of Knowledge."
50. Stockton, "Mathematical Competitions in Hungary." On Leningrad High School Olympiad and Moscow Math Olympiad, see Grozman and Shapovalov, *60 Odd Years of Moscow,* 1–3.
51. Kuratowski, *A Half Century of Polish Mathematics,* 108.
52. Durnová and Bušková, "Československá matematická olympiáda."
53. Pavlíček, "Mathematical Olympiad in People's Poland."
54. In some countries the competition had a different name but a very similar format (for example, the Romanian Competition in 1950, the Yugoslav Federal Mathematical Competition in 1960). In the German Democratic Republic, mathematical competitions began after the first International MO, as was the case in Bulgaria (where the first national MO was held in 1962, Nacionalna olimpiada po matematika).
55. In Moscow, for example, A. Kolmogorov. See Zdravkovska and Duren, *Golden Years of Moscow Mathematicians,* for Czechoslovakia *Květy.* For Poland a new journal on didactics *Matematyka.*
56. He was invited by Wacław Zawadowski. Tomáš W. Pavlíček, interview with Wiesław Żelazko and Hanna Stande Żelazko, 27 June 2022.
57. Klich-Kluczewska, "Kobieta wobec rodziny," 235.
58. Pavlíček, Interview with Adamowicz, 14 July 2022. Her grandfather, Professor Stanisław Straszewicz, helped organize the Polish MO for several years.
59. See Concretely in Kulawiaková and Pavlíček, "Zofia Adamowicz".
60. Klánský, "Olympiáda, o níž se nepíše," 9.
61. Prečan and Slavíčková, *Bez Františka Janoucha.*

62. The circle was organized by Semadeni's classmate Przemysław Zbijewski (1933–1966), who selected books that were not discussed in basic courses and not available in the Polish language: *Theory of the Integral* (1937) by S. Saks (victim of war), *Moderne Algebra* (1930, 1931) by B. van der Waerden, *Topologie I* (1933) by K. Kuratowski. See also Naskręcki, *Prof. Dr hab. Józef Witkowski*, 14.
63. Instead of Orlicz, he was supervised by junior professor Andrzej Alexiewicz. Pavlíček, interview with Semadeni, 12 October 2023.
64. The interviewed scholars repeatedly said they did not want to do probability calculations or teach math. Tomáš W. Pavlíček, interview with Aleš Pultr, 10 January 2023.
65. There was less experience with repatriation in Czechoslovakia, mostly limited to peripheral regions. Nevertheless, there emotions were connected with deterritorialization from the anthropological point of view. Deleuze and Guattari, *A Thousand Plateaus*.
66. E. Marczewski, B. Knaster, H. Steinhaus and assistant S. Hartman plus Professor W. Ślebodziński (from Poznań). Duda, *Historia matematyki*, 321.
67. In their memoirs, however, they never mentioned Lviv during the war because of its association with conflicting loyalties. Tomáš W. Pavlíček, interview with Roman Duda, 21 March 2024.
68. Monument in Memory of Polish Scientists—Victims of Hitlerism. See Mierzecki, "Budowa wrocławskiego pomnika".
69. Archiv MUNI, collection Otakar Borůvka, inv. nr 230, correspondence with Jur Hronec, 1946–58.
70. The impression of exclusivity is reinforced by the collection of short memoirs of graduates. Netuka and Stiborová, *Univerzita Karlova v Praze*. Eduard Čech pushed for the creation of an autonomous Institute of Mathematics at Charles University within the new Faculty (Matematický ústav Univerzity Karlovy), which would function as a creative environment for a professorial sabbatical.
71. It happened after Anton Kotzig joined the Institute in 1959. It was an example of socialist reciprocity, though Bratislava later fell behind, when Kotzig moved to Canada in 1969, at which point graphs began to be taught in Prague as well. Hudek and Kováč, *Dejiny Slovenskej Akadémie vied*, 505; Tomáš W. Pavlíček, interview with Václav Chvátal, 2 May 2023.
72. See passus "I to były bardzo ciekawe seminaria." Pavlíček, interview with Duda, 21 March 2024.
73. Kulawiaková and Pavlíček, "Zofia Adamowicz".
74. See passus "Niestety myśmy tego nie widzieli." Pavlíček, interview with Duda, 21 March 2024. From the researcher's perspective, the interview with Roman Duda was extraordinarily useful for two reasons. First, as the author of a monograph on the history of mathematics in the Polish lands, he reviewed my transcripts, and second, during the interview he was able to describe his subjective student experience in the topology seminar.
75. Later, for the Congress of Polish Science in 1972, Duda sent a critical assessment of the state of topology in the country. In Warsaw, in the PAN Committee of Topology, he was given credit for his statement, but on an official note, the Committee diplomatically softened his remarks. Archiwum Uniwersytetu Wrocławskiego, Roman Duda collection, box 6 and 10, correspondence, unsorted.

76. Brown, "The Mathematical Work of R H Bing," 3.
77. Tomáš W. Pavlíček, interview with Roman Duda 4 June 2025.
78. The conference organized by P. Alexandrov concentrated on homology theory, with the participation of the following invited speakers: Čech from Czechoslovakia, Borsuk from Poland, Solomon Lefschetz (Russian origin, educated in Paris) from USA and Izrail I. Gordon from USSR. Each conference talk in the proceedings was in one of four languages: English, French, German, or Russian. See Mardešić, "Topology in Eastern Europe"; Apushkinskaya, Nazarov, and Sinkevich, "In Search of Shadows."
79. Tomáš W. Pavlíček and Barbora Kulawiaková, interview with Roman Duda, 19 April 2024.
80. Pavlíček, interview with Semadeni, 12 October 2023.
81. While Czechoslovak mathematicians had few emigrant facilitators in the West, the Poles had S. Eilenberg, S. Ulam, A. Zygmund, and A. Tarski who established new scientific schools in the United States. See Eilenberg and MacLane, "General Theory of Natural Equivalences."
82. When he was preparing to present his master's thesis in Prague, he discovered that he had made a mistake in his initial manuscript that the reviewers had overlooked because they had not read it all the way through. That is why Duda remembers the successful publication in Prague with gratitude.
83. Pavlíček, interview with Duda, 21 March 2024. Under the influence of Knaster, Duda was tasked with developing the problem of homomorphisms: Duda, "Sur Le Prolongement."
84. Duda, "Bronisław Knaster (1893–1980)."
85. "When I came back from Cambridge, I took it upon myself to promote this new topology in our environment. And it was successful, because this Wrocław topology was revived, refreshed, so to speak, and brought about some important work and good reputations." Pavlíček, interview with Duda, 21 March 2024.
86. *Theory of Shape* (Warsaw: PWN, 1975) was translated into Russian without a translation of the title (Карол Борсук, Теория шейпов. [Teoria sheipov] Moscow: Mir, 1976). With this Cyrillic transcription, the new concept became part of Russian mathematics.
87. Pavlíček, interview with Duda, 4 June 2025. Zdravkovska and Duren, *Golden Years of Moscow Mathematicians*, 232–33.
88. Duda, *Lwowska Szkoła Matematyczna*.
89. See Toposym, "First Symposium on General Topology."
90. Netuka and Stiborová, *Univerzita Karlova v Praze*, 161. Trnková's written recollection was similar to Pultr's oral narrative in the interview.
91. The school was organized by Semadeni's former classmate, Kazimierz Gęba. See Pavlíček, interview with Semadeni, 12 October 2023.
92. Although Katětov was the rector of Charles University in the early 1950s, he was only permitted to hold a research position at the university after 1968. Netuka and Stiborová, *Univerzita Karlova v Praze*, 161.
93. "Katětov's area was—well, apart from topology, which he liked—functional analysis, which bored me terribly. And conversely, he was terribly 'bored' by algebraic topology, which I really wanted. But he just gave me free rein. And in the end, it was all very different—with [my choice of] category theory, and he enjoyed that." Pavlíček, interview with Pultr, 10 January 2023.

94. See Toposym, "Second Symposium on General Topology."
95. "It meant a lot to me because I met Jürgen Schmidt and Bernhard Banaschewski there. With Bernhard it was a lifelong collaboration." Pavlíček, interview with Pultr, 10 January 2023.
96. Ibid.
97. Pavlíček, interview with Semadeni, 12 October 2023. On Orlicz's students, see Mathematics Genealogy Project, "Władysław Roman Orlicz."
98. The importance of the Second Department is disproportionately marginalized in the UAM 50th Anniversary Handbook (compared to Orlicz's First Department), and the name of its head, Alexiewicz, is not mentioned at all. See Grot, *50 lat Uniwersytetu*, 235–36.
99. Because of this particular lack of even-handedness, scholars from Poznań began to look elsewhere for new intellectual insights. Semadeni transferred to Warsaw. Kazimierz Gęba was not admitted to *aspirantura*. He received an internship outside Poznań and went on to make a career in Gdańsk: "In short, of the three of us [Zbijewski, Semadeni, and Gęba] he was the best beyond compare, but at the time of graduation he did not shine." Pavlíček, interview with Semadeni, 12 October 2023.
100. Kulawiaková and Pavlíček, "Andrzej Sołtysiak."
101. Bourdieu, *Homo Academicus.*
102. Netuka and Stiborová, *Univerzita Karlova v Praze*, 160. Čech's friendly approach was a way of alleviating the sentiments of mistrust and fear in the late 1950s. See Pavlíček, interview with Pultr, 10 January 2023.
103. Masarykův ústav a Archiv AV ČR, collection MÚ ČSAV; Personal archive collections of Jaroslav Kurzweil (held by his family), Wiesław Żelazko.
104. Archiwum PAN Poznań branch, Władysław Orlicz collection, IV B) korespondencja wpływająca, files 151–74.
105. Archiv MUNI, Brno, Otakar Borůvka collection, inv. nr 117, correspondence with Bielecki, 1957–59.
106. Zahradníček, *Polské poučení z pražského jara.*
107. Duda, *Historia matematyki*, 345.
108. Surman and Pavlíček, "The Formal and the Informal."
109. Kulawiaková and Pavlíček, "Andrzej Sołtysiak."
110. Translated from Polish and quoted by Kulawiaková and Pavlíček, "Andrzej Sołtysiak," 210.
111. Pták exploited his interest in the spectral theory of Banach algebras and together they published two papers on continuity of the spectrum and the spectral radius.
112. Surman, "Re-Connecting Central European Science."
113. Although, in practice, he often failed to deal with correspondence in a timely manner. Tomáš W. Pavlíček, interview with Wiesław Żelazko and Hanna Stande Żelazko, 27 June 2022.
114. Surman and Pavlíček, "The Formal and the Informal." For Prague, see Netuka and Stiborová, *Univerzita Karlova v Praze.*
115. Pavlíček, interview with Żelazko and Stande-Żelazko 27 June 2022, 12.
116. Gerovitch, "The Kitchen and the Dacha"; Zdravkovska and Duren, *Golden Years of Moscow Mathematicians.*
117. Duda, "Paweł Siergiejewicz Aleksandrow"; Jakimowicz and Szczepański, *Matematyka na Pomorzu*, 125–26.

118. Oreskes and Krige, *Science and Technology*, 435–8.
119. Těšínská, Československá jaderná a částicová fyzika; Hladký, *Paměti kosmika*; Pavlíček, Šolc, and Hyklová, *Astronomers behind the Iron Curtain*.
120. Patiniotis and Gavroglu, "The Sciences in Europe."
121. Kraft, "Spaces of Knowledge and Gender Regimes."
122. Švejdar, "Modern Czech Logic."
123. Thanks to his tutors Kuratowski, A. Tarski, and K. Gödel, he spent several years as a host at the IAS in Princeton, Berkeley, and Oxford. Duda, *Słownik biograficzny*, 271–73. see MacTutor Index, "Andrzej Mostowski."
124. See Murawski, *Z historii logiki*, 251–61.
125. Pavlíček, interview with Adamowicz, 14 July 2022. Similarly, Dina Štěrbová recalls the community of mathematical analysis in Olomouc, praising the opportunity to attend seminars in Prague, where she even met Żelazko and invited him to Olomouc. Pavlíček, interview with Dina Štěrbová, 10 March 2023.
126. Pavlíček, interview with Adamowicz, 14 July 2022.
127. Hájek and Petr Vopěnka knew Mostowski, as he had helped them by writing an English circle review on their articles published in Russian and Czech. In 1972, their monograph *Theory of Semisets* was translated into English and published by North Holland Publishing Company (Hájek was extremely proficient in languages).
128. Haniková, "Petr Hájek," 24–25.
129. Ibid. Along with Hájek, the team included Ivan B. Havel, Metoděj Chytil, and the statistician Tomáš Havránek.
130. At one point, Hájek held the seminar in his apartment so that his sick wife could attend. Pavlíček, interview with Pavel Pudlák, 3 April 2023.
131. However, the informal nature of minority research runs the risk of overlooking the genesis of the team. See Haniková, "Petr Hájek," 24–25.
132. Benda and Bendová, *Dopisy přes mříže.*
133. Pavlíček, interview with Pudlák, 3 April 2023. Pultr, Bendová, Chvátal, and Jaroslav Nešetřil described similar experiences with Hedrlín. Pavlíček, interview with Pultr, 10 January 2023, interview with Bendová, 9 September 2024, interview with Chvátal, 2 May 2023.
134. Tomáš W. Pavlíček, interview with Zofia Adamowicz, 26 March 2024.
135. Bendová worked at MÚ ČSAV as a doctoral assistant while managing a household with six children. Benda and Bendová, *Dopisy přes mříže.*
136. Švejdar, "Modern Czech Logic."
137. In 1963, Paul Cohen published a seminal paper demonstrating that Gödel's famous theorem, the continuum hypothesis, could not be proven, introducing the method of forcing as a way to construct alternative set-theoretical models. See Cohen, "*The Independence of the Continuum Hypothesis.*"
138. Pavlíček, interview with Adamowicz, 14 July 2022.
139. Haniková, "Petr Hájek," 27.
140. For Vencovská's postgraduate studies at the Institute, they negotiated an exceptional supervision with Professor of Differential Equations, Jaroslav Kurzweil, who sympathized with transnational couples and helped the postgraduate to publish three joint papers. Pavlíček, "Rozhovor s prof. Jaroslavem Kurzweilem."
141. The following list of accomplishments is not complete but nevertheless impressive: metamathematics (Pudlák and Hájek), applied logic (GUHA method: Havel,

Hájek and Chytil), fuzzy logic (Hájek), computational complexity class (Pudlák and Krajíček), bounded arithmetic (Krajíček and Adamowicz), information storage and retrieval systems (Marek and Pawlak), rough set (Pawlak), model theory (East Germans). It is necessary to mention two significant achievements in Slovakia: number theory (Štefan Schwarz) and graph theory (Anton Kotzig). Hudek and Kováč, *Dejiny Slovenskej Akadémie vied*, 505.

142. From 1974 to 2015 the proceedings were published by Springer. Traditionally, the conference moved between Poland, the Czech lands and Slovakia, and since 2013, the conference has traveled around Europe.
143. The first conference took place in Jabłonne near Warsaw, the second in Štrbské Pleso in the Tatra Mountains, the third in Poland once more, and the fourth in Mariánské lázně in the Bohemian part of Czechoslovakia. The parity was represented by three key scholars who organized the conferences: Andrzej Blikle from Warsaw, Ivan Korec from Bratislava (obtained his PhD after a stay at the Banach Center in Warsaw), and Ivan M. Havel (brother of the well-known Czech dissident and playwright, Václav Havel). Pavlíček, interview with Bendová, 9 September 2024.
144. Pavlíček, interview with Semadeni, 12 October 2023, interview with Żelazko and Stande Żelazko, 27 June 2022.
145. Duda, "Emigration of Mathematicians"; Duda, *Historia matematyki*, 342–45.

Bibliography

Apushkinskaya, D. E., A. I. Nazarov, and G. I. Sinkevich. 2019. "In Search of Shadows: The First Topological Conference, Moscow 1935." *The Mathematical Intelligencer* 41: 37–42. https://doi.org/10.1007/s00283-019-09907-6.

Arnason, J. P. 2000. "Communism and Modernity." *Daedalus* 129: 71.

Babuška, Ivo. 1956. "Úvodem" [Introduction]. *Aplikace matematiky* 1, no. 1: 1.

Benda, Václav, and Kamila Bendová. 2007. *Dopisy přes mříže* [Letters through bars]. Kostelní Vydří: Karmelitánské nakladatelství.

Borsuk, Karol. 1975. *Theory of Shape*. Warsaw: PWN.

Bourdieu, Pierre. 1988. *Homo Academicus*. Stanford, CA: Stanford University Press.

Brenner, Christiane. 2015. *Mezi Východem a Západem: české politické diskurzy 1945–1948* [Between East and West: Czech political discourses 1945–1948]. Prague: Argo.

Brown, Morton. 2011. "The Mathematical Work of R H Bing." *Celebratio Mathematica*: 1–11.

Candido, Leandro. 2022. "On the Semadeni Derivative of Banach Spaces C(K, K)." *Studia Mathematica* 266: 225–40.

Cohen, Paul, 1963. "The Independence of the Continuum Hypothesis." *Proceedings of the National Academy of Sciences of the United States of America* 50, no. 6: 1143–48.

Connelly, John. 2000. *Captive University: The Sovietization of East German, Czech, and Polish Higher Education, 1945–1956*. Chapel Hill, NC: The University of North Carolina Press.

Deleuze, Gilles, and Félix Guattari. 1987. *A Thousand Plateaus*. Minneapolis: University of Minnesota Press.

Duda, Roman. 1983. "Bronisław Knaster (1893–1980)." *Wiadomości Matematyczne* 25, no. 1: 99–116. https://doi.org/10.14708/wm.v25i01.4004.

———. 2012. "Emigration of Mathematicians from Poland in the 20th Century (Roughly 1919–1989)." *Organon* 44: 95–125.

———. 2019. *Historia matematyki w Polsce na tle dziejów nauki i kultury* [The History of mathematics in Poland in the context of the history of science and culture]. Warsaw: Instytut Historii Nauki PAN—Oficyna Wydawnicza ASPRA-JR.

———. 2007. *Lwowska Szkoła Matematyczna* [Lviv school of mathematics]. Wrocław: Wydawnictwo Uniwersytetu Wrocławskiego.

———. 1986. "Paweł Siergiejewicz Aleksandrow." *Wiadomości Matematyczne* 27, no. 1: 59–74. https://doi.org/10.14708/wm.v27i01.4128.

———. 1958. "Sur Le Prolongement Des Homéomorphies" [On the Extension of Homeomorphisms]. *Fundamenta Mathematicae* 58, no. 2: 175–86.

———, ed. 2024. *Słownik biograficzny matematyków polskich. Słowniki biograficzne uczonych polskich* [Biographical Dictionary of Polish Mathematicians. Biographical Dictionaries of Polish Scholars]. Kraków: Polska Akademia Umiejętności.

Dupré, Sven, and Geert Somsen. 2019. "The History of Knowledge and the Future of Knowledge Societies." *Berichte zur Wissenschaftsgeschichte* 42, no. 2–3: 186–99.

Durnová, Helena, and Petra Bušková. 2023. "Československá matematická olympiáda v 50. letech 20. století" [Czechoslovak Mathematical Olympiad in the 1950s]. *Učitel matematiky* 31, no. 2: 135–47.

Eilenberg, Samuel, and Saunders Mac Lane. 1945. "General Theory of Natural Equivalences." *Transactions of the American Mathematical Society* 58, no. 2: 231–94.

Entanglements. 2025. "Asymmetries in Mathematical Knowledge." Retrieved 25 June 2025 from https://entanglements.ihpan.edu.pl/interview/testowy-wywiad/.

Furinghetti, Fulvia, and Livia Giacardi, eds. 2022. *The International Commission on Mathematical Instruction, 1908–2008: People, Events, and Challenges in Mathematics Education*. Cham: Springer Nature.

Fürst, Juliane. 2010. *Stalin's Last Generation: Soviet Post-War Youth and the Emergence of Mature Socialism*. Oxford: Oxford University Press.

Gerovitch, Slava. 2024. "The Kitchen and the Dacha: Productive Spaces of Soviet Mathematics." *Studia Historiae Scientiarum* 23: 657–83. https://doi.org/10.4467/2543702XSHS.24.018.19591.

Górny, Maciej. 2024. "The Decade of Reciprocity: The Debate Over Socialist Planning in Poland and Czechoslovakia, 1956–1968." *History of Economy Ideas* 32, no. 2: 61–86.

Grot, Zdzisław. 1969. *50 lat Uniwersytetu im. Adama Mickiewicza: 1919–1969* [50 Years of Adam Mickiewicz University: 1919–1969]. Poznań: UAM.

Grozman, Pavel, and Alexander Shapovalov. 1997. *60 Odd Years of Moscow Mathematical Olympiads (Manuscript)*, edited by Dimitry Leites. Stockholm. Retrieved 30 June 2025 from https://www.scribd.com/document/522561886/60-Odd-Years-of-Moscow-Mathematical-Olympiads-by-D-Leites-Ed-G-Galperin-A-Tolpygo-P-Grozman-A-Shapovalov-V-Prasolov-A-Fomenko-Z-lib-o.

Hamilton, Paula, and Linda Shoeps. 2008. "Building Partnerships between Oral History and Memory Studies." In *Oral History and Public Memories*, edited by Paula Hamilton and Linda Shoeps, vii–xvii. Philadelphia, PA: Temple University Press.

Haniková, Zuzana. 2015. "Petr Hájek: A Scientific Biography." In *Petr Hájek on Mathematical Fuzzy Logic, Outstanding Contributions to Logic*, edited by F. Montagna, 21–38. Cham: Springer Nature. https://doi.org/10.1007/978-3-319-06233-4_2.

Havránek, Jan. 2000. "Kontakty s vědou Východního bloku a první styk se Západem" [Contacts with the scientific community of the Eastern Bloc and the first interaction with the West]. In *Věda v Československu v letech 1953–1963* [Science in Czechoslovakia, 1953–1963], edited by Hana Barvíková, 46–47. Prague: Archiv Akademie věd ČR.

Herczyński, Ryszard. 2008. *Spętana nauka: opozycja intelektualna w Polsce 1945–1970* [The Shackled academy: intellectual opposition in Poland, 1945–1970]. Warsaw: Wydawnictwo Naukowe Semper.

Hladký, Jan. 2018. *Paměti kosmika Fyzikálního ústavu ČSAV* [Memoirs of a cosmic physicist from the Institute of Physics of the Czechoslovak Academy of Sciences]. Prague: Academia.

Hollings, Christopher. 2016. *Scientific Communication across the Iron Curtain*. Cham: Springer.

Hollings, Christopher, Reinhard Siegmund-Schultze, and Henrik Kragh Sørensen. 2020. *Meeting under the Integral Sign? The Oslo Congress of Mathematicians on the Eve of the Second World War*. Providence, RI: American Mathematical Society.

Hudek, Adam, and Dušan Kováč, eds. 2023. *Dejiny Slovenskej Akadémie vied* [History of the Slovak Academy of Sciences]. Bratislava: VEDA.

Jakimowicz, Emilia, and Andrzej Szczepański, eds. 2017. *Matematyka na Pomorzu Gdańskim 1945–2015* [Mathematics in the Gdańsk region, 1945–2015]. Gdańsk: Wydawnictwo Uniwersytetu Gdańskiego.

Jareš, Jakub, and Martin Franc. 2018. *Mezi konkurencí a spoluprací: Univerzita Karlova a Československá akademie věd 1945–1969* [Between competition and cooperation: Charles University and the Czechoslovak Academy of Sciences, 1945–1969]. Prague: Karolinum.

Kałwa, Dobrochna. 2021. "Herstoria mówiona w Polsce: Kilka uwag o feministycznych projektach oral history" [Oral herstory in Poland: a few thoughts on feminist oral history projects]. *Wrocławski Rocznik Historii Mówionej* 11: 8–21. https://doi.org/10.26774/wrhm.283.

Klánský, Mojmír. 1957. "Olympiáda, o níž se nepíše" [Olympiad that nobody writes about]. *Květy* 7, no. 22: 8–9.

Klich-Kluczewska, Barbara. 2020 "Kobieta wobec rodziny" [Woman and the family]. In *Kobiety w Polsce 1945–1989: Nowoczesność—równouprawnienie—komunizm* [Women in Poland 1945–1989: modernity—equality—communism], edited by Małgorzata Fidelis, Piotr Perkowski, Barbara Klich-Kluczewska and Katarzyna Stańczak-Wiślicz, 291–336. Kraków: Universitas.

Kotarbiński, Tadeusz. 1965. *Praxiology: An Introduction to the Science of Efficient Action*. New York: Pergamon Press.

Kraft, Claudia. 2018. "Spaces of Knowledge and Gender Regimes: From Double Marginalization to a Gendered History of Knowledge in Central and Eastern Europe." *Acta Poloniae Historica* 117: 7–26. https://doi.org/10.12775/APH.2018.117.01

Kreuder-Sonnen, Katharina. 2016. "From Transnationalism to Olympic Internationalism: Polish Medical Experts and International Scientific Exchange, 1885–1939."

Contemporary European History 25, no. 2: 207–31. https://doi.org/10.1017/S0960777316000047.

Krige, John. 2019. "Introduction: Writing the Transnational History of Science and Technology." In *How Knowledge Moves: Writing the Transnational History of Science and Technology*, edited by John Krige, 1–31. Chicago: University of Chicago Press.

Kulawiaková, Barbora, and Tomáš W. Pavlíček. 2023. "Andrzej Sołtysiak: 'The Best Way for You Is to Study Math, You Just Put Some Chalk in Your Pocket, Go to the Classroom and Don't Care.' Choosing the Field of Studies—the Entangled Cooperation between Polish and Czech Mathematicians." *Wrocławski Rocznik Historii Mówionej* 13: 198–222. https://doi.org/10.26774/wrhm.372.

———. 2021. "The Training of the Czech Mathematician Jaroslav Kurzweil with Władysław Orlicz in Poland." *Antiquitates Mathematicae* 15, no. 2: 188–206. https://doi.org/10.14708/am.v15i1.7078.

———. 2026. "Zofia Adamowicz: 'You Know, Women Didn't Have It Easy at All. I Needed to Be Three Times than a Man to Be Justly Appreciated Fairly.' Stress and Time between Scientific Work and Family." *Wrocławski Rocznik Historii Mówionej* 16 (in print).

Kuratowski, Kazimierz. 1980. *A Half Century of Polish Mathematics: Remembrances and Reflections*, trans. Andrzej Kirkor. Oxford: Pergamon Press.

Kurkowska-Budzan, Marta, Emilia Soroko, and Marcin Stasiak. 2020. "Wywiad historyczny: Propozycja metody badawczej" [Historical interview: a proposed research method]. *Wrocławski Rocznik Historii Mówionej* 10: 8–36. https://doi.org/10.26774/wrhm.269.

Kurzweil, Jaroslav. 1953. "A Characterization of Analytic Operations in Real Banach Spaces." *Studia Mathematica* 14: 82–83.

Lochmatow, Alexej. 2023. *Public Knowledge in Cold War Poland: Scholarly Battles and the Clash of Virtues, 1945–1956*. New York: Routledge.

MacTutor Index. "Andrzej Mostowski." Retrieved 12 July 2024 from https://mathshistory.st-andrews.ac.uk/Biographies/Mostowski/.

Mardešić, Sibe. 2000. "Topology in Eastern Europe 1900–1950." *Topology Proceeding* 25 (Spring): 397–430.

Maruszewski, Tomasz. 2005. *Pamięć autobiograficzna* [Autobiographical memory]. Gdańsk: Gdańskie Wydawnictwo Psychologiczne.

Mathematics Genealogy Project. "Władysław Roman Orlicz." Retrieved 31 March 2023 from https://www.mathgenealogy.org/id.php?id=51907.

Mazur, Stanisław. 1951. "Walka o postępową naukę w dziedzinie matematyki" [The struggle for progressive education in mathematics]. *Matematyka. Czasopismo dla nauczycieli* 4, no. 1: 1–4.

McAllister, James W., and Diedel J. Kornet. 2002. "Historical and Structural Approaches in the Natural and Human Sciences." In *The Future of the Sciences and Humanities: Four Analytical Essays and a Critical Debate on the Future of Scholastic Endeavour*, edited by Arie Rip and Aan Elzinga, 19–62. Amsterdam: Amsterdam University Press.

Mierzecki, Roman. 2007. "Budowa wrocławskiego pomnika w latach 1956–1964: Ku czci polskich profesorów zamordowanych we Lwowie w 1941 roku" [Construction of the Wrocław monument between 1956 and 1964: in memory of the Polish professors murdered in Lviv in 1941]. *Analecta* 16(1–2). Retrieved 12 June 2024 from https://www.lwow.com.pl/mierzecki/analecta.html.

Murawski, Roman. 2019. *Z historii logiki i filozofii matematyki* [From the history of logic and the philosophy of mathematics]. Poznań: Wydawnictwo Naukowe UAM.

Naskręcki, Władysław. 1995. *Prof. Dr hab. Józef Witkowski.* Poznań: Obserwatorium astronomiczne Uniwersytetu im. Adama Mickiewicza.

Nečasová, Denisa. 2017. *Nový socialistický člověk: Československo 1948–1956* [New socialist man: Czechoslovakia 1948–1956]. Brno: Host.

Netuka, Ivan, and Milena Stiborová, eds. 2002. *Univerzita Karlova v Praze: Matematicko-fyzikální fakulta. Jubilejní almanach* [Charles university in Prague: Faculty of mathematics and physics. Bicentennial almanac]. Prague: Matfyzpress.

Norman, Donald A., and David E. Rumelhart. 1975. *Exploration in Cognition.* San Francisco: W. H. Freeman.

Olšáková, Doubravka. 2018. "Pugwash in Eastern Europe: The Limits of International Cooperation under Soviet Control in the 1950s and 1960s." *Journal of Cold War Studies* 20, no. 1: 210–40. https://doi.org/10.1162/jcws_a_00805.

Oreskes, Naomi, and John Krige, eds. 2014. *Science and Technology in the Global Cold War.* Transformations: Studies in the History of Science and Technology. Cambridge, MA: MIT Press.

Patiniotis, Manolis, and Kostas Gavroglu. 2012. "The Sciences in Europe: Transmitting Centers and the Appropriating Peripheries." In *The Globalization of Knowledge in History,* edited by Jürgen Renn, 321–43. Berlin: Max-Planck-Gesellschaft zur Förderung der Wissenschaften.

Pavlíček, Tomáš W. 2023. "Czech and Polish Mathematicians Facing the Choice between Transnational and Slavic Contacts." *Historická sociologie* no. 1: 11–25. https://doi.org/10.14712/23363525.2023.2.

———. n.d. "The Mathematical Olympiad in People's Poland and its Neighbours: Youth Encounters and the Search for New Cadres." In *Entanglements and Disentanglements: Towards a Transnational History of Poland and Its Neighbours,* edited by Stanisław Witecki. Leiden: Brill (in print).

———. 2022. "Memento obětí totalitních režimů jako motor kariéry, nebo závazek kontinuity vědecké práce" [Memento of victims of totalitarian regimes as a career boost or a commitment to the continuity of scholarly work]. *Kultúrne dejiny* 13, Supplementum: 128–52. https://doi.org/10.54937/kd.2022.13.Supp.128-152.

———. 2020. "Rozhovor s prof. Jaroslavem Kurzweilem" [Interview with prof. Jaroslav Kurzweil]. *Pokroky matematiky, fyziky a astronomie* 65, no. 2: 90–117.

Pavlíček, Tomáš W., Martin Šolc, and Petra Hyklová. 2024. *Astronomers behind the Iron Curtain: The First Postwar Generation in Czechoslovakia.* Prague: Matfyzpress.

Prečan, Vilém, and Miloslava Slavíčková, eds. 2016. *Bez Františka Janoucha ani moře nešplouchá: Františku Janouchovi k narozeninám* [František Janouch, influential and essential: In honor of František Janouch on his birthday]. Prague: Československé dokumentační středisko.

Semadeni, Zbigniew. 1960. "Banach Spaces Non-Isomorphic to Their Cartesian Squares. II." *Bull. Acad. Polon. Sci. Sér. Sci. Math. Astr. Phys.* 8: 81–84.

Steinhaus, Hugo. 1959. "Matematyka wczoraj i dziś" [Mathematics then and now]. *Zeszyty Naukowe Uniwersytetu Wrocławskiego im. B. Bieruta, Seria B. Zeszyty Naukowe* no. 2: 5–19.

———. 2002. *Wspomnienia i zapiski* [Memoirs and notes]. 2. Wrocław: Centrum Steinhausa—Hugo Steinhaus Center (HSC) and Oficyna Wydawnicza ATUT.

Stockton, Julianna Connelly. 2012. "Mathematical Competitions in Hungary: Promoting a Tradition of Excellence & Creativity." *The Mathematics Enthusiast* 9, no. 1: 37–58.

Surman, Jan. 2022. "Re-Connecting Central European Science: An Introduction." In *Science Interconnected: German–Polish Scholarly Entanglements in Modern History*, edited by Jan Surman, in cooperation with Friedrich Cain, Marcin Dolecki, Maciej Górny, Peter Haslinger, Adam Kożuchowski, Katharina Kreuder-Sonnen, Ewa Manikowska, and Thomas Strobel, 1–36. Marburg: Verlag Herder Institut.

Surman, Jan. 2022. "Discussion with Dietlind Hüchtker, Katrin Steffen, Jan Surman, Joanna Wawrzyniak: Stereotypes, Imageries and Hegemonies: Race, Gender and History of Science in East Central Europe." In *Science Interconnected: German–Polish Scholarly Entanglements in Modern History*, edited by Jan Surman, in cooperation with Friedrich Cain, Marcin Dolecki, Maciej Górny, Peter Haslinger, Adam Kożuchowski, Katharina Kreuder-Sonnen, Ewa Manikowska, and Thomas Strobel, 327–40. Marburg: Verlag Herder Institut.

Surman, Jan, and Tomáš W. Pavlíček. 2024. "The Formal and the Informal in the History of Socialist Scholarly Interconnectedness in East Central Europe." *NTM Zeitschrift Für Geschichte Der Wissenschaften, Technik Und Medizin* 32, no. 2: 137–66. https://doi.org/10.1007/s00048-024-00387-0.

Švejdar, Vladimír. 2018. "Modern Czech Logic: Vopěnka and Hájek, History and Background." *Journal of Applied Logics—IfCoLoG Journal of Logics and Their Applications* 5, no. 6: 1261–71.

Těšínská, Emilie, ed. 2019. *Československá jaderná a částicová fyzika: Mezi SÚJV a CERN. Pamětnická ohlédnutí za českou a slovenskou spoluprací se Spojeným ústavem jaderných výzkumů Dubna (SJÚV) a Evropskou organizací pro jaderný výzkum (CERN)* [Czechoslovak nuclear and particle physics: between SÚJV and CERN. Memoirs reflecting on Czech and Slovak cooperation with the Joint Institute for Nuclear Research in Dubna (SÚJV) and the European Organization for Nuclear Research (CERN)]. Prague: Ústav pro soudobé dějiny AV ČR, v. v. i. a České vysoké učení technické, Ústav technické a experimentální fyziky.

Thompson, Paul. 2003. "The Voice of the Past: Oral History." In *The Oral History Reader*, edited by Alistair Thomson and Robert Perks, 21–22. New York: Routledge.

Thomson, Alistair. 2006. "Four Paradigm Transformations in Oral History." *Oral History Review* 34, no. 1: 49–70.

Toposym. "First Symposium on General Topology." Retrieved 12 July 2024 from http://www.toposym.cz/1961.html.

———. "Second Symposium on General Topology." Retrieved 12 July 2024 from http://www.toposym.cz/1966.html.

Wikipedia. "Zbigniew Semadeni." Retrieved 31 March 2023 from https://pl.wikipedia.org/wiki/Zbigniew_Semadeni.

Zahradníček, Tomáš. 2011. *Polské poučení z pražského jara: Tři studie z dějin politického myšlení 1968–1981* [Poland's learning from the Prague Spring: three studies from the history of political thought, 1968–1981]. Prague: Ústav pro soudobé dějiny AV ČR.

Zdravkovska, Smilka, and Peter L. Duren, eds. 2007. *Golden Years of Moscow Mathematicians*, 2nd ed. History of Mathematics 6. Rhode Island: American Mathematical Society.

Zrull, Lindsay Smith. 2021. "Women in Glass: Women at the Harvard Observatory during the Era of Astronomical Glass Plate Photography, 1875–1975." *Journal for the History of Astronomy* 52, no. 2: 115–46. https://doi.org/10.1177/00218286211000470.

Zysiak, Agata. 2015. "Modernizing Science: Between a Liberal, Social, and Socialistic University—the Case of Poland and the University of Łódź (1945–1953)." *Science in Context* 28, no. 2: 215–36.

———. 2016. *Punkty za pochodzenie: Powojenna modernizacja i uniwersytet w robotniczym mieście* [Points for social origin: postwar modernization and the university in the working class city]. Kraków: Nomos.

Oral History Sources (Interviews with Scientists Conducted by Tomáš W. Pavlíček)

Interview with Zbigniew Semadeni, 12 October 2023. Retrieved 25 January 2026 from https://entanglements.ihpan.edu.pl/interview/mama-byla-jedynym-lekarzem-miala-sanitariuszki-ktore-byly-harcerkami-zbigniew-semadeni/.

Interview with Wiesław Żelazko and Hanna Stande Żelazko, 27 June 2022. Retrieved 25 January 2026 from https://entanglements.ihpan.edu.pl/interview/prof-wieslaw-zelazko-i-dr-hanna-stande-zelazko-wywiad-1/.

Interviews with Zofia Adamowicz, 14 July 2022, 26 March 2024. Retrieved 25 January 2026 from https://entanglements.ihpan.edu.pl/interview/kobiety-w-ogole-nie-mialy-latwo-rozmowa-z-matematycka-zofia-adamowicz/.

Interviews with Roman Duda, 21 March 2024, 19 April 2024, 4 June 2025. Retrieved 25 January 2026 from https://entanglements.ihpan.edu.pl/interview/dzisiaj-zamiast-wspolpracy-obserwuje-konkurencje-o-granty-o-punkty-o-stypendia-wyjazdy-rozmowy-z-matematykiem-romanem-duda/.

Interview with Dina Štěrbová, 10 March 2023. Retrieved 25 January 2026 from https://entanglements.ihpan.edu.pl/cs/interview/dina-sterbova-a-nesnaze-v-horach-i-vede/.

Interview with Jaroslav Kurzweil—see Bibliography (Kulawiaková and Pavlíček).

Interview with Andrzej Sołtysiak—see Bibliography (Kulawiaková and Pavlíček).

Interview with Václav Chvátal, 2 May 2023. Retrieved 25 January 2026 from https://entanglements.ihpan.edu.pl/interview/vasek-chvatal-graphs-and-postcards-from-erdos/.

Interview with Kamila Bendová, 9 September 2024. Retrieved 25 January 2026 from https://entanglements.ihpan.edu.pl/cs/interview/do-polska-jsem-nikdy-nemohla/.

Interview with Pavel Pudlák, 3 April 2023. Retrieved 25 January 2026 from https://entanglements.ihpan.edu.pl/interview/learn-more-about-philosophy/.

Interview with Aleš Pultr, 10 January 2023. Retrieved 25 January 2026 from https://entanglements.ihpan.edu.pl/cs/kdyz-byste-z-matematiky-zapomnel-vsechno-co-jste-mel-k-maturite-ale-naucilo-by-vas-to-ze-se-veci-museji-nejakym-zpusobem-dokazovat-ze-si-nedate-neco-jen-tak-nabulikovat-tak-je-to-dobry/.

CHAPTER 6

Roads (Not) Taken

Practices of Cooperation between Polish and Czechoslovak Sociologists under State Socialism

Kornelia Kończal

In both Poland and Czechoslovakia, the history of sociology between 1945 and 1989 was marked by similar processes. The discipline was institutionalized in both countries after World War I, drawing on the intellectual capital that its leading figures had accumulated while studying and researching outside their native realm. Ironically, the cosmopolitan outlook and shared intellectual inspirations of sociologists from Poland and Czechoslovakia had little impact on bilateral cooperation in the interwar period: national duty proved more important than neighborly or any form of international cooperation. During World War II, sociology in the Polish and Czech lands was discontinued due to the closure of universities, a fate that did not befall Slovakia, where academic institutions remained open. Although it was revived in Poland and Czechoslovakia after the end of the war, sociology was officially banned under Stalinism for being a bourgeois pseudoscience—as it was in the entire Eastern Bloc. In Poland, the revival of the discipline began in 1956, whereas the corresponding process was only launched in Czechoslovakia in 1964 and then curtailed again after the suppression of the Prague Spring in 1968. As a result, relations between Polish and Czechoslovak sociologists under state socialism were shaped by a structural asymmetry: while sociologists in Czechoslovakia regarded the work of their Polish colleagues as an important intellectual stimulus and a source of access to Western research, Polish scholars showed little interest in the work of their southern neighbors.[1]

By paying particular attention to the interplay of individuals, institutions and ideology, I address the trope of asymmetry in three steps. The first section shows that before the emergence of Polish–Czechoslovak asymmetry, relations between Polish and Czechoslovak sociologists were actually characterized by an inverse dynamic. A closer look at the well-known but little-studied case of Jiří Kolaja (1919–2003) reveals how knowledge of both Polish society and Polish sociology was produced by a Czech-born scholar who pursued an academic career in North America. The second section focuses on the institutional and intellectual setting in which cooperation between Polish and Czechoslovak sociologists took place during the golden age of the discipline in both countries (and beyond), that is, in the 1960s. In doing so, this section sets the stage for the case studies discussed in the final section, which revolve around the different trajectories of the most popular Polish sociologists in Czechoslovakia: Jerzy Wiatr (*1931), Zygmunt Bauman (1925–2017), and Jan Szczepański (1913–2004). While the second section is devoted to bilateral cooperation in the strict sense of the word, sections one and three go beyond the Iron Curtain. Taken together, examples of Polish–Czechoslovak encounters examined in this chapter demonstrate that the most fertile frameworks for cross-border cooperation were to be found in settings driven by both individual and institutional receptivity.

Methodologically, this chapter opens a praxeological perspective on Polish–Czechoslovak encounters in sociology. Rather than treating the history of the social sciences as a linear progression of ideas, the praxeological lens focuses on the everyday practices of researchers, including data collection, theorizing, collaborating, and publishing, by taking into account the wider institutional, cultural and material contexts that enable or constrain these practices. In other words, the praxeological approach highlights the multiple and situated activities through which concepts, tools, and theories are produced, transmitted and transformed.[2]

Typologically, encounters between Polish and Czechoslovak sociologists can be divided into two types of practices: apparent activities and inspiring interactions. While apparent activities, as defined by Jan Lutyński, refer to formalized actions that serve symbolic or performative functions rather than substantive, practical outcomes,[3] inspiring interactions highlight the (often informal) exchange of ideas and experiences that foster genuine intellectual interaction and innovation. Since formal activities can sometimes provide the framework for meaningful informal exchanges, and informal interactions may carry symbolic weight or be influenced by institutional contexts, the distinction between these two types of practice is at best ideal-typical. Nevertheless, a closer look at the interplay between structure and agency in shaping cooperation between Polish and Czechoslovak sociologists helps to reveal the limits of institutional initiatives and the crucial role of personal relationships in overcoming material and ideological barriers to enable impactful intellectual exchange.

Empirically, this chapter draws on a growing body of literature on the history of sociology in Central and Eastern Europe, personal documents written by sociologists, and archival material produced by institutional actors in People's Poland and Czechoslovakia.[4] Before entering into the core analysis, it is worth reflecting briefly on the challenges and limitations that operated behind the scenes of my own research practice. First, for whatever reason, Czechoslovak sociologists documented their work much more thoroughly than their Polish counterparts. As a result, although the field of sociology in People's Poland was much broader than in Czechoslovakia, it is far less well documented.[5] Second, as in many other countries, the private papers of sociologists from Poland and Czechoslovakia were rarely transferred to proper archives. For example, Jan Szczepański's papers are easily accessible in a university archive, while those of Jiří Kolaja are in private hands (and have remained inaccessible to me),[6] and the fate of the personal papers of many other sociologists mentioned in this chapter is unknown. Third, while certain aspects of sociological practice in postwar Poland and Czechoslovakia were closely surveilled and thoroughly documented by security forces, others remain inaccessible to historians. For example, in contrast to the prewar period, much cross-border communication after 1945 was conducted by telephone rather than by mail, leaving fewer written records and creating significant gaps in the archival trace of the individual and institutional interactions.[7] Fourth, some Polish–Czechoslovak encounters that took place outside of official channels remain largely undocumented because their initiators invested their limited resources in the practice of cooperation rather than in that of documentation. A series of Polish–Czechoslovak seminars on urban sociology organized in the 1980s by Ewa Kaltenberg-Kwiatkowska (1935–2015), who was recognized as a facilitator of Polish–Czechoslovak encounters as early as 1956,[8] is a case in point.[9] Finally, there were a number of common sociological interests in Poland and Czechoslovakia that, as far as one can tell, did not produce any relevant cross-border encounters. The most puzzling example is the socioeconomic transformation that took place in both countries as a result of the expulsion of the Germans around 1945 and the repopulation of the post-German territories in the second half of the 1940s. Did Polish and Czechoslovak sociologists working on these politically sensitive issues not share their experiences? Or did they do so, but did not bother to document it? Or did they not document it because they were not allowed to? The uneven availability of sources, the fragmentary nature of personal documents, and the selective preservation of archival material have shaped both the scope of this chapter and the interpretations it offers, highlighting the challenges of reconstructing transnational academic practices in any context in which much of the historical record remains incomplete or inaccessible.

Transatlantic Encounters

In the summer of 1957, Jiří Kolaja, a sociologist from Talladega College in Alabama, the oldest black liberal arts college in the United States, spent eight weeks in Łódź, popularly known as the Polish Manchester, conducting sociological research in a textile factory. Kolaja may have been making an implicit comparison with his native Brno, known as the Moravian (or Austrian) Manchester, when he wrote: "The city of Lodz is not beautiful; being located on a plain and without any major river, the city does not have any dominant landmark." The Marchlewski factory employed around 8,000 people, mostly women, and was "surrounded by rather shabby red-brick houses and fields. Earlier, at one side of the factory block there was a Jewish part of Lodz, partially destroyed by Nazis. On the whole, a newcomer from the West is depressed by the neighborhood."[10] Indeed, the multiethnic and vibrant community that had characterized prewar Łódź was irrevocably gone. Once home to the second largest Jewish community in Poland and a vibrant German minority, Łódź had become a grim reminder of the devastating effects of war, the Holocaust, and expulsion.[11]

Kolaja's interest, however, was not in the socioeconomic transformation and postwar reconstruction of the destroyed city but in the functioning of its workers' councils (*rady robotnicze*), a new institution introduced in October 1956, which related to the shop committees (*rady oddziałów*) that had existed in Poland immediately after the war and had some similarities with the workers' councils introduced in Yugoslavia in 1950. The main aims of the new body, which would become one of the symbols of the destalinization of Poland, were to increase the participation of Polish workers in the management of factories, to improve productivity, to redress workers' grievances and to resolve industrial disputes. In studying the Polish case, Kolaja hoped to contribute to the new field of industrial sociology: "For with its Western cultural ties, coupled with its present involvement in the Soviet political and economic system, Poland's situation may furnish us with clues to the effectiveness of communist theory in improving worker happiness and efficiency in particular, and to the problem of workers' participation in management in general."[12] Kolaja's book *A Polish Factory: A Case Study of Workers' Participation in Decision-Making* was published by the University of Kentucky Press in 1960. In his foreword, William Foote Whyte (1914–2000), Kolaja's advisor at the New York State School of Industry and Labor Relations at Cornell University and the pioneer of participant observation in sociology, wrote: "this is the first time we have had a field research report on human relations in a factory in a communist society."[13] For twenty years, Kolaja's book was the only original English-language monograph based on firsthand research shaping "the

American sociological imagination about Poland" (as defined by Michael Kennedy).[14] Despite the transatlantic distance between Kolaja's object of study in Łódź and his doctoral dissertation defended at Cornell University in 1959, the research for and reception of *A Polish Factory* provides interesting insights into the history of Czechoslovak–Polish encounters in sociology—both in person and on paper.

Jiří Kolaja was born in 1919 to a Czech family in the Czechoslovak city of Brno,[15] where he studied sociology, psychology, and philosophy. In 1947, he defended his PhD on "Sociological and Psychological Problems of the Film." Kolaja's academic advisor was Inocenc Arnošt Bláha (1879–1960), one of the founders of academic sociology in interwar Czechoslovakia. His research stay in Sweden in early 1948 coincided with the communist coup d'état.[16] Kolaja decided not to return to Czechoslovakia. In moving to the United States, he was one of 50,000 Czechoslovak citizens who left the country. Initially, Kolaja studied sociology at the University of Chicago. In the early 1950s, he worked as an assistant to the sociologist Pitirim Sorokin (1889–1968) and was a research fellow at the Russian Research Center at Harvard University (1951–1952). Kolaja's journey from Brno to Harvard can be considered part of the international circulation of Durkheimian sociology: Kolaja's mentor Bláha had studied with Emile Durkheim in Paris in the years 1908–9 and later promoted his sociology in interwar Czechoslovakia, while Sorokin advocated Durkheimian ideas in Russia and subsequently in the United States.[17] In the following years Kolaja benefited from Sorokin's letters of recommendation when successfully applying for positions in sociology across North America, including Cornell University (1952–53), Talladega College in Alabama (1954–58), the University of Kentucky (1958–66), McMaster University in Hamilton, Canada (1966–68), the State University of New York (1968–70), and the West Virginia University in Morgantown, West Virginia, which he joined in 1970.[18] Thus, when arriving in Poland in 1957, Kolaja was not a Czechoslovak but an American citizen. As he put it: "As one could expect, my role as a researcher from America was unusual and produced some distrust. Under Stalin, as Poles would say, nobody but the secret police had ever interviewed the people in the shop. So far, they had never been exposed to a sociological inquiry. Also, my United States passport and my Czechoslovak accent in my Polish speech contributed to general surprise and wonder as to my identity."[19]

Kolaja's engagement with Poland went far beyond academic pursuits, and his research trip to Łódź in 1957 was not his first visit to the country. Reflecting on his lifelong connection to Polish culture, he wrote: "Since my childhood, I have been much interested in Poland, its history, literature, and people."[20] This deep-seated interest was shaped by his father, a lecturer in Polish at Masaryk University in Brno, which allowed Kolaja to learn Polish during his formative years.[21] In 1946, while studying sociology in Brno, he traveled to Poland, where he established connections with prominent sociologists, including Stanisław Ossowski (1897–1963).[22] The following year, he spent more than two months in Poland, aiding the World Student Relief in distributing assistance to Polish universities.[23] These experiences also had an intellectual impact, inspiring Kolaja to write reviews of Polish books for *Sociologická revue*, Czechoslovakia's leading pre-Marxist sociological journal.[24]

When he arrived in Poland in the summer of 1957 from the United States, he found a radically different reality. The Cold War division of the continent was in full swing, Poland's postwar reconstruction was largely accomplished, and the death of Joseph Stalin in 1953 and the following Khrushchev Thaw had enabled a partial liberalization throughout the entire Eastern Bloc. In the second half of 1956, Poland was at the forefront of change. Importantly for Kolaja, sociology, which had been banned from Polish universities as "bourgeois pseudoscience" in 1948, was re-established and restrictions on foreigners visiting the country were lifted. Kolaja applied to the Polish Ministry of Light Industry for permission to undertake research in Łódź. His application was sponsored by the newly created Institute of Philosophy and Sociology of the Polish Academy of Sciences in Warsaw and accepted by the ministry. "The management had already been informed of our coming by a letter from the ministry in Warsaw, and was consequently most cooperative."[25]

Upon arrival in the factory on 11 July 1957, Kolaja noted: "I immediately became an object of great attention. Those who stood around me, mostly women, started to express their complaints to me."[26] Most criticisms revolved around low pay, bad working conditions, and disrespect or distrust of the management. For forty-two working days or nights, Kolaja made personal observations and gathered data through interviews and questionnaires—both in the factory and beyond. However, his sudden appearance and constant presence on the shop floor defied easy categorization: "The roles ascribed to me at first were 'good' or 'neutral,' such as 'American,' 'Professor,' or 'Czech,' as well as 'bad,' such as 'Western Spy,' 'American Communist,' or 'Management's Ally.'"[27] Kolaja cleared up some misunderstandings about his identity and work by writing personal letters to the workers and by asking a Polish colleague who assisted his research for help. When doubts about him had been dispelled, it paved the way for deeper interactions: "As soon as I had become accepted by the workers and management personnel, I also met them outside the factory, frequently being invited to their homes."[28]

According to Kolaja's observations, the workers of Łódź showed little interest in official ideology and production problems and saw in their superiors an opposing class. The creation of workers' councils did little to change this state of affairs. The real reason for this failure was threefold. First, there was a dichotomy between the management and the weavers: while the management tended to define the function of the factory in terms of the collective purpose of production, the weavers saw the factory from an individualistic point of view, that is, they worked in order to earn a living. In other words, the transformation of the means of production into collective property did not automatically imply that, as suggested by Marxian theory, an individual would identify with the factory. Second, there was a great deal of workers' distrust of their superiors: although management asked workers to participate in decisions, in fact, no decisions were made with their participation, and the memory of Stalinist suppression was still fresh. At the same time, the discrepancy between the verbal definition of the situation and the real situation weakened the messages sent by the management to the weavers. Third, representatives of the management, the party, the union, and increasingly also the workers' council spoke the same language and addressed the weavers in the same way. From the workers' point of view, it was not their own organization.

Kolaja was right, insofar as Łódź was not beautiful. And yet it had a magnetic effect on left-wing intellectuals and academics in postwar Poland, who were drawn to the newly founded and rapidly growing university, which became a hub of progressive thought and a center for the intellectual reconstruction of the country.[29] The Department of Sociology was part of this dynamic. In choosing Łódź for his fieldwork, Kolaja was able to consult with some of Poland's leading sociologists, or at least with their students. Kolaja's research in the factory was assisted by Józef Kądzielski (1929–1990), a twenty-eight-year-old sociologist and student of Józef Chałasiński (1904–1997), a major figure in the Łódź school of sociology and rector of the university (1949–1952), who had possibly known Kolaja since the early postwar period.[30] Working with Kądzielski, Kolaja was not only able to discuss sociology, but also benefit from his partner's connections. Kądzielski had worked in the factory as a weaver before entering the university, as did his parents: "his mother was still living in one of the red-brick apartment houses. Therefore, it was to our advantage to settle at this plant . . . because of my friend's familiarity with the job and with the people."[31] Kolaja contends in the book that his research was supported by "other Polish sociologists in Lodz" who gave him "spontaneous help and advice"[32] without mentioning names. It is safe to assume that one of the local sociologists advising him was Jan Lutyński (1921–1988), who, at that time, was Chałasiński's assistant.[33] According to the Polish secret service, in the summer of 1957 Kolaja lived in Lutyński's apartment in a well-known townhouse on Uniwersytecka Street, a hub for university faculty at all career stages and an ideal setting for socializing and networking.[34]

As Kolaja pointed out, his interpretation of the workers' behavior was his own, and "neither Mr. Kądzielski nor anyone else in Poland should be held responsible for it. In fact, there was some disagreement in the definition of the problem of inquiry between my Polish colleague and myself during our cooperation in research." He considered this tension "natural because of the difference between the a priori dialectic Marxist framework on the one hand and the empirical structural-functional sociological approach on the other."[35] Was this statement just a description of belonging to two completely different communities of sociological practice? Was it an attempt to shield Kądzielski from potential trouble at the university? Or was it a subtle hint that the relationship between Kolaja and Kądzielski had deteriorated after or (perhaps) during their fieldwork in the factory? Whatever happened, Kądzielski's field report, which was to be published in Polish, did not appear,[36] and his review of Kolaja's book, written for *The Polish Sociological Bulletin*, a periodical published in Warsaw in the English language and thus aimed at a foreign audience, proved to be highly critical.[37]

Kądzielski's review was not an exception: Whereas Western reviewers (including American expatriates from Poland and Czechoslovakia) were enthusiastic about Kolaja's book and referred explicitly to his linguistic versatility and close ties to Poland,[38] sociologists from Poland (and Czechoslovakia) criticized the author's "methodological failures"[39] and the "questionable interpretation of facts and conclusions"[40] of the "American sociologist."[41]

The review of Kolaja's book written for *Social Problems* by the main protagonist of the third section of this chapter, Jan Szczepański, from 1952 professor of sociology and between 1952 and 1956 rector of the University of Łódź, is particularly insightful.[42] With regard to the "a priori dialectic Marxist framework" allegedly shaping the minds of Polish sociologists, as opposed to his own "structural-functional sociological approach," as Kolaja put it, Szczepański replied: "I believe that the Polish sociologists in Lodz are fully aware of the methodological requirements of empirical studies, and we would ask rather for recognition of the broader context of the studied phenomena (precisely Malinowski's functional approach) than for some a priori framework."[43] Szczepański further explained: "I think that if Dr. Kolaja had been less concerned to show the inadequacy of Marxist theory and had concentrated more on a sociological interpretation of the gathered materials, his book would be more valuable."[44] Missing from Kolaja's work was not only reflection upon the postwar reconstruction of Poland and deep changes in the composition of the working class in the early postwar period but also the comparative perspective. Szczepański asked in a slightly whataboutist tone: "Why did the *Mitbestimmung** fail in Western Germany if Marxist theory should be held the factor mainly responsible for failure of the workers' council in Poland?"[45] The virtual conversation between Szczepański and Kolaja testifies to two of the main problems that Polish (and Czechoslovak) sociologists had with *A Polish Factory*: the bigger picture was missing, including the destructive effects of wartime occupation, and they were apparently offended by Kolaja's general suspicion of being captured by the Hegelian bite. Needless to add that while *A Polish Factory* was reprinted in the United States in 1973, it has not been translated into Polish (or Czech or Slovak) to this day.

Ironically, Szczepański wrote his critical review of Kolaja's book during his seven-month stay at the Center for Advanced Study in the Behavioral Sciences at Stanford University in late 1960 and early 1961.[46] The author and the reviewer met in the United States, and it is safe to assume that they continued to meet in the following years at international conferences.[47] In 1961, Kolaja wrote a highly appreciative review of Szczepański's books that had been published at that time in Polish,[48] and continued reviewing his publications in the years to follow.[49]

Finally, whereas Kolaja sent a reply to the editor of the *American Journal of Sociology* that published a generally positive review of this book authored by a US sociologist who only criticized its minor shortcomings in order to explain his methodological decisions,[50] he did not engage in public conversation with the much harsher disapproval voiced in US and Central and Eastern European journals by sociologists based in the Eastern Bloc—including Szczepański. What is more, Kolaja wrote a very polite and personal letter to Szczepański thanking him for the review.[51] The conclusion that might be drawn from this overview of the reception of Kolaja's book is twofold. First, there is no doubt that despite extensive networks in Central and Eastern Europe, Kolaja's main community of practice was the field of academic sociology in the United States. Second, it seems that the ideologically inspired criticism of Kolaja's book voiced by his Polish colleague may have been more ritual than real. Both figures, the US social scientist treating Eastern Europe as a mere object of study and the Eastern European Marxist paying lip service to ideology, were not isolated cases but rather representative of broader trends and patterns in the intellectual dynamics of the Cold War era.

On the institutional level, however, Kolaja's anti-Marxist interpretation of the Polish workers' councils had serious ramifications. All his requests for an extended research stay in Poland were refused and his short visits to Poland—both personal and private—were closely monitored by Polish security forces.[52] Strikingly, the assessment of the threat that Kolaja apparently posed to public security in Poland did not reflect his research output but internal developments in the country. For example, in 1968, when the authorities launched an antisemitic campaign and forced thousands of Polish citizens of Jewish origin to leave the country, the security forces char-

Figure 6.1. United they stand: Jan Szczepański (in the middle) and Jiří Kolaja (on the right) in the United States, late 1960/early 1961. Jan Szczepański's papers, Archives of Polish Sociology at the Nicolaus Copernicus University Library in Toruń, published with permission.

acterized Kolaja as "probably of Jewish nationality."[53] In the early 1970s, when the government faced mounting pressure to address the growing discontent among the working class after a wave of mass strikes against low wages, poor working conditions, and rising prices, the security forces argued that *A Polish Factory* approaches the workers' self-government "in a fragmentary way," "promotes American sociology," and "does not take into account the social and historical context, which results in a false interpretation."[54] In 1976, as Poland's opening to the West gained momentum, the number of potential external enemies of the socialist order continued to grow. Polish security forces followed Kolaja at the 6th International Congress of Rural Sociology in Toruń and his extended networking activities caught the eye of the Polish agents. In tune with the "tonnage ideology," they noted that Kolaja "established contact with 150 congress participants from socialist countries. He handed out business cards to the delegates he selected, asking for correspondence. He recorded the characteristics of the researchers and activists he met."[55] In 1978, Polish security forces had no doubts about Kolaja's alleged "ties to secret services" in the United States[56] and declared him an "unwanted person," which equated to a ban on entering Polish territory (and publishing in the Polish language).[57]

Deprived of the opportunity to conduct research in Poland, Kolaja continued to write sociological books on Eastern Europe but did not produce any further studies of Polish society. A copy of *A Polish Factory* somehow found its way into the library of the University of Łódź, but it is difficult to say whether it was part of the sociology syllabus.[58] At the same time, Kolaja continued to shape the global perception of Polish sociology by reviewing the work of his Polish colleagues for the leading sociological journals in English (and Spanish), publishing on the state of the field in Poland, and writing about the legacy of selected Polish sociologists.[59]

Kolaja's case is both exceptional and emblematic. It is exceptional because it does not fit into the asymmetrical relationship between Polish and Czechoslovak sociology, and it is safe to assume that this is because Kolaja left Czechoslovakia in 1948. This departure allowed him to establish an uninterrupted academic career, freed him from having to reconcile intellectual curiosity with ideological constraints, and enabled him to engage more freely with international scholarly communities and research agendas. At the same time, Kolaja's trajectory is emblematic of many sociologists born in Central and Eastern Europe who left or did not return to their countries of origin for political reasons. In Poland, this was the case of Feliks Gross (1906–2006), Aleksander Hertz (1895–1983), and Florian Znaniecki (1882–1958), who remained in the United States after the end of the war, Stanisław Andrzejewski (or Stanislav Andreski) (1919–2007), who did not return from Great Britain, Alicja Iwańska (1918–1996) and Jerzy Zubrzycki (1920–2009), who settled in the United States and Australia respectively, Zygmunt Bauman and Maria Hirszowicz (1925–2007), who were forced to emigrate after the antisemitic campaign of 1968 and continued their work at British universities, and a number of established sociologists who left the country in the 1970s including Aleksander Matejko (1924–1999) and Adam Podgórecki (1925–1998). In Czechoslovakia, Kolaja was part of a larger group of sociologists who either left the country or stayed abroad after the communist coup in 1948. Among them were Ivan Gaďourek (1923–2013), Richard Jung (1926–2014), Ferdinand Kolegar (1929–2007), Otakar Machotka (1899–1970), Jiří Nehněvajsa (1925–1966), and Zdeněk Ullrich (1901–1955), who became professors of sociology across the globe, while Josef Macek (1887–1972), Karel Maiwald (1902–1979), Emil Schoenbaum (1882–1967), and Zdeněk Suda (1920–2005) shifted their focus to applied sociology or journalism. In the late 1960s and 1970s, the list of sociologists from Czechoslovakia who decided to continue their careers abroad included, among others, Bedřich Baumann (1922–1984) and Miroslav Disman (1925–1996) who pursued their academic careers in Canada, and Václav Lamser (1924–1986) and Ilja Šrubař (*1946) who became professors at West German universities in Bielefeld and Nürnberg-Erlangen respectively. The dual nature of Kolaja's experience not only illustrates the costs of political exile and the opportunities it might offer,[60] but also shows that the geography of Polish–Czechoslovak encounters in sociology during socialism went far beyond the borders of both countries.

Unequal Neighbors

The 1960s were the golden age of sociology in both Poland and Czechoslovakia, though its scale and scope differed between the two countries. In Czechoslovakia, sociology reentered the sphere of academic disciplines in 1964, marking what some scholars described as a revival, renaissance, and revitalization,[61] while others referred to it as a sociological explosion.[62] In Poland, the 1960s represented the peak of a continued development that had been relaunched in 1956. In both countries, the year 1968 brought an essential caesura, albeit for different reasons and with different consequences. In Poland, the effects of the antisemitic campaign were felt primarily on the individual level. In Czechoslovakia, the "normalization" that followed the suppression of the Prague Spring led to systematic purges, with some sociological institutions being closed, resulting in a severe setback for the discipline as a whole.[63]

Beyond these major differences, the golden age of sociology in both countries was marked by some structural similarities typical of the entire Eastern Bloc. In institutional terms, the field of sociology consisted of four partially overlapping settings: the Institute of Sociology at the respective academies of sciences, sociology departments at universities, and the Sociological Association (which in Czechoslovakia comprised the Czech Sociological Society and the Slovak Sociological Society). The intensity of political control and public funding varied between these subsystems. The Academy had the highest degree of political control, but also the most financial support, which enabled it to carry out large-scale research projects. The sociological associations, on the other hand, enjoyed the greatest political independence, but also received the least funding. The universities, somewhere in between, were under moderate political control, but also received limited financial resources because their main role was teaching. Lastly, a certain number of (applied) sociologists worked—in a relatively unrestricted way—in public service and state enterprises.[64] In ideological terms, most Polish and Czechoslovak sociologists adhered to Marxism, which was more "superficial, fragmented and hybrid" in Poland than in

Czechoslovakia,[65] and considered Western sociologies as their significant other.

And yet, to put it in praxis–theoretical terms: Polish and Czechoslovak sociologists belonged to very different communities of practice. To begin with, the identification with historical materialism was much higher among Czechoslovak than among Polish sociologists. The reasons for this difference were twofold. In contrast to Poland, there was virtually no continuity between prewar and postwar sociology in Czechoslovakia because—as was shown in the previous section—a significant proportion of critical academics left the country in 1948 and those who remained in the country were evicted from universities or were simply too old to have a real impact on the development of the discipline. The sociological landscape in postwar Czechoslovakia was thus shaped by the "reform generation" (Michael Voříšek) of scholars who were largely aligned with the communist regime and Marxist ideology, which left little room for the continuation of the more diverse, critical sociological traditions. This rupture hindered the development of a diverse sociological community and resulted in a more rigid, state-controlled intellectual environment, in stark contrast to Poland, where prewar traditions were more successfully maintained and integrated into the postwar context. Second, unlike in Poland, the development of sociology in Czechoslovakia was more closely supervised by the Communist Party and primarily carried out by its members. Suffice it to recall that the Scientific Board of Philosophy and Sociology at the Czechoslovak Academy of Sciences, responsible for coordinating the development of sociology, had no counterpart in Poland, and while the Polish Sociological Association was an organization autonomous from the government, its Czechoslovak counterpart was affiliated with the Academy of Sciences.[66] Lastly, there was a great imbalance in the composition of the two sociological communities. Whereas sociologists in Czechoslovakia were mainly concerned with teaching, Polish sociologists invested much more intellectual energy and resources in research.

As a result, there was a striking Polish–Czechoslovak asymmetry in sociological relations in the book market. Most handbooks for sociology published in Czechoslovakia during the 1960s were translations from Polish. With five books translated into Czech and Slovak, Jerzy J. Wiatr[67] was the most successful Polish sociologist in Czechoslovakia, followed by Zygmunt Bauman[68] and Jan Szczepański[69] with four and three translations respectively. In addition, manuals by Stefan Szostkiewicz (1924–1986) and the aforementioned Maria Hirszowicz were translated into Slovak,[70] as well as Czech translations of a number of sociological studies on more specific topics by Antonina Kłoskowska (1919–2001), Mikołaj Kozakiewicz (1923–1998), Aleksander Matejko (1924–1999), Adam Podgórecki (1925–1998),

Włodzimierz Wesołowski (1929–2020) and Jan Zieleniewski (1901–1973). In all, twenty-one books by Polish sociologists were translated into Czech and Slovak in the 1960s, more than into Russian and all Western languages combined (not a single one of them, however, was authored by Stanisław Ossowski [1897–1963], Józef Chałasiński [1904–1979] or any other prewar sociologist).[71] As a Czech sociologist put it in retrospect: "The translation of Polish works and their study not only resulted in a knowledge of Polish sociology, but also introduced representatives of sociology from the West whose works were not easily accessible in our country."[72]

In contrast, the first monograph by a Czechoslovak sociologist to be translated into Polish was the comparative study *Urbanization in Socialist Countries* (1977) by Jiří Musil (1928–2012), which appeared in Polish only in 1984, that is, four years after its translation into English.[73] Not a single Polish translation was published of works by the most prominent sociologists in socialist Czechoslovakia and authors of several monographs, namely, Jaroslav Klofáč (1921–1981), head of the Czechoslovak Sociological Association, and Pavel Machonin (1927–2008), head of the Scientific Board of Philosophy and Sociology. The only Czechoslovak exception was the sociologist and philosopher Radovan Richta (1924–1983), whose two acclaimed books—but not single-authored monographs—were translated from Czech and Russian into Polish in 1971 and 1981 respectively.[74]

Perhaps the most striking manifestation of the asymmetry between Polish and Czechoslovak sociology was the prominent role played by Polish scholars in the International Sociological Association (ISA). Stanisław Ossowski served as vice president of the ISA from 1959 to 1962, Jan Szczepański held the presidency from 1966 to 1970, and Magdalena Sokołowska (1922–1989) later assumed the vice presidency from 1978 to 1982—distinctions that no sociologist from Czechoslovakia achieved. Ironically, the strong Polish presence in the ISA under socialism was not merely a reflection of individual accomplishments or the international prestige of Polish sociology as a whole; it also resulted from the relative weakness of sociology in Czechoslovakia and other socialist countries. In other words: the pool of non-Western candidates for such distinguished roles was notably limited.[75]

As a result, encounters between Polish and Czechoslovak sociologists were limited in several ways. First of all, throughout the 1960s the Polish population was twice that of Czechoslovakia, and the same ratio applied to the number of academics in both countries. In addition, the period in which substantial cooperation could take place between sociologists based in Poland and Czechoslovakia was limited to the brief window between 1964 and 1968. Finally, the relations between Polish and Czechoslovak sociologists during this period were shaped by a clear division of labor. The development of a discipline on the periphery typically requires strong models to emulate, and this need becomes even more critical in the context of generational rupture, as was the case in Czechoslovakia. Paradoxically, the Soviet Union could not serve as a model for the reconstruction of sociology in Czechoslovakia because the state of the discipline in both countries was quite similar. Western sociologies could not serve as models for ideological reasons. Under these circumstances, the semi-peripheral Polish sociology was the only model worth following for the peripheral sociology of Czechoslovakia.[76]

In the second half of the 1960s, sociologists in both countries were acutely aware of the fundamental differences between them. Only English-language literature garnered more reviews in sociological journals published in Czechoslovakia than works by Polish authors.[77] Czech and Slovak sociologists closely followed debates among their Polish counterparts and wrote about them extensively.[78] The number of Czechoslovak sociologists claiming knowledge of Polish was far greater than the reverse; the desire to learn from Polish sociology became a recurring theme in academic mission requests submitted by Czech and Slovak scholars to the Czechoslovak Academy of Sciences and their reports from stays in Poland often read like detailed descriptions of educational trips, whereas those by Polish sociologists visiting Czechoslovakia tended to present a picture of cultural enrichment.[79]

Despite the real existing asymmetry between Polish and Czechoslovak sociology, communication between the two academies of sciences in the second half of the 1960s was imbued with the spirit of socialist reciprocity. Although the interest of Polish sociologists to visit Czechoslovakia was much less pronounced than the eagerness of Czechoslovak sociologists to go to Poland, the problem was not addressed in any of the many bilateral research plans. On the contrary, research exchange based on an equal amount of time being spent by Polish and Czechoslovak sociologists in the neighboring country was considered the core of the cooperation.[80] The outcomes of the mutual visits were underwhelming. Despite shared interests, a common theoretical framework and similar methods in empirical studies, comparative research projects on urbanization, industrialization, and other socioeconomic processes unfolding in both countries were often proposed but ultimately left unrealized.[81] This was mainly due to the Polish lack of interest and the general lack of time.[82] As a result, the complex and time-consuming process of drafting, consulting, and approving joint research plans was a typical example of an apparent activity.

The highlight of cooperation between Polish and Czechoslovak sociologists was arguably a joint conference held in Smolenice, a small town situated in the Little Carpathians in western Slovakia, in April 1967. Organized by both academies of sciences and sociological associations, the event brought together eighteen Polish and fifty Czechoslovak sociologists to discuss potential collaborations on methodological issues and comparative studies of social structure, industrial work, and leisure. However, the organization of the conference was not without complications. One of the Polish participants, Stefan Nowak (1925–1989), had signed a letter in defense of Adam Michnik, a student at Warsaw University at the time, who was already well-known for his revisionist activities. As a result, Nowak was denied permission to travel to Czechoslovakia, and the participation of much of the Polish delegation remained uncertain until the last moment, as colleagues from Warsaw University expressed solidarity with Nowak.[83] Ultimately, the issue was resolved, and the conference went ahead as planned. For some

Figure 6.2. Smolenice Castle, postcard from the 1960s. Author's private collection.

The Polish–Czechoslovak conference took place "in the idyllic setting of a stylishly restored castle complex in the foothills of the Little Carpathians."[84]

After the first day of the Polish–Czechoslovak conference in Smolenice, Jan Szczepański observed that breakfast conversations were dominated by participants "sharing stories of the previous night's drinking and singing." The Polish delegation took particular pride in having "out-sung" their Slovak counterparts, celebrating it as an "international success." The day after, Szczepański wrote: "Dinner in the Franciscan cellar, red wine, gypsy orchestra, singing. Until midnight." However, after the conference concluded, Szczepański adopted a more reserved tone, remarking, "The behavior of some delegation members under the influence of alcohol was not always appropriate."[85]

participants, the three-day gathering in Slovakia was a chance to reconnect with familiar colleagues; for others, it was their first encounter with sociologists from the neighboring country. Despite some "difficulties in communicating between people who speak three languages" noted by one of the participants,[86] the conference was widely regarded as a success, fostering both intellectual exchange and social connections. Polish sociologists were impressed by the advancements their Czech and Slovak counterparts had made in sociological research and noted their revolutionary spirit in political terms. Plans were made to publish the conference papers and host a follow-up event in Poland in 1969.[87]

In the months that followed, a significant number of Czechoslovak sociologists actively participated in the reformist movement. Zygmunt Bauman recalled a conversation with a Czech intellectual who visited Warsaw at that time assuring him "warmly but honestly that Dubček was not Gomułka, and the Czechs were not the Poles, and they would manage perfectly all that the Poles had tried to do but messed up, and what the Hungarians wanted to do but weren't allowed to."[88] This optimism was widely shared among Czechoslovak social scientists who viewed the Prague Spring as a unique opportunity to implement a more humanistic version of socialism. When Soviet and its allied forces, including Polish troops, invaded Czechoslovakia in August 1968, members of the Institute of Sociology at the Academy of Sciences openly condemned the invasion and expressed support for the Prague Spring reforms. During the subsequent period of normalization, the Institute of Sociology was dissolved and merged with other departments, resulting in the dismissal of about half its staff.[89]

Consequently, the papers presented in Smolenice were never published, and the planned follow-up conference, after multiple postponements, ultimately only took place in 1973. In hindsight, one of the Czechoslovak participants of the follow-up conference in Jabłonna, near Warsaw, observed that it was "probably driven more by external and formal circumstances than by genuine interest on our part."[90] It is therefore hardly surprising that no conference report documenting the meeting was published, bringing the brief tradition of bilateral sociological conferences to an end.[91]

Ironically, during the golden age of sociology in Poland and Czechoslovakia, the only research projects that advanced beyond the planning stage and actively involved sociologists from both countries were initiated by foreign scholars. Between 1964 and 1966, Hungarian sociologist Sándor Szalai (1912–1983) led the "Multinational Comparative Time-Budget Research Project," a UNESCO-funded cross-national study on time use in twelve countries across the Americas and Europe, including Poland and Czechoslovakia.[92] In 1969, the Bureau of Applied Social Research at Columbia University in New York, directed by Paul F. Lazarsfeld (1901–1976) and Robert K. Merton (1910–2003), initiated "The International Study of Opinion Makers," mapping networks among elites in politics, economy, and culture in several countries worldwide, including Poland and Czechoslovakia.[93] Both initiatives were part of a broader wave of theoretical and methodological innovation during that period: the rise of empirical sociology based on standardized research tools, including surveys, sampling, panel studies, and mathematical modeling. This shift was significantly influenced by Austrian American sociologist and institutional innovator Paul F. Lazarsfeld, exemplifying the transatlantic circulation of sociological ideas from pre-Nazi Vienna to New York and back to Europe. After 1956, Lazarsfeld's methodologies began to influence Polish sociology; a decade later, similar trends emerged in Czechoslovakia.[94] More broadly, the roads (not) taken discussed in this section offer valuable insights into the limitations and possibilities of cross-border intellectual exchange during the Cold War era.

Knowledge Brokers

In contrast to the institutional attempts at launching Polish–Czechoslovak research projects mentioned in the previous section, which rarely advanced beyond planning stages or formal agreements, individual initiatives proved significantly more effective. This section explores interpersonal encounters and their impact on the sociological communities in both countries, focusing on Jerzy Wiatr, Zygmunt Bauman, and Jan Szczepański, who were the most widely read and recognized Polish sociologists in Czechoslovakia. While their efforts were undoubtedly shaped by institutional frameworks, they were embedded within these contexts in distinct ways. Ultimately, it was their individual commitment and intellectual originality that made the difference.

The success of Wiatr and Bauman in Czechoslovakia, who were close friends both intellectually and personally, had diverse foundations. Both were Marxists, members of the influential circle around Julian Hochfeld (1911–1966), and among the best-connected Polish sociologists internationally, but their intellectual backgrounds, research interests and status in Polish academia in the 1960s were different. While Wiatr belonged to the non-Jewish Warsaw intelligentsia, Bauman was born into a Jewish family in Poznań and was a first-generation academic. Wiatr held important positions both at the Academy of Sciences and at Warsaw University. He founded the new sub-discipline of military sociology (and later the new discipline of political science) and had a chair in military sociology in the Military Political Academy in Warsaw. His work reflected his engagement with the ideological and practical dimensions of the socialist state. Bauman lectured at Warsaw University and was a member of the board of the Sociological Association. His relationship with the political establishment was more strained and his broader theoretical explorations foreshadowed the path of his career abroad and global recognition.[95] In the 1960s, books by both Wiatr and Bauman perfectly aligned with the needs of Czechoslovak sociology: Wiatr provided practical frameworks for understanding political and military structures in socialist states, while more theoretical contributions by Bauman offered insights into ideology and modernity that resonated with broader intellectual debates.

Figure 6.3. Sociologists of the military in Prague 1967: Jerzy Wiatr (left) and Colonel Jaromír Cvrček. Jerzy J. Wiatr's family archive. Reproduced with permission.

Wiatr recalled:

> In the USSR and other Warsaw Pact countries, the sociology of the military simply did not exist. In the mid-sixties, some interest in this subject emerged in Czechoslovakia, where a translation of my book on the American army was published (in 1963). In 1967, an international conference on the sociology of the military was organized at the Klement Gottwald Military Political Academy in Prague, with the participation of military officers from the USSR and other socialist countries. I was the only civilian in this group, and I was treated as the father of sociology in our part of Europe.[96]

The sociologist who maintained the closest and most personal relationships with colleagues from Czechoslovakia—and arguably even stronger ties with Slovak sociologists than with Czech ones—was Jan Szczepański. The reasons for his special relationship with scholars from Czechoslovakia were twofold: his personal background, shaped by his upbringing in the multiethnic and multilingual region of Teschen Silesia, had fostered an affinity for Slovak culture and language and his professional standing made him a key figure in fostering cross-border sociological exchanges.

Born in 1913 to a family of Polish peasants in what was then Austrian Silesia, Szczepański grew up in a region repeatedly contested and divided between Poland and Czechoslovakia throughout the twentieth century. These shifting borders and cultural interactions left a profound mark on Szczepański, as did his fluency in both Polish and Slovak, a skill that later proved invaluable in his engagement with Czechoslovak sociologists. After earning his doctorate in sociology in prewar Poznań with Znaniecki, Szczepański quickly established himself as one of Poland's leading sociologists after 1945, holding key positions at Łódź University and at the Polish Academy of Sciences. His work blended theoretical insights with empirical studies on industrialization, urbanization, and class dynamics, reflecting a deep understanding of the sociopolitical realities of his time. By the 1960s, Szczepański was not only a towering figure in Polish sociology but also an influential voice internationally. His ability to navigate ideological and cultural boundaries, possibly rooted in his borderland heritage, made him an essential interlocutor in Polish–Czechoslovak sociological exchanges, in which his leadership and insights left a lasting impact.[97]

In the 1960s, no other Polish sociologist visited Czechoslovakia as frequently as Jan Szczepański. His first invitation to lecture in both Prague and Bratislava came in December 1964,[98] and in the years that followed, he returned several times. Particularly noteworthy was his participation in the first postwar congress of Czechoslovak sociologists in Špindlerův Mlýn, a mountain resort located in the northeastern part of Bohemia, in November 1966, where he delivered a lecture on the prospects of Marxist sociology,[99] as well as his co-chairing of the aforementioned Polish–Czechoslovak conference in Smolenice in April 1967. Yet, Szczepański did not confine himself to high-profile events. He willingly accepted invitations to smaller venues and less prestigious events, making him the most engaged and effective Polish sociologist in fostering cross-border collaborations with Czech and Slovak colleagues.

In the second half of the 1960s, Szczepański's prominent standing in Czechoslovakia was formally recognized on two significant occasions. In April 1967, he was awarded honorary membership in the Slovak Sociological Society, underscoring his close ties to Slovak sociology.[100] In May 1969,

Szczepański received an honorary doctorate from the University of Brno, reflecting his influence and the high regard in which he was held within the Czechoslovak academic community.[101] While the political context of the Prague Spring's suppression suggests that the latter event, barely recorded in archives, was a rather formal affair, the earlier ceremony in Bratislava offers a more vivid glimpse of Szczepański's standing and the rapport he shared with Slovak sociologists. Along with the diploma and flowers, he was presented with a traditional Carpathian shepherd's axe called *ciupaga* in Polish and *valaška* in Slovak.[102] Often ornately carved, it was used as both a walking stick and a light weapon. The axe is therefore associated with the traditional pastoral culture, embodying the values of self-reliance, hard work, and a deep respect for the local tradition. In ceremonies or gifts, it is a gesture of respect and admiration, signifying honor, loyalty, and the strength of the highland traditions. As a ceremonial gift, it conveys admiration and devotion, though in this case, it may have been offered with a lighthearted touch, possibly referencing Szczepański's previous visit to Bratislava in early 1966.

Szczepański's diary from that earlier trip offers a vivid account of his experiences, blending professional engagements with rich cultural immersion. He detailed his stay at the legendary Zochova Chata near Bratislava, describing it as "furnished with Slovakian style taste, having a modern wine cellar with a large baker's oven from which the baker was just taking out fresh, fragrant loaves." He also recounted a visit to the village of Svätý Jur, where he attended a *zabíjačka* at the cooperative pub—a traditional event involving the slaughter of a pig, accompanied by a communal feast and the preparation of various pork products. There, he enjoyed "a variety of sausages, crackling, and drank white wine, semi-sweet, fragrant." Another memorable stop was at a local restaurant named after the legendary premodern rebel called Jánošík in Slovak and Janosik in Polish, which he described as "a mountain hut from the outside, decorated inside in the American style, with a big hearth, walls covered with skins, and tables made of blocks." [103] Szczepański's reflections leave little doubt that he thoroughly enjoyed his time in Slovakia, both professionally and personally, forging connections that would later be reflected in the warmth and recognition he received.

It might also be, however, that the axe gifted to Szczepański in Bratislava was imbued with yet another layer of meaning. Deeply rooted in the highlander (Goral) culture of the Carpathian Mountains, which stretches across parts of Poland, Slovakia, and Ukraine, the shepherd's axe holds relatively little cultural significance in Czechia compared to Slovakia and Poland, where it is more integral to folk traditions and identity. Associated with Jánošík/Janosik, who was known for taking property from the rich and giving it to the poor, the shepherd's axe also symbolizes the fight for justice.[104] The choice of this gift might therefore reflect the effort made by Slovak sociologists to distinguish themselves within the Czechoslovak academic community and perhaps even to demonstrate their equal standing—or ambition to achieve parity—with their Czech counterparts, which was indeed a leitmotif in Czech–Slovak academic relations at that time.[105] In this context, the presentation of the axe could also be interpreted as a strategic gesture, leveraging Szczepański's recognition to bolster their own position and affirm the distinctiveness of Slovak sociology.

The warm feelings were clearly mutual. Szczepański's fondness for socializing with Slovaks is evident throughout his accounts. During the aforementioned conference in Špindlerův Mlýn in late 1966, he noted in his diary: "After dinner, I sat and drank wine with the Slovaks, and then I had serious conversations with the Germans."[106] A few months later, he sent a deeply personal letter to the organizers of the conference in Smolenice, writing that Slovakia had come to feel like a second home to him. In the letter, he expressed not only a sense of belonging but also a profound emotional connection to the country, emphasizing the warmth and hospitality he had encountered. His words reflected the strong bonds he had formed with the Slovak people and the intellectual and cultural ties that had flourished during the conference, underscoring the lasting impact that his time in Slovakia had had on him.[107]

During his travels to Czechoslovakia, Szczepański observed the dynamics within and between Czech and Slovak sociology with keen interest. Already his first visit to Czechoslovakia in late 1964 presented him with a puzzle. After arriving in Prague from Warsaw on an overnight train, Szczepański engaged in several conversations with Czech colleagues, delivered a lecture at the Czech Sociological Society, and continued the exchange over dinner. To his surprise, however, he was left to his own devices afterward:

> I have a train to Bratislava at midnight. I think if a member of the Czech Academy of Sciences came to Poland, they wouldn't let him lug his suitcase alone to the station around midnight. But at the end of the day, I'm supposedly a guest of the Slovak Sociological Society and my status in Prague is unclear, and I don't know if I've committed some kind of faux pas by coming here.[108]

It seems that the question of whether international guests should be accompanied to train stations or not may have pointed to a lot of subtle cultural differences and unspoken expectations. In some circles, such gestures were seen as a mark of respect and hospitality, reflecting the host institution's attentiveness; in others, they were considered unnecessary formalities, leaving guests to navigate on their own. For Szczepański, the absence of such courtesies in Prague seemed to highlight an ambiguity about his status and perhaps reflected underlying tensions between Czech and Slovak sociologists,[109] if not even between the Czechoslovak and Slovak Academies of Sciences (the latter, established in 1953 to modernize Slovakia's academic landscape, was officially incorporated into the Czechoslovak Academy in 1960 but remained largely independent in practice).[110] Unfamiliar with the local train connections, Szczepański learned at the last minute which station to head to. The journey itself—with the company of a "nice Slovak" in the compartment—was uneventful, and his welcome in Bratislava proved to be exceptionally warm. Before delivering his lecture at the Slovak Sociological Society, he reflected: "I don't really know how to structure today's lecture for them, what audience it will be for. What kind of audience it will be compared to Prague. The Bratislavans have the ambition to create a faster and better intellectual center, and especially to create a better sociology than Prague."[111] Szczepański's diary entries from his first visit to Czechoslovakia reveal both the distance between Polish, Czech, and Slovak sociologies at the time—even for someone fluent in two of the three languages involved, and the difference between the state of sociology in Czech lands and Slovakia.

Importantly, Szczepański's frequent visits to Czechoslovakia were not merely a one-way exchange of knowledge; they were pivotal in shaping his understanding of his own position at home and abroad. While Czech and Slovak sociologists benefited from his professional expertise and reflections on the state of their field, Szczepański gained valuable insights into his influence within the Socialist Bloc. During his first visit in late 1964, he noted in his diary: "In socialist countries, as the head of an institute, I am considered something of an institutional leader of sociology in Poland, and by extension, within the Bloc as well. This realization has implications, but I only became fully aware of it during my recent visit to Prague and Bratislava."[112] In early 1966, after another extended visit, he wrote: "My stay in Czechoslovakia was successful. I found that in Prague, Brno, Bratislava, Opava, Ostrava and Olomouc people were reading my books, learning from them, referring to them. I discussed with professors and young enthusiasts. I have helped with the ferment that has swept the Czech universities. I myself also benefited."[113] These experiences, which combined intellectual exchange with personal growth, can be seen as stepping stones to greater responsibilities on the global stage. At the 6th World Congress of Sociology in Évian, France, in September 1966, Szczepański was elected President of the International Sociological Association (ISA). Despite his initial reluctance, he accepted the role with a full understanding of its significance and the responsibilities it entailed.[114] His writings suggest a keen awareness of the symbolic and practical importance of representing both the Socialist

Bloc and the broader discipline of sociology on the global stage, as well as the challenges of balancing competing aspirations within the organization.[115] Reflecting on this milestone, he wrote: "I have a big job ahead of me in this position, manipulating the aspirations of socialist countries, countries of three continents, reorganizing the ISA, revitalizing international contacts, revitalizing the work of research committees."[116] Szczepański's time in Czechoslovakia undoubtedly shaped the vision and confidence he brought to this influential global role.

Ironically, it was a reference to Czechoslovakia that brought Szczepański to fully grasp the shift in his standing in 1968. In late March of that year, deeply critical of the antisemitic campaign in Poland and disillusioned by the silence of his colleagues, he reflected in his diary: "The current situation in Poland, the witch hunt against Polish antisemitism, the protests against Polish policies—all of this may render my position as ISA president very difficult or even untenable. What is happening is a horrifying bankruptcy of communism in Poland. I have always tried to remain loyal to the state and government, to the system—but now? How can one defend a system that, in the name of some unknown principles and undisclosed interests, brutally suppresses its best youth and leads a campaign of lies?" Drawing a sharp contrast with the international perception of Czechoslovakia under Dubček's reformist leadership, he wrote: "Unlike Czechoslovakia, we currently have the worst press coverage in the world, while the Czechs and Dubček are garnering enormously positive attention."[117] Highlighting Szczepański's moral and professional crisis, these entries also reveal how the comparison to Czechoslovakia deepened his awareness of the challenges he faced in balancing his role as a global sociological leader with the political upheaval in his homeland. On 21 August 1968, Szczepański's diary tersely records: "Today, the armies of Bulgaria, East Germany, Hungary, the USSR, and Poland entered the territory of Czechoslovakia. Damn it!"[118] A few months later, he ceased writing in his diary altogether.

The year 1968 disrupted the Czechoslovak careers of all three sociologists discussed in this section. In late August 1968, Jerzy Wiatr, who was "firmly opposed to the intervention and ashamed of the role Poland played in it," resigned from his position at the Military Political Academy. He later described this step as "a very modest gesture of disapproval at the use of the Polish forces in the intervention."[119] Wiatr also refused to sign a public letter endorsing the Polish intervention in Czechoslovakia.[120] His last book was translated into Czech in 1970, but his promising collaboration with military sociologists in Czechoslovakia came to a halt as the field itself was stifled during the period of normalization.[121] In the years that followed, his only contact with Czechoslovakia came through Zdeněk Mlynář (1930–1997)—a sociologist, political scientist and key architect of the Prague

Spring reforms—whom Wiatr occasionally visited during Mlynář's exile in Austria, where he settled in 1977.[122]

Zygmunt Bauman was forced to leave Poland in the wake of the antisemitic campaign of 1968. He initially settled in Israel before eventually establishing himself at the University of Leeds in the United Kingdom. His Polish works were not republished, and his English writings were banned from translation, isolating him from his Czech and Slovak readers as well as audiences across the Eastern Bloc. Yet, his career might have unfolded differently. After his dismissal from Warsaw University in March 1968, Bauman and Maria Hirszowicz received offers of academic positions at Charles University in Prague and the Czechoslovak Academy of Sciences.[123] Despite his admiration for the Prague Spring, his deep affinity for Czech culture and his knowledge of the Czech language,[124] Bauman chose not to settle in Czechoslovakia, possibly sensing the fragility of its reformist moment. Reflecting on this choice in later years, Bauman explained:

> I had many friends in the Czech lands, I highly valued the Czech sociologists I knew personally, I read Czech literature, and I was fascinated by Czech films, painting, music, and theater . . . The offer was therefore tempting (although I could not know to what extent the Polish authorities would allow me to take advantage of it). However, I must honestly admit that I was a person of little faith and did not believe in the success of the Prague Spring.[125]

Unlike Jerzy Wiatr, whose collaboration with Czechoslovak sociologists ceased entirely, Bauman eventually regained a readership in Czechoslovakia. After retiring in 1990, he (co-)authored over forty books, many of which found enthusiastic audiences among Czech and Slovak readers. His lectures and writings on modernity, postmodernity, and global social transformations resonated deeply in post-communist societies, marking a significant revival of his influence in the region.[126]

The case of Szczepański lies somewhere in between. While he did not face the professional ruptures that Bauman and Wiatr experienced, the events of 1968 profoundly shaped his intellectual and personal relationship with Czechoslovakia.[127] As president of the ISA, he maintained a prominent position on the global sociological stage, but the Warsaw Pact invasion of Czechoslovakia deeply affected his sense of loyalty to both his country and the broader Socialist Bloc. Reflecting the broader chilling of cross-border collaboration in the wake of normalization, Szczepański's previously close connections with Czech and Slovak sociologists gradually withered away because most of his intellectual partners were forced to leave academia. The few exceptions to this rule were Jiří Musil (1928–2012), who pursued comparative research including the developments in Poland, Miloslav Petrusek (1936–2012), who had a genuine interest in Polish sociology, and

Jan Sedláček (1930–2021), who—despite the fundamental distrust of his Czechoslovak colleagues defending orthodox Marxism against sociological pluralism in Poland—continued translating sociological publications from Polish into Czech during the 1970s.[128] Ironically, when the Slovak dream of federalization began to materialize in 1968, granting the Slovak Academy of Sciences greater autonomy, this newfound institutional agency did little to improve professional relations between Slovak and Polish sociologists.[129] Though Szczepański's works remained accessible in Czechoslovakia's academic libraries, they were neither recommended to students nor cited, effectively falling into oblivion.[130] Unlike Bauman, Szczepański did not find a later resurgence among Czech and Slovak audiences, leaving his career and legacy intertwined with the political and intellectual fractures of the era.

A remarkable effort to revive Szczepański's cross-border legacy has taken root in his hometown of Ustroń, where the Free School of Philosophical and Social Sciences, named in his honor, was established in 2014. This initiative directly draws inspiration from a reflection in Szczepański's wartime diary, written in early 1940. Imagining the future of his hometown, he noted: "And perhaps one day, we will manage to fulfill another dream: a free school of philosophical and social sciences in Ustroń."[131] Motivated by this vision, the Free School has evolved into a dynamic and interdisciplinary platform, providing young researchers from Poland and Czechia with the opportunity to engage in dialogue and showcase their ongoing work.[132]

The Czechoslovak trajectories of Wiatr, Bauman, and Szczepański, though distinct, underscore the profound rupture of 1968 in Polish–Czechoslovak sociological cooperation. Following this turning point, four interrelated developments shaped the nature of this relationship during the 1970s and 1980s. The overall cooperation diminished significantly, reflecting the broader political and intellectual constraints of the era. At the same time, apparent activities largely overshadowed inspiring interactions. In more general terms, sociologists in both countries concentrated primarily on analyzing developments within their own borders, with limited efforts to engage with or understand processes on the other side of the border.[133] Lastly, adherence to Marxism within the sociological communities continued to diverge: while most Czechoslovak sociologists upheld a relatively orthodox interpretation, their Polish counterparts adopted a more diverse and functional approach.[134] A telling illustration of this Polish–Czechoslovak difference is the institutional rise of chairs of "Marxist–Leninist sociology" at Czechoslovak universities, positions that were notably absent in Poland.[135]

One particularly striking missed opportunity for collaboration emerged within the dissident realm. *Sociologický obzor* (Sociological horizon), cofounded and edited by Miloslav Petrusek in 1987—and the only samizdat sociological journal in Eastern Europe—neither reached Polish audiences nor featured contributions from Polish sociologists, despite Petrusek's strong connections to Poland, the active involvement of many Polish sociologists in the dissident movement, and shared Polish–Czechoslovak interests.[136] Notably, dissident social scientists from both countries who lived in exile seemed to cooperate more closely with one another than their counterparts who remained in Poland and Czechoslovakia.[137] This fragmentation further limited the potential for cross-border collaboration and deprived both sociological traditions of a deeper, mutually enriching exchange during a pivotal period in their histories.

Conclusion and Outlook

The history of sociological cooperation between Poland and Czechoslovakia under state socialism was shaped by two layers of asymmetry: one between Poland and Czechoslovakia and another within Czechoslovakia itself, between Czech and Slovak sociology. Polish sociology, benefiting from earlier institutional development and a stronger international presence, often served as a model for its Czechoslovak counterpart. In contrast, Czechoslovak sociology was shaped by more centralized state control, a narrower intellectual base, and internal disparities between its Czech and Slovak branches. Czech sociology, which was anchored in the more developed academic infrastructure of Prague, often dominated the field, while Slovak sociology faced challenges of marginalization and sought to assert its distinct identity.

Despite these structural imbalances, efforts were made to foster cooperation. Joint initiatives, such as collaborations between the Institutes of Sociology at the Polish and Czechoslovak Academies of Sciences, were emblematic of the spirit of socialist solidarity. However, these initiatives were often more symbolic than substantive, reflecting the limitations imposed by political constraints and ideological frameworks. While these institutional efforts produced limited practical outcomes, they nonetheless underscored the difficulties of achieving equitable bilateral and internal collaboration in a highly politicized academic environment.

At the same time, the individual trajectories of sociologists played a pivotal role in shaping cross-border interactions. The chapter underscored the significance of figures like Jiří Kolaja, Jerzy Wiatr, Zygmunt Bauman, and Jan Szczepański, who each contributed uniquely to the intellectual exchange between the two sociological communities. Kolaja's *A Polish Factory* demonstrated the potential of qualitative research across borders but also revealed the limits of such endeavors within an ideologically divided academic landscape. Meanwhile, the personal connections and professional commitments of Wiatr, Bauman, and Szczepański testify to the crucial role of agency in overcoming institutional and political barriers. Szczepański, in particular, stood out for his consistent engagement with Czech and (even more so) Slovak sociologists, forging meaningful ties despite the broader chilling of relations after 1968.

Indeed, the Prague Spring and its suppression in 1968 marked a turning point in Polish–Czechoslovak sociological relations. The normalization period that followed not only curtailed the autonomy of Czechoslovak sociology but also led to the erosion of cross-border collaborations. The subsequent decades were characterized by a shift toward apparent activities—formalized and largely unproductive exchanges that overshadowed the more inspiring and meaningful interactions of the 1960s. While dissident social scientists from both countries in exile occasionally bridged this gap, those who remained within Poland and Czechoslovakia often turned inward, focusing on their own national contexts rather than engaging in sustained comparative or cooperative research.

The legacy of Polish–Czechoslovak sociological relations invites further reflection on the intersections of politics, ideology, and the practice of intellectual work. The role of Marxism, as both a unifying framework and a source of tension, presents a compelling avenue for further exploration, particularly in how it influenced the development of sociological thought and practice in Poland and Czechoslovakia. In addition, the parallels and divergences between the two traditions offer a rich field for comparative analysis of developments that reconfigured the field of sociology in both countries. Institutionally, these included sociology's proximity to philosophy and the separation of political science from sociology—processes that reshaped disciplinary boundaries and priorities. Intellectually, the emergence and evolution of applied sociology and public sociology reflected efforts by sociologists to navigate political constraints, address societal needs, and engage with broader audiences. These under-researched developments highlight the adaptive strategies of sociologists in both countries, as they sought to influence public policy and contribute to discourse despite challenging circumstances.

The transnational dimension of Polish–Czechoslovak sociological interactions remains an underexplored area of inquiry. The involvement of sociologists from both countries in international settings—such as the international projects mentioned above, but also international conferences, venues facilitating the exchange of ideas like the Frankfurt International Book Fair or Europe Talks in Vienna, and more or less institutionalized diaspora networks—points to opportunities for a broader understanding of the dynamics of knowledge transfer and adaptation beyond the Socialist Bloc. These contexts, often shaped by shared fellowship programs and participation in transnational academic communities, offer valuable insights into the possibilities and limitations of cross-border intellectual exchange during the Cold War era and serve as a reminder of the enduring challenges and possibilities inherent in fostering meaningful intellectual exchange across borders.

This chapter also raises critical questions about representation within the sociological field. Is it purely coincidental that all the individuals discussed and depicted here in detail are men? Were women truly absent from these ostensibly egalitarian societies, or have their contributions simply been overlooked? More specifically, did Jiří Kolaja fail to consult with sociologist Krystyna Lutyńska (1931–2016), the wife of Jan Lutyński, during his work in Łódź in the 1950s? Did Jerzy Wiatr not engage in research discussions with sociologist Irena Dubská (1924–2010), the first editor-in-chief of the *Czech Sociological Review* (*Sociologický časopis*) between 1965 and 1969, while visiting her husband Zdeněk Mlynář in their Vienna exile in the 1970s? Or did they do so, but the traces of their exchanges have remained less visible in the historical record—as was the case of the aforementioned Ewa Kaltenberg-Kwiatkowska? These questions highlight a crucial avenue for further investigation: the role of women (and other marginalized voices) in shaping the sociological landscape of Poland and Czechoslovakia.

In conclusion, the history of Polish–Czechoslovak sociological cooperation offers fertile ground for speculation on the roads not taken, particularly in relation to alternative intellectual paths that could have redesigned the field. What if Zygmunt Bauman and Maria Hirszowicz had chosen to settle in Prague in 1968? To what extent might Bauman's theoretical insights and Hirszowicz's expertise in applied sociology have provided Czechoslovak sociology with a new intellectual direction, thereby altering the asymmetries within the Polish–Czech–Slovak triangle? Similarly, had Jiří Kolaja not been declared persona non grata in Poland and Czechoslovakia, would he have authored the much-needed comparative book on Polish and Czechoslovak societies under socialism or perhaps even inspired a new line of cross-national studies that were frequently advocated in programmatic papers by both academies of sciences? Reflecting on these scenarios underscores the complex interplay between sociology, politics, and society in East-Central Europe, highlighting how political decisions and ideological climates can significantly influence academic collaboration and the development of the social sciences.

Author Biography

Kornelia Kończal holds a PhD in History from the European University Institute in Florence, Italy (2017) and serves as Assistant Professor of Public History at Bielefeld University, Germany. She specializes in contemporary

European history, with a particular focus on Central and Eastern Europe, France, and Germany. She is currently completing her first monograph, *Politics of Plunder: How Expropriating German Expellees Helped and Harmed the Postwar Reconstruction of East Central Europe*, and editing a volume on *Doing Heritage without Heirs? Practices of Cultural Succession in Post-German Spaces*. co-editing a Polish–French volume of ego documents on the international history of sociology in the first half of the twentieth century.

List of Abbreviations

AAN	Archiwum Akt Nowych (Central Archives of Modern Records)
ABS	Archiv bezpečnostních složek (Archive of the Security Services)
A PAN	Archiwum Polskiej Akademii Nauk (Archives of the Polish Academy of Sciences)
AP Ł	Archiwum Państwowe w Łodzi (State Archive in Łódź)
ASP UMK	Archiwum Socjologii Polskiej w Bibliotece Uniwersyteckiej w Toruniu (Archives of Polish Sociology at the Nicolaus Copernicus University Library)
A UŁ	Archiwum Uniwersytetu Łódzkiego (University of Łódź Archives)
IH PAN	Instytut Historii Polskiej Akademii Nauk (Institute of History, Polish Academy of Sciences)
IPN	Instytut Pamięci Narodowej (Institute of National Remembrance)
MÚA AV ČR	Masarykův ústav a Archiv Akademie věd České republiky (Masaryk Institute and Archives of the Czech Academy of Sciences)

Notes

I would like to express my sincere gratitude to Jan Chodějovský (MÚA AV ČR) and Oleksandr Pestrykov (IH PAN) for their invaluable assistance in collecting archival material. I am also deeply grateful to the contributors to this volume—particularly Michaela Šmidrkalová—for her insightful suggestions on earlier drafts of this chapter. Additionally, I would like to thank my external critical readers, Marcin Jarząbek, Jarosław Kilias, Zdeněk R. Nešpor, and Agata Zysiak, whose thoughtful comments greatly contributed to improving the chapter's clarity and argumentation.

1. Bucholc, *Sociology in Poland*; Skovajsa and Balon, *Sociology in the Czech Republic*; Kilias, "An Older Brother."

2. Platt, *A History of Sociological Research Methods in America*; Oakes and Vidich, *Collaboration, Reputation and Ethics in American Academic Life*; Schatzki, Cetina, and von Savigny, *The Practice Turn in Contemporary Theory*; Igo, *The Average American*; Bert and Mauss, *Henri Hubert et la sociologie des religions*.
3. Lutyński, "Apparent Activities."
4. The archival material consulted for the purposes of this chapter includes collections held at the Archives of the Polish Academy of Sciences (A PAN) in Warsaw (Institute of Philosophy and Sociology), the Archives of Polish Sociology at the Nicolaus Copernicus University Library (ASP UMK) in Toruń (Jan Szczepański's papers; Polish Sociological Association), the University of Łódź Archives (A UŁ) (Jan Lutyński's papers), the Central Archives of Modern Records (AAN) in Warsaw (Ministry of Foreign Affairs; Ministry of Light Industry; Ministry of Science and Higher Education), the Institute of National Remembrance (IPN) in Warsaw and Katowice (Jiří Kolaja's files), the Masaryk Institute and Archives of the Czech Academy of Sciences (MÚA AV ČR) in Prague (Foreign Department of the Presidium; Institute of Philosophy and Sociology; Institute of Sociology; Scientific Board of Philosophy and Sociology), the Archive of the Security Services (ABS) in Prague (Jiří Kolaja's files), the State Archive in Łódź (AP Ł) (Julian Marchlewski's cotton mill), the University of Saskatchewan Special Collections, Canada (Pitirim A. Sorokin Collection).
5. This difference is particularly striking for documents produced by sociologists at both Academies of Sciences (Szczepański, *Dzienniki z lat 1945–1968*, 322). The only exceptions to this rule are Sociological Associations in both countries: the archives of the Czechoslovak Sociological Associations have been lost.
6. Despite several attempts and promising conversations, I did not manage to gain access to Kolaja's family archives in the United States.
7. Interesting exceptions to this rule are phone calls that were bugged and recorded by security forces. For example, Kolaja's file compiled by the Czechoslovak security forces contains a transcript of a short telephone conversation between him and his academic teacher conducted during Kolaja's stay in Czechoslovakia in 1956 (ABS, Prague: Jiří Kolaja's files, 10027 I.S | 10027_309, p. 81–90).
8. Sedláček, "Nazdar," esp. 111.
9. However, evidence of these events remains largely anecdotal, with little formal documentation available.
10. Kolaja, *A Polish Factory*, 14. The name of the factory is not explicitly referred to in Kolaja's book. His description suggests, however, that it must have been the Marchlewski factory. Built in the late nineteenth century by the Polish Jewish businessman and philanthropist Izaak Poznański (1833–1900), the factory was a major textile enterprise, contributing to the city's industrial growth. After 1945, it was nationalized, renamed after the revolutionary activist Julian Marchlewski (1866–1925) and became a symbol of socialist industrialization. With the fall of communism and the decline of the textile industry in the 1990s, the factory eventually closed and its facilities were largely dismantled or repurposed.
11. Bömelburg, *Lodz*; Chu, "From Łódź to Litzmannstadt," 240–67.
12. Kolaja, "Acknowledgments," esp. 18.
13. Whyte, "Forward," esp. 5.

14. Kennedy, "Poland in the American Sociological Imagination." The 1980s saw the publication of, among others, Goldfarb, *The Persistence of Freedom*; Goldfarb, *On Cultural Freedom*; Touraine, *Solidarity*; Wedel, *The Private Poland.*
15. During the war, Kolaja worked as a film scriptwriter and director in Zlín. Those interested in the history of the animated film might know him as co-author of the iconic *Ferda the Ant* (*Ferda mravenec*) (1944).
16. Kolaja's stay in Sweden was arguably his last attempt to reconnect with film. In early 1948, he worked on a study of Swedish feature film in Stockholm. Kolaja, "Swedish Feature Films and Swedish Society," esp. 189.
17. Heilbron, "Les métamorphoses du durkheimisme"; Marcel, *Le durkheimisme dans l'entre-deux-guerres.* Kolaja's interest in Durkheimian approach is best documented in the following publications: Kolaja, *Social System and Time and Space;* Kolaja, "The Social and Time."
18. University of Saskatchewan Special Collections, Pitirim A. Sorokin Collection: box 17, K
19. Kolaja, *A Polish Factory,* 12.
20. Kolaja, "Acknowledgments," esp. 18.
21. In 1925, Maxmilián Kolaja (1883–1966) co-founded the Czechoslovak–Polish Club in Brno. His later life is a good illustration of the uneasy Polish–Czechoslovak neighborhood in the 1930s and 1940s. In 1929 he was awarded the Polish Order of Polonia Restituta, the second-highest civilian state award, in recognition of his efforts on behalf of the Polish cause in Czechoslovakia. He returned it in 1938, however, when Poland occupied part of the Trans-Olza region, which had been disputed between Poland and Czechoslovakia for several years. Because of his Polonophile views and activities, Maxmilián Kolaja was arrested by the Gestapo on 1 September 1939, the day of the German invasion of Poland, and imprisoned for several weeks. Gregor, "Za prof. dr. Maxmiliánem Kolajou."
22. Kolaja, "Stanislaw Ossowski."
23. Kolaja, "Acknowledgments," esp. 18. Extensive research in the Central Archives of Modern Records (AAN) in Warsaw, including relevant collections produced by the Ministry of Foreign Affairs, the Ministry of Light Industry and the Ministry of Science and Higher Education did not yield any further information on Kolaja's request.
24. *Sociologická revue* was founded in 1930 and ceased publication in 1949. Jiří Kolaja reviewed several works on Polish sociology and philosophy for this journal, see Kolaja, "Review of *Prusy Wschodnie*"; Kolaja, "Review of *Polskie tradycje demokratyczne*."
25. Kolaja, *A Polish Factory,* 16.
26. Ibid., 17.
27. Ibid., 12.
28. Ibid., 11.
29. Zysiak, *Limiting Privilege.*
30. Wincławski, "Józef Kądzielski." In 1948, when Kolaja studied in Stockholm, his review of a Swedish book was published in *Przegląd Socjologiczny,* a sociological journal edited by Chałasiński. Kolaja, "Review of *De Svenska*."
31. Kolaja, *A Polish Factory,* 19.

32. Ibid., xiii–xviii, esp. xviii.
33. It is safe to assume that Lutyński and Kolaja kept in touch. Lutyński's unclassified papers include his diary from 1959, when he spent some time in the United States as a Ford Foundation fellow, in which Kolaja's telephone number is noted (A UŁ: Jan Lutyński's papers).
34. Notatka, Warszawa, 3 May 1968 (IPN Katowice: Jiří Kolaja's files, 04/2439, p. 7); Szczerba, *Dom uczonych*, 110–36.
35. Kolaja, "Acknowledgments," esp. 18.
36. Kądzielski, "Informacja o badaniach nad małą grupą w przemyśle."
37. Kądzielski, "Review of *A Polish Factory* by J. Kolaja."
38. Bazant, "Review of *A Polish Factory* by J. Kolaja,"esp. 355; Gross, "Review of *A Polish Factory* by J. Kolaja," esp. 236; Rodnick, "Review of *A Polish Factory* by J. Kolaja"; Wandycz, "Review of *A Polish Factory* by J. Kolaja," esp. 165; Taborsky, "Review of *A Polish Factory* by J. Kolaja," esp. 172; Loucks, "Review of *A Polish Factory* by J. Kolaja," esp. 308; Noland, "Review of *A Polish Factory* by J. Kolaja," esp. 363. One of the Czechoslovak newspapers in exile noted: "Dr. Kolaja signed his book with his Czech first name 'Jiří,' which in itself shows that he did not succumb (like many others) to the desire to adapt to the point of being indistinguishable from the American environment. It may seem like a small detail at first glance, but it suggests a great deal." ("Český sociolog v polské továrně"; see also "Socialistický člověk zatím neexistuje.")
39. Obrdlíková, "Sociolog o dělnických radách," esp. 735.
40. Sarapata, "Review of *A Polish Factory* by J. Kolaja," esp. 433.
41. Szczepanski, "Review of *A Polish Factory* by J. Kolaja," esp. 98; Kądzielski, "Review of *A Polish Factory* by J. Kolaja," esp. 117; Sarapata, "Review of *A Polish Factory* by J. Kolaja," esp. 433.
42. In the summer of 1957, Szczepański did not keep a diary so that it is difficult to learn more about his possible encounters with Kolaja during the latter's fieldwork in Łódź (Szczepański, *Dzienniki z lat 1945–1968*).
43. Szczepanski, "Review of *A Polish Factory* by J. Kolaja," esp. 99.
44. Ibid., 98.

* *Mitbestimmung* (German): co-determination, involvement of employees in management decisions.

45. Ibid., 99.
46. Szczepański, *Dzienniki z lat 1945–1968*, 103–20.
47. Kolaja's later requests for research stays in Poland lists several names of Polish sociologists he knew well including Jan Szczepański (IPN, teczka osobowa Kolaji, 1218/25023, p. 148, 172).
48. Kolaja, "Review of *Wykształcenie a Pozycja*."
49. Kolaja, "Review of *Sociologia, teoria, metody, problem*"; Kolaja "Review of *Social Problems of Work and Production*"; Kolaja, "Review of *Polish Society*"; Kolaja, "Review of *Odmiany czasu teraźniejszego*"; Kolaja, "A Polish Survey of Human Issues" (the original reviews contain spelling errors that have been intentionally retained).
50. Perrow, "Review of *A Polish Factory*"; Kolaja, "*A Polish Factory*: Letters to the Editor."
51. Szczepański to Kolaja, 19 November 1961 (ASP UMK in Toruń: Jan Szczepański's papers, 1/130).

52. In 1963, Kolaja married a Polish woman from Silesia and visited her family. As he was not allowed to enter Czechoslovak territory, stays in Poland offered him an opportunity to arrange meetings with members of his Czech family (IPN Katowice 04/2439, p. 7).
53. In original, "prawdopodobnie narodowości żydowskiej" (IPN, Akta osobowe Kolaji, 1218/25023, p. 121).
54. Berger, Notatka służbowa, Warszawa, May 1974 (IPN, teczka osobowa Kolaji, 1218/25023, p. 134–35: 134).
55. IPN, Akta osobowe Kolaji, 1218/25023, p. 176–78).
56. IPN, teczka osobowa Kolaji, 1218/25023, p. 159, 167–68).
57. IPN, teczka osobowa Kolaji, 1218/25023, p. 182–85).
58. The catalog of the University Library does not contain information about how and when the book was included in the collection.
59. Kolaja, "The Contemporary Polish Sociological Review"; Kolaja, "Review of *Robotnicy Warszawskiej Fabryki Motocykli*"; Kolaja, "Review of *Sytuacja społeczna dziennikarzy polskich*"; Kolaja, "To the Editor"; Kolaja, "Review of *Praca i koleżeństwo*"; Kolaja, "Review of *Social Processes of Industrialization*"; Kolaja, "Review of *Occupational Choice and Prestige*"; Kolaja, "Dos revistas demográficas"; Kolaja, "Review of *Kultura masowa: krytyka i obrana*"; Kolaja, "Review of *Film na wsi i w miastach powiatowych*"; Kolaja, "Review of *Sociology of a Work Enterprise*"; Kolaja, "Review of *Silesian Workers*"; Kolaja, "Review of *The Formation of a Modern Labor Force*"; Kolaja, "Review of *Valores y pautas culturales*"; Kolaja, "Review of *Cultural Values and Patterns*"; Kolaja, "Review of *Wright Mills si Marxismul*"; Kolaja and Das, *Glimpses of Sociology in Eastern Europe.*
60. Petrusek, *České sociální vědy v exilu 1948 až 1989*; Nešpor et al., *Dějiny české sociologie*, esp. 263–274.
61. Havelka, "Tschechische Soziologie im gesellschaftlichen Wandel," esp. 243–44; Musil, "Poznámky o české sociologii za komunistického režimu," esp. 580; Petrusek, "Výuka sociologie v čase tání a v časech normalizace (1964–1989)," esp. 599.
62. Voříšek, *The Reform Generation*, 11.
63. Brokl et al. "Česká sociologie v letech 1965–1989."
64. Ibid., 725, 729.
65. Kurczewska, "Socjologia polska w latach 1944–1989 w służbie społeczeństwa," esp. 116.
66. Voříšek, *The Reform Generation*, 237–75; Sułek and Kraśko, "The Multifarious and Changing Functions"; Nešpor, "Československá sociologická společnost při ČSAV."
67. Wiatr, *Wojsko, społeczeństwo i polityka w Stanach Zjednoczonych*; Wiatr, *Szkice o materializmie historycznym i socjologii*; Wiatr, *Czy zmierzch ery ideologii?*; Wiatr, *Społeczeństwo*; Wiatr, *Socjologia zaangażowana.*
68. Bauman, *Zarys socjologii*; Bauman, *Kariera*; Bauman, *Socjologia na co dzień*; Bauman, *Wizje ludzkiego świata.*
69. Szczepański, *Socjologia: rozwój problematyki i metod*; Szczepański, *Elementarne pojęcia socjologii*; Szczepański, *Zagadnienia socjologii współczesnej.*
70. Szostkiewicz, *Procedury i techniki badań socjologicznych*; Hirszowicz: *Konfrontacje socjologiczne.*
71. Kilias, "An Older Brother," 193–95.

72. Urbanek, "Ups and Downs in Czech Sociology," esp. 81.
73. Musil, *Urbanizace v socialistických zemích.*
74. See the chapter by Jan Surman in this volume.
75. Platt, *A Brief History*; Markiewicz, *Sto lat przeciw głupocie*, 243–44.
76. Kilias, "Socjologia polska i czeska"; Warczok, "Globalne pole nauk społecznych a socjologia Polska."
77. Kilias, "Socjologia polska i czeska," 292–93.
78. For example, Katriak: "Niektoré aspekty rozvoja poľskej sociológie"; Sedláček, "Další hlasy v polské diskusi o sociologii."
79. MÚA AV ČR, Vědecké kolegium filosofie (a sociologie) ČSAV, inv. č. 79, sign. 60: Ústav pro filosofii a sociologii ČSAV, inv. č. 69, sign. 655.
80. MÚA AV ČR, Vědecké kolegium filosofie (a sociologie) ČSAV, inv. č. 90, sign. 646.
81. MÚA AV ČR, Vědecké kolegium filosofie (a sociologie) ČSAV, inv. č. 90, sign. 646.
82. MÚA AV ČR, Vědecké kolegium filosofie (a sociologie) ČSAV, inv. č. 90, sign. 646.
83. Stola, *Kampania antysyjonistyczna*, 111–12; Szczepański, *Dzienniki z lat 1945–1968*, 328.
84. Battěk, "Společné rokování polských," 625.
85. Szczepański, *Dzienniki z lat*, 330, 332, 333.
86. Szczepański, *Dzienniki z lat 1945–1968*, 330.
87. Bauman, "Pierwsza ogólnokrajowa konferencja"; Battěk, "Společné rokování polských"; Szostkiewicz, "First Meeting"; Szczepański, *Dzienniki z lat*, 329–33.
88. Bauman, *My Life in Fragments*, 144.
89. Prokůpek, "Normalizace Ústavu po filosofii a sociologii ČSAV."
90. Sedláček, "Nazdar," 110.
91. Kilias, *Goście*, 114.
92. Szalai, *The Use of Time.*
93. International Studies of Values in Politics, *Values and the Active Community*; see also Lyons, *Theory, Data and Analysis*, 181–84; Szczepański, *Dzienniki z lat*, 125–27.
94. Barton, "Paul Lazarsfeld as Institutional Inventor"; Jeřábek, *Lazarsfeld's Methodology and Its Influence*, 61–69, 122–48; Sułek, "To America!"; Sułek, *Sondaż Polski*, 21–36.
95. Wagner, *Bauman*; Wiatr and Wysocki, "Zygmunt powinien był jasno powiedzieć."
96. Wiatr and Kozłowski, *O socjologii w Polsce Ludowej rozmów jedenaście*, 73.
97. Nešpor, Jan Szczepański v dějinách české sociologie."
98. MÚA AV ČR, Vědecké kolegium filosofie (a sociologie), inv. č. 173, sg. 91; Szczepański, *Dzienniki z lat*, 231–34.
99. Szczepański, "Některé aspekty situace, v níž se nachází marxistická sociologie"; Szczepański, *Dzienniki z lat*, 314–15; MÚA AV ČR, Vědecké kolegium filosofie (a sociologie) ČSAV, inv. č. 123, sign. 44.
100. MÚA AV ČR, Vědecké kolegium filosofie (a sociologie) ČSAV, inv. č. 166, sign. 82; Szczepański, *Dzienniki z lat*, 327.
101. ASP UMK in Toruń: Jan Szczepański's papers, 2/126.
102. Szczepański, *Dzienniki z lat*, 332–33.
103. Ibid., 281.
104. Votruba, "Hang Him High."
105. Hudek, "The Development of Relations," 36, 40–41.

106. In original, "Po kolacji siedziałem i piłem wino ze Słowakami, a potem prowadziłem *seriozne* rozmowy z Niemcami." (Szczepański, *Dzienniki z lat,* 315.)
107. Undated letters of Szczepański to his Slovak colleagues (ASP UMK in Toruń: Jan Szczepański's papers, 1/30).
108. Szczepański, *Dzienniki z lat,* 232.
109. The question of how hospitality—and by extension, professional acknowledgment—should be handled was often addressed during visits of international guests (for example, Szczepański, *Dzienniki z lat,* 245).
110. Hudek, "The Development of Relations."
111. Szczepański, *Dzienniki z lat,* 233–34.
112. Ibid., 235.
113. Ibid., 282.
114. Markiewicz, *Sto lat przeciw głupocie,* 243–44.
115. Three entries in Szczepański's diary provide a compelling glimpse into his self-perception and the broader implications of his election as ISA president. The first, written on 9 September 1966, captures his characteristic self-irony and humility as he reflected on the momentous event with a casual tone that downplayed its significance: "This afternoon, at the ISA Council [*sic*] meeting at 7:30 p.m., Johnny Szczepański from the hamlet Brzeziny in Zawodzie was elected *the President* of the International Sociological Association." (In original, "Dzisiaj po południu na zebraniu Counsil [sic] ISA o godz. 19.30 Jaś Szczepański z Brzeziny na Zawodziu został wybrany *the President* tejże International Sociological Association."). The second entry, from 3 October, reveals a more reflective and analytical side. Szczepański pondered the potential ramifications of his election and contextualized his role within the dynamics of global sociology and Cold War geopolitics: "I wonder whether being elected as the president of the ISA might have any further consequences. Just like the positions of rector or member of parliament, this one came about by chance, without any effort or desire on my part. There are hundreds of international organizations like the ISA. Sociology is, admittedly, gaining prominence in the contemporary world, and the fact that I am a president from Eastern Europe also carries some significance. Additionally, it matters that, while not being a communist, I have the support of the entire bloc of communist states; that, while not being a Catholic, I represent an ultra-Catholic country; and that, at the same time, I have sympathies and friendships within the Protestant ecumenical movement." The third entry, from December 1967, captures a turning point in Szczepański's self-understanding as a leader within the sociological community. Reflecting on his evolving role, he wrote: "I decided, under the circumstances as they were, to take on the role of organizer and leading sociologist in Eastern European countries, with some influence on sociology in the USSR." (Szczepański, *Dzienniki z lat,* 309, 311, 365.) These entries showcase Szczepański's ability to balance self-deprecating humor with profound insights into the symbolic weight of his position, while also tracing the process through which he gradually recognized and embraced the responsibilities of leadership within both the ISA and the broader sociological community of the Socialist Bloc.
116. Szczepański, *Dzienniki z lat,* 309–10.
117. Ibid., 391.
118. Ibid., 412.

119. Wiatr, "Hopes and Tragedy," esp. 427.
120. Szczepański, *Dzienniki z lat*, 418.
121. Wiatr, "Hopes and Tragedy," esp. 426.
122. Ibid., 427; see also Catalano, "Zdeněk Mlynář and the Search for a Socialist Opposition."
123. MÚA AV ČR, Vědecké kolegium filosofie (a sociologie) ČSAV, inv. č. 90, sign. 646.
124. Wagner, *Bauman*, 357.
125. Bauman, Petrusek, and Reifová, "Deset otázek profesoru Zygmuntu Baumanovi," esp. 210; see also J. Bauman, *Nigdzie na ziemi*, 27; Wagner, *Bauman*, 288.
126. Petrusek, "Zygmunt Bauman and Czech Sociology (1964–2010)."
127. In 1968, Szczepański accepted the role of head of the Institute of Philosophy and Sociology at the Polish Academy of Sciences following the request of its founder, Adam Schaff (1913–2006), who was forced to step down because of his Jewish origins.
128. Sedláček, "Socjologia polska"; Sedláček, Burian, and Šubrt, "Mezigenerační rozprava o vědě a životě"; Petrusek, "Jan Sedláček: náš současník"; Kaltenberg-Kwiatkowska, "Jan Sedláček (1935–2008)"; Červinková, "Miloslav (Milan) Petrusek (1936–2012)."
129. Hudek, "The Development of Relations," 36, 40–44.
130. Petrusek [pseud. J. Polehňa], "Z čeho studují adepti sociologické vědy?"
131. Szczepański, *Dzienniki z lat*, 108.
132. Kasperek and Böhm, *Spotkania filozoficzno-socjologiczne*.
133. The interdisciplinary field of futurology, discussed in another chapter of this book, is a case in point.
134. Building upon Jan Strzelecki's concept of "lyric Marxism" (*marksizm liryczny*)—which refers to the practice of including scholarly references to Marxist classics, typically in book introductions, to demonstrate ideological reliability to authorities—Joanna Kurczewska identifies two additional forms of Marxism in late socialist Poland: "epic Marxism," aligning closely with official propaganda when addressing Poland's situation, and "dramatic Marxism," applied in analyses of capitalist and Third World countries. (Kurczewska, "Socjologia polska," esp. 116–117.)
135. Petrusek, "Sociologie marxisticko-leninská."
136. Nešpor, "Miloslav Petrusek and the Samizdat *Sociologický obzor*"; Kopeček, "Dissident Legalism."
137. This topic warrants further investigation. Notably, in the early 1980s, Zdeněk Mlynář initiated comparative projects examining sociopolitical crises in Eastern Bloc countries. These projects involved the participation of several prominent Polish researchers, including sociologists Maria Hirszowicz and Krzysztof Mreła (*1950), political scientists Aleksander Smolar (*1940) and Jan Zielonka (*1955), and economist Włodzimierz Brus (1921–2007) (Catalano, "Zdeněk Mlynář and the Search for a Socialist Opposition," 328–39).

Bibliography

Barton, Allen H. 2001. "Paul Lazarsfeld as Institutional Inventor." *International Journal of Public Opinion Research* 13, no. 3: 245–69.

Battěk, Rudolf. 1967. "Společné rokování polských a československých sociologů" [Joint conference of Polish and Czechoslovak sociologists]. *Sociologický časopis / Czech Sociological Review* 3, no. 5: 625–27.

Bauman, Janina. 2011. *Nigdzie na ziemi: Powroty. Opowiadania* [Nowhere on earth: homecomings. A collection of short stories]. Łódź: Wydawnictwo Oficyna.

Bauman, Zygmunt. 1960. *Kariera: cztery szkice socjologiczne* [Career: four sociological essays]. Warsaw: Iskry.

———. 1962. *Socjologia na co dzień* [Sociology in everyday life]. Warsaw: Iskry.

———. 1962. *Zarys socjologii: Zagadnienia i pojęcia* [Outline of sociology: issues and concepts]. Warsaw: PWN.

———. 1964. *Wizje ludzkiego świata: Studia nad społeczną genezą i funkcją socjologii* [Visions of the human world: studies on the social genesis and function of sociology]. Warsaw: Książka i Wiedza.

———. 1965. *Sociologie* [Sociology], transl. Antonín Měšťan. Prague: Orbis.

———. 1965. *Sociologie pro každý den* [Sociology for everyday life], transl. Jan Sedláček. Prague: Mladá fronta.

———. 1967. *Kariéra: Sociologické črty* [Career: sociological sketches], transl. Jan Sedláček. Prague: Mladá fronta.

———. 1967. *Vízie ľudského sveta: Štúdia o spoločenskej genéze a funkcii sociológie* [Visions of the human world: studies on the social genesis and function of sociology], transl. Otakar Mošner and Jozef Marušiak. Bratislava: Obzor.

———. 1967. "Pierwsza ogólnokrajowa konferencja socjologów czechosłowackich" [First nationwide conference of Czechoslovak sociologists]. *Studia Socjologiczne* 2: 323–24.

———. 2023. *My Life in Fragments*, edited by Izabela Wagner, with translations by Katarzyna Bartoszyńska. London: Polity.

Bauman, Zygmunt, Miloslav Petrusek, and Irena Reifová. 1997. "Deset otázek profesoru Zygmuntu Baumanovi" [Ten questions for professor Zygmunt Bauman]. *Sociologický časopis / Czech Sociological Review* 33, no. 2: 203–16.

Bazant, Jan. 1961. "Review of *A Polish Factory* by J. Kolaja." *El Trimestre Económico* 28, no. 110/2: 355–58.

Bert, Jean-François, and Marcel Mauss. 2012. *Henri Hubert et la sociologie des religions: Penser et écrire à deux* [Henri Hubert and the sociology of religions: thinking and writing together]. Paris: La Cause des Livres.

Bömelburg, Hans-Jürgen. 2022. *Lodz: Geschichte einer multikulturellen Industriestadt im 20. Jahrhundert* [Łódź: history of a multicultural industrial city in the 20th century]. Paderborn: Brill / Schöningh.

Brokl, Lubomír et al. 2004. "Česká sociologie v letech 1965–1989" [Czech sociology in the years 1965–1989]. *Sociologický časopis / Czech Sociological Review* 40, no. 5: 695–740.

Bucholc, Marta. 2016. *Sociology in Poland: To Be Continued?* London: Palgrave Pivot.

Catalano, Alessandro. 2013. "Zdeněk Mlynář and the Search for a Socialist Opposition: From Active Politician to Dissident to Editorial Work in Exile." *Soudobé dějiny / Contemporary History* 20, no. 3: 261–344.

Červinková, Hana. 2012. "Miloslav (Milan) Petrusek (1936–2012)." *Kultura i Społeczeństwo* 56, no. 3: 165–69.

"Český sociolog v polské továrně" [A Czech sociologist in a Polish factory]. *Zprávy SVU* 2, no. 10 (December 1960): 87.

Chu, Winson. 2017. "From Łódź to Litzmannstadt: German Pasts and Holocaust Sites in Post-Communist Poland." *Holocaust and Genocide Studies* 31, no. 2: 240–67.

Goldfarb, Jeffrey C. 1980. *The Persistence of Freedom: The Sociological Implications of Polish Student Theatre*. Boulder, CO: Westview Press.

———. 1982. *On Cultural Freedom: An Exploration of Public Life in Poland and America*. Chicago: University of Chicago Press.

Gregor, Alois. 1967. "Za prof. dr. Maxmiliánem Kolajou" [In memory of prof. dr. Maxmilián Kolaja]. *Vlastivědný věstník moravský* 1: 117–18.

Gross, Edward. 1961. "Review of *A Polish Factory* by J. Kolaja." *The Sociological Quarterly* 2, no. 3: 235–36.

Havelka, Miloš. 1995. "Tschechische Soziologie im gesellschaftlichen Wandel: Zur Sozial- und Geistesgeschichte der tschechischen Soziologie bis 1989" [Czech sociology in social transformation: on the social and intellectual history of Czech sociology up to 1989]. *Jahrbuch für Soziologiegeschichte* 1995: 223–53.

Hirszowicz, Maria. 1964. *Konfrontacje socjologiczne: marksizm i socjologia współczesna* [sociological confrontations: Marxism and contemporary sociology]. Warsaw: Książka i Wiedza.

———. 1966. *Sociologické konfrontácie: marxizmus a súčasná sociológia* [Sociological confrontations: Marxism and contemporary sociology], transl. Anna Varsíková, and Verona Chorváthová. Bratislava: Vydavateľstvo politickej literatúry.

Heilbron, Johan. 1985. "Les métamorphoses du durkheimisme, 1920–1940" [The metamorphoses of Durkheimianism, 1920–1940]. *Revue française de sociologie* 26, no. 2: 203–37.

Hudek, Adam. 2022. "The Development of Relations between the Slovak and the Czechoslovak Academy of Sciences until 1970." *Práce z dějin Akademie věd* 14, no. 2: 27–47.

Igo, Sarah E. 2007. *The Average American: Surveys, Citizens, and the Making of Mass Public*. Cambridge, MA: Harvard University Press.

International Studies of Values in Politics. 1971. *Values and the Active Community: A Cross-National Study of the Influence of Local Leadership*. New York: Free Press.

Jeřábek, Hynek. 2025. *Lazarsfeld's Methodology and Its Influence on Postwar Sociology in Europe: The Rise of the Columbia Model in Sociology*. London: Routledge.

Kądzielski, Józef. 1958. "Informacja o badaniach nad małą grupą w przemyśle]" [Information on research on a small group in industry]. *Przegląd Socjologiczny* 12: 323–24.

———. 1961. "Review of *A Polish Factory* by J. Kolaja." *The Polish Sociological Bulletin* 1/2: 117–18.

Kaltenberg-Kwiatkowska, Ewa. 2008. "Jan Sedláček (1935–2008): Wspomnienie o naszym czeskim przyjacielu" [Jan Sedláček (1935–2008): in memory of our friend]. *Studia Socjologiczne* 4: 147–52.

Kasperek, Andrzej, and Hynek Böhm, eds. 2018. *Spotkania filozoficzno-socjologiczne na pograniczu w ramach Polsko-Czeskiej Wolnej Szkoły Nauk Filozoficznych i Społecznych im. Profesora Jana Szczepańskiego* [Philosophical and sociological encounters in the borderland within the Polish-Czech Free School of Philosophical and Social Sciences named after Professor Jan Szczepański]. Katowice: Polska Akademia

Nauk. Oddział Katowice; Cieszyn: Stowarzyszenie Rozwoju i Współpracy Regionalnej "Olza."

Katriak, Martin. 1966. "Niektoré aspekty rozvoja poľskej sociológie" [Some aspects of the development of Polish sociology]. *Sociologický časopis / Czech Sociological Review* 5: 691–97.

Kennedy, Michael D. 2004. "Poland in the American Sociological Imagination." *Polish Sociological Review* 4: 361–83.

Kilias, Jarosław. 2016. "Socjologia polska i czeska: wzajemne stosunki, ich charakter i konteksty" [Polish and Czech sociology: mutual relations, their character and contexts]. *Stan Rzeczy* 2: 382–315.

———. 2017. "An Older Brother: Polish and Czechoslovak Sociology in the 1960s." In *The Social Sciences in the Other Europe since 1945*, edited by Adela Hîncu and Victor Karady, 185–208. Budapest: CEU Press.

———. 2017. *Goście ze wschodu Socjologia polska lat sześćdziesiątych XX wieku a nauka światowa* [Guests from the East: Polish sociology of the 1960s and world scholarship]. Cracow: Nomos.

Kolaja, Jiří. 1947. "Review of *Prusy Wschodnie: Studium geograficzne, gospodarcze i społeczne* [East Prussia: a geographical, economic, and social study by Stanisław Srokowski, Instytut Bałtycki 1945.]" *Sociologická revue* 13, no. 1/2: 47.

———. 1947. "Review of *Polskie tradycje demokratyczne* [Polish democratic traditions], by Bogdan Suchodolski, Wrocław 1946; *Kultura* [Culture], by Stefan Czarnowski, Warsaw-Bydgoszcz 1946; *Opinia parlament prasa* [Opinion, parliament, press], by Henryk Jabłoński, Warsaw 1947; *Uspołecznienie kultury* [The socialization of culture], by Bogdan Suchodolski, Warsaw 1947." *Sociologická revue* 13, no. 4: 273–78.

———. 1948. "Review of *De Svenska privatanställda* [Swedish private-sector employees] by Fritz Croner, Stockholm 1939." *Przegląd Socjologiczny: Sociological Review* 10: 731–32.

———. 1950. "Swedish Feature Films and Swedish Society." *Hollywood Quarterly* 5, no. 2: 189–94.

———. 1960. "The Contemporary Polish Sociological Review." *American Sociological Review* 25, no. 1: 105–7.

———. 1960. *A Polish Factory: A Case Study of Workers' Participation in Decision Making*. Lexington: University of Kentucky Press.

———. 1960. "Acknowledgments." In *A Polish Factory: A Case Study of Workers' Participation in Decision Making,* by Jiří Kolaja, 7–18. Lexington: University of Kentucky Press.

———. 1961. "Review of *Wykształcenie a Pozycja Spoleczna Inteligencji* [Education and the social position of the intelligentsia]; *Socjologia: Rozwoj Problematyki i Metod,* by J. Szczepanski." *Social Forces* 40, no. 2: 198–99.

———. 1961. "*A Polish Factory*: Letters to the Editor." *American Journal of Sociology* 67, no. 2: 200.

———. 1962. "Review of *Robotnicy Warszawskiej Fabryki Motocykli / Workers of the Warsaw Motorcycles Factory,* by J. Malanowski." *American Journal of Sociology,* 68, no. 3: 365–365.

———. 1962. "Review of *Sytuacja społeczna dziennikarzy polskich* [The social situation of Polish journalists], by S. Dziecielska." *The Public Opinion Quarterly* 26, no. 4: 686–88.

———. 1963. "To the Editor." *American Journal of Sociology* 68, no. 4: 490.
———. 1963. "Review of *Praca i koleżeństwo* [Work and collegiality], by A. Matejko." *American Sociological Review* 28, no. 6: 1045–46.
———. 1964. "Stanislaw Ossowski." In *Revista Mexicana de Sociología* 26, no. 1: 287–88.
———. 1965. "Review of *Social Processes of Industrialization*, by W. Markiewicz." *American Sociological Review* 30, no. 2: 302–3.
———. 1965. "Review of *Occupational Choice and Prestige in Silesia; Social Function of Urbanists and Their Training; Socio-Psychological Productivity Factors in the Profession of Accountancy*, by K. Żygulski, & F. Michon." *Social Forces* 44, no. 1: 133–34.
———. 1965. "Dos revistas demográficas: Studia Demograficzne Polaca y Demografie Checoeslovaca [Two demographic journals: Polish demographic studies and Czechoslovak demography]." *Revista Mexicana De Sociología* 27, no. 2: 1073–75.
———. 1966. "Review of *Kultura masowa: krytyka i obrana / Mass Culture: Critique and Defense*, by A. Kłoskowska." *The Public Opinion Quarterly* 30, no. 3: 522–23.
———. 1967. "Review of *Sociologia, teoria, metody, problemy* [Sociology, theory, methods, problems] by Andrej Siracky; *Drogi rozwoju kultury masowej* [Paths of development of mass culture] by Kazimierz Zygulski; *Zakladní sociologické pojmy* [Basic sociological concepts] by J. Szczepanski." *Slavic Review* 26, no. 3: 500–1.
———. 1969 (reprint in 1980). *Social System and Time and Space: An Introduction to the Theory of Recurrent Behavior.* Pittsburgh, PA.: Duquesne University Press.
———. 1969. "The Social and Time." In *Revue Internationale de Sociologie: Actes du XXII Congrès de l'Institut International de Sociologie, Rome, 15–21 Septembre 1969* 2, 5, no. 3: 258–61.
———. 1969. "Review of *Film na wsi i w miastach powiatowych, Zagadnienia socjologiczne filmu* [Film in the countryside and in district towns], by K. Zygulski." *The Public Opinion Quarterly* 33, no. 4: 657–58.
———. 1970. "Review of *Silesian Workers: Their Working and Living Conditions*, by J. Raba." *The Polish Review* 15, no. 4: 117–18.
———. 1970. "Review of *Sociology of a Work Enterprise*, by A. Matejko." *The Polish Review* 15, no. 3: 92–93.
———. 1971. "Review of *Social Problems of Work and Production: A Soviet-Polish Comparative Research*, edited by G. V. Osipov & J. Szczepanski," *Social Forces* 49, no. 3: 514–15.
———. 1971. "Review of *Polish Society*, by J. Szczepanski." *The Sociological Quarterly* 12, no. 4: 554–55.
———. 1971. "Review of *Odmiany czasu teraźniejszego* [Variations of the present time], by J. Szczepanski." *The Kansas Journal of Sociology* 7, no. 4: 166.
———. 1976. "Review of *The Formation of a Modern Labor Force: Upper Silesia, 1865–1914*, by L. Schafer." *The Polish Review* 21, no. 1/2: 160.
———. 1976. "Review of *Valores y pautas culturale: Reflexiones de un sociólogo*, by K. Zygulski [Cultural values and patterns: reflections of a sociologist]." *Revista Española de La Opinión Pública* 45: 210.
———. 1978. "Review of *Cultural Values and Patterns: Reflections of a Sociologist*, by K. Zygulski." *International Review of Modern Sociology* 8, no. 2: 275–76.
———. 1980. "Review of *Wright Mills si Marxismul* [Wright Mills and Marxism]; *Šetření populačního klimatu: metodika, mezinárodní přehled a šetření venkovských žen v*

Čechách [Survey of population climate: methodology, international overview, and survey of rural women in Bohemia]; *Anton Štefánek, 1877–1977: Ľud a národ očami sociológa* [Anton Štefánek, 1877–1977: the people and the nation through the eyes of a sociologist], by A. Mihu, Z. Pavlík, and B. Štefánek." *Contemporary Sociology* 9, no. 1: 161–62.

———. 1982. "A Polish Survey of Human Issues: Review of *Sprawy ludzkie (Human Issues)*, by J. Szczepanski." *The American Journal of Economics and Sociology* 41, no. 1: 94–96.

Kolaja, Jiří, and Man Singh Das, eds. 1990. *Glimpses of Sociology in Eastern Europe.* New Delhi: Prints India.

Kopeček, Michal. 2022. "Dissident Legalism: Human Rights, Socialist Legality, and the Birth of Legal Resistance in the 1970s Democratic Opposition in Czechoslovakia and Poland." In *Making Sense of Dictatorship: Domination and Everyday Life in East Central Europe after 1945*, edited by Celia Donert, Ana Kladnik, and Martin Sabrow, 241–70. Budapest: CEU Press.

Kurczewska, Joanna. 2006. "Socjologia polska w latach 1944–1989 w służbie społeczeństw, siebie samej i/lub władzy" [Polish sociology in the years 1944–1989 in the service of society, itself, and/or power]. In *Humanistyka polska w latach 1945–1990* [Polish humanities in the years 1945–1990], edited by U. Jakubowska and J. Myśliński, 96–126. Warsaw: Fundacja Akademia Humanistyczna i Instytut Badań Literackich PAN.

Loucks, William N. 1961. "Review of *A Polish Factory* by J. Kolaja." *Current History* 40, no. 237: 308–9.

Lutyński, Jan. 1978. "Apparent Activities." *The Polish Sociological Bulletin* 41: 47–57.

Lyons, Pat. 2012. *Theory, Data and Analysis: Data Resources for the Study of Politics in the Czech Republic.* Prague: Institute of Sociology, Academy of Sciences of the Czech Republic.

Marcel, Jean-Christophe. 2001. *Le durkheimisme dans l'entre-deux-guerres* [Durkheimianism in the interwar period]. Paris: Presses Universitaires de France.

Markiewicz, Władysław. 2016. *Sto lat przeciw głupocie: Rozmowy z przyjaciółmi Pawłem Kozłowskim i Jerzym Słabickim* [A hundred years against stupidity: conversations with friends Paweł Kozłowski and Jerzy Słabicki]. Cracow: Universitas.

Musil, Jiří. 1977. *Urbanizace v socialistických zemích* [Urbanization in socialist countries]. Prague: Svoboda.

———. 1980. *Urbanization in Socialist Countries*, with a foreword by Simona Ganassi Agger. New York: Sharpe; Croom Helm.

———. 1984. *Urbanizacja w krajach socjalistycznych* [Urbanization in socialist countries], transl. by Kwieta Król. Warsaw: Książka i Wiedza.

———. 2004. "Poznámky o české sociologii za komunistického režimu" [Notes on Czech sociology under the communist regime]. *Sociologický časopis / Czech Sociological Review* 40, no. 5: 573–96.

Nešpor, Zdeněk R. 2014. "Miloslav Petrusek and the Samizdat *Sociologický obzor*: (New) Insight into a Sociological Thinker." *Sociologický časopis / Czech Sociological Review* 50, no. 2: 189–210.

———. 2014. "Jan Szczepański v dějinách české sociologie" [Jan Szczepański in the history of Czech sociology]. In *Profesorowi w hołdzie: publikacja poświęcona setnej rocznicy urodzin Profesora Jana Szczepańskiego* [In homage to the professor: a publication dedicated to the centenary of Professor Jan Szczepański's birth], edited by Daniel Kadłubiec, 52–61. Cieszyn: Oficyna Drukarsko-Wydawnicza Akant.

———. 2019. "Československá sociologická společnost při ČSAV jako institucionální báze obnoveného oboru" [The Czechoslovak Sociological Society at the Czechoslovak Academy of Sciences as the institutional basis of the restored discipline]. *Práce z dějin Akademie věd* 11: 19–30.

Nešpor, Zdeněk R. et al. 2014. *Dějiny české sociologie* [History of Czech sociology]. Prague: Academia.

Noland, E. William. 1961. "Review of *A Polish Factory* by J. Kolaja." *Social Forces* 39, no. 4: 362–63.

Obrdlíková, Juliána. 1968. "Sociolog o dělnických radách [A sociologist on workers' councils]: Review of *A Polish Factory* and *Workers' Councils* by J. Kolaja." *Sociologický časopis: Czech Sociological Review* 4, no. 6: 735–38.

Oakes, Guy, and Arthur J. Vidich. 1999. *Collaboration, Reputation and Ethics in American Academic Life: Hans H. Gerth and C. Wright Mills.* Urbana: University of Illinois Press.

Perrow, Charles. 1961. "Review of *A Polish Factory: A Case Study of Workers' Participation in Decision-Making,* by J. Kolaja." *American Journal of Sociology* 66, no. 6: 651–52.

Petrusek, Miloslav [pseud. J. Polehňa]. 1988. "Z čeho studují adepti sociologické vědy? [What do students of sociology study from?]" *Sociologický obzor* 2, no. 3: 54–59.

Petrusek, Miloslav. 2004. "Výuka sociologie v čase tání a v časech normalizace (1964–1989) (Kapitola o vztahu vědění a moci, vědy a politiky, reality a mýtotvorby)" [Teaching sociology in the thaw period and in the normalization era (1964–1989): a chapter on the relationship between knowledge and power, science and politics, reality and mythmaking]. *Sociologický časopis / Czech Sociological Review* 40, no. 5: 597–607.

———. 2010. "Zygmunt Bauman and Czech Sociology (1964–2010)." *Sociologický časopis / Czech Sociological Review* 46, no. 6: 1035–45.

———. 2011. *České sociální vědy v exilu 1948 až 1989* [Czech social sciences in exile 1948 to 1989]. Prague: Sociologické nakladatelství.

———. 2015. "Jan Sedláček: náš současník" [Jan Sedláček: our contemporary]. In *"Prima to bývalo a prima je to i teď!": memoáry k 50. výročí založení katedry sociologie na FF UK* ["It used to be great, and it is great even now!": memoirs for the 50th anniversary of the founding of the department of sociology at the Faculty of Arts, Charles University], edited by Markéta Sedláčková and Jakub Mlynář, 34–37. Prague: Filozofická fakulta Univerzity Karlovy.

———. "Sociologie marxisticko-leninská" [Marxist-Leninist sociology]. In *Sociologická encycklopedie* [Sociological encyclopedia], edited by Zdeněk R. Nešpor. Retrieved 26 March 2025 from https://encyklopedie.soc.cas.cz/w/Sociologie_marxisticko-leninská.

Platt, Jennifer. 1997. *A History of Sociological Research Methods in America, 1920–1960.* Cambridge: Cambridge University Press.

———. 1998. *A Brief History of the International Sociological Association, 1949–1987.* Montreal: ISA.

Prokůpek, Ladislav. 2002. "Normalizace Ústavu po filosofii a sociologii ČSAV " [Normalization of the Institute of Philosophy and Sociology of the Czechoslovak Academy of Sciences]. In *Věda v Československu v období normalizace (1970–1975)* [Science in Czechoslovakia during the normalization period (1970–1975)], edited by Antonín Kostlán, 201–17. Prague: Výzkumné centrum pro dějiny vědy.

Rodnick, David. 1961. "Review of *A Polish Factory* by J. Kolaja." *American Sociological Review* 26, no. 2: 308.

Sarapata, Adam. 1962. "Review of *A Polish Factory* by J. Kolaja." *Industrial and Labor Relations Review* 15, no. 3: 433–34.

Schatzki, Theodore R., Karin Knorr Cetina, and Eike von Savigny, eds. 2001. *The Practice Turn in Contemporary Theory*. London: Routledge and Taylor & Francis Group.

Sedláček, Jan. 1965. "Socjologia polska w oczach socjologa czeskiego" [Polish sociology through the eyes of a Czech sociologist]. *Studia Socjologiczne* 4: 317–21.

———. 1968. "Další hlasy v polské diskusi o sociologii" [Further voices in the Polish discussion on sociology]. *Sociologický časopis* 4: 741–44.

———. 2005. "Nazdar" [Cheers]. In *Jan Szczepański: Humanista—uczony—państwowiec: Księga wspomnień* [Jan Szczepański: humanist—scholar—statesman: book of reminiscences], edited by J. Kulpińska, 107–13. Warsaw: Wydawnictwo Sejmowe.

Sedláček, Jan, Jiří Burian, and Jiří Šubrt. 2000. "Mezigenerační rozprava o vědě a životě" [An intergenerational conversation about science and life]. In *Jedinec a společnost. Sborník k 65. narozeninám doc. PhDr. Jana Sedláčka, CSc.* [The individual and society: volume for the 65th birthday of doc. PhDr. Jan Sedláček, CSc.], 13–31. Prague: Charles University, Faculty of Philosophy.

Skovajsa, Marek, and Jan Balon. 2016. *Sociology in the Czech Republic: Between East and West*. London: Palgrave Pivot.

"Socialistický člověk zatím neexistuje" [The socialist human does not yet exist]. *České slovo* 7, no. 5 (1961): 6.

Staniszkis, Jadwiga. 1984. *Poland's Self-Limiting Revolution*. Princeton, NJ: Princeton University Press.

Stola, Dariusz. 2000. *Kampania antysyjonistyczna w Polsce 1967–1968* [The anti-Zionist campaign in Poland 1967–1968]. Warsaw: Instytut Studiów Politycznych PAN.

Sułek, Antoni. 2006. *Sondaż Polski: Przygarść rozpraw o badaniach ankietowych* [The Polish survey: a handful of essays on survey research]. Warsaw: Wydawnictwo Instytutu Filozofii i Socjologii PAN.

———. 2010. "'To America!': Polish Sociologists in the United States after 1956 and the Development of Empirical Sociology in Poland." *East European Politics and Societies* 24, no. 3: 327–52.

Sułek, Antoni, and Nina Kraśko. 2002. "The Multifarious and Changing Functions of the Polish Sociological Association." *International Sociology* 17: 213–31.

Szalai, Alexander, eds. 1972. *The Use of Time: Daily Activities of Urban and Suburban Populations in Twelve Countries*. The Hague: Mouton.

Szczepanski, Jan. 1961. "Review of *A Polish Factory* by J. Kolaja." *Social Problems* 9, no. 1: 98–99.

———. 1961. *Socjologia: rozwój problematyki i metod* [Sociology: the development of problems and methods]. Warsaw: PWN.

———. 1961. *Sociológia: vývin problematiky a metód* [Sociology: the development of problems and methods], transl. Vladimír Košnar, Bohuslava Vajdová, and Anna Vasilová. Bratislava: Vydavateľstvo politickej literatury.

———. 1963. *Elementarne pojęcia socjologii* [Elementary concepts of sociology]. Warsaw: PWN.
———. 1966. *Základní sociologické pojmy* [Basic sociological concepts], transl. Miroslav Disman. Prague: NPL.
———. 1965. *Zagadnienia socjologii współczesnej* [Issues of contemporary sociology]. Warsaw: PWN.
———. 1968. *Spektrum společnosti: Otázky soudobé sociologie* [Spectrum of society: questions of contemporary sociology], transl. by Jan Sedláček. Prague: Mladá fronta.
———. 1968. "Některé aspekty situace, v níž se nachází marxistická sociologie" [Some aspects of the situation in which Marxist sociology finds itself]. In *Sociologie ′66. Sborník z I. celostátní konference* československých *sociologů* [Sociology '66: proceedings of the first nationwide conference of Czechoslovak sociologists], 49–51. Prague: Svoboda.
———. 2013. *Dzienniki z lat 1945–1968* [Diaries from the years 1945–1968], edited by Daniel Kadłubiec. Ustroń: Muzeum Ustrońskie im. Jana Jarockiego.
Szczerba, Adrianna. 2019. *Dom uczonych: Kamienica przy ul. Uniwersyteckiej 3 w Łodzi w latach 1945–2017* [House of scholars: the building at 3 Uniwersytecka Street in Łódź in the years 1945–2017]. Łódź: Wydawnictwo Uniwersytetu Łódzkiego.
Szostkiewicz, Stefan. 1960. *Procedury i techniki badań socjologicznych: skrypt* [Procedures and techniques of sociological research: course script]. Warsaw: Wojskowa Akademia Polityczna im. F. Dzierżyńskiego.
———. 1964. *Sociologický výskum: Procedúry a techniky* [Sociological research: procedures and techniques], transl. Ján Pašiak. Bratislava: Vydavateľstvo politickej literatúry.
———. 1967. "First Meeting of Czechoslovak and Polish Sociologists." *The Polish Sociological Bulletin* 16: 130–31.
Taborsky, Edward. 1961. "Review of *A Polish Factory* by J. Kolaja." *The Annals of the American Academy of Political and Social Science* 338: 172–73.
Touraine, Alain. 1983. *Solidarity: The Analysis of a Social Movement: Poland, 1980–1981.* Cambridge: Cambridge University Press.
Urbanek, Eduard. 1994. "Ups and Downs in Czech Sociology." In *Eastern Europe in Transformation: The Impact on Sociology*, edited by M. F. Keen and J. Mucha, 79–87. Westport, CT: Greenwood.
Voříšek, Michael. 2012. *The Reform Generation: 1960s Czechoslovak Sociology from a Comparative Perspective.* Prague: Kalich.
Votruba, Martin. 2006. "Hang Him High: The Elevation of Jánošík to an Ethnic Icon." *Slavic Review* 65, no. 1: 24–44.
Wagner, Izabela. 2020. *Bauman: A Biography.* Cambridge: Polity.
Wandycz, Piotr S. 1961. "Review of *A Polish Factory* by J. Kolaja." *The Polish Review* 6, no. 1/2: 165–67.
Warczok, Tomasz. 2016. "Globalne pole nauk społecznych a socjologia Polska: Zarys centro-peryferyjnego przepływu idei" [The global field of social sciences and Polish sociology: an outline of centre-periphery flows of ideas]. In *Polska jako peryferie*, edited by Tomasz Zarycki, 170–86. Warsaw: Wydawnictwo Naukowe Scholar.
Wedel, Janine. 1986. *The Private Poland: An Anthropologist's Look at Everyday Life.* New York: Facts on File.

Whyte, William Foote. 1960. "Forward." In *A Polish Factory: A Case Study of Workers' Participation in Decision Making*, by Jiří Kolaja, 5–6. Lexington: University of Kentucky Press.
Wiatr, Jerzy J. 1962. *Wojsko, społeczeństwo i polityka w Stanach Zjednoczonych* [The military, society and politics in the United States]. Warsaw: MON.
———. 1962. *Szkice o materializmie historycznym i socjologii* [Essays on historical materialism and sociology]. Warsaw: Książka i Wiedza.
———. 1963. *Armáda, společnost, politika v USA* [The army, society and politics in the USA], transl. Josef R. Burian. Prague: Naše vojsko.
———. 1964. *O sociológii* [On sociology], transl. Šimon Ondráš. Bratislava: Osveta.
———. 1964. *Społeczeństwo: wstęp do socjologii systematycznej* [Society: an introduction to systematic sociology]. Warsaw: PWN.
———. 1968. *Společnost: Úvod do systematické sociologie* [Society: an introduction to systematic sociology], transl. Eda Sládková. Prague: SPN.
———. 1965. *Socjologia zaangażowana* [Engaged sociology]. Warsaw: Książka i Wiedza.
———. 1966. *Czy zmierzch ery ideologii* [The end of the era of ideology]? Warsaw: Książka i Wiedza.
———. 1968. *Zánik ideológie? Problémy politiky a ideológie v súčasnom svete* [The end of ideology? Problems of politics and ideology in the contemporary world], transl. Andrej Kopčok. Bratislava: Vydavateľstvo politickej literatúry.
———. 1970. *Angažovaná sociologie: črty o sociologii a politice* [Engaged sociology: essays on sociology and politics], transl. Peluška Bendlová. Prague: Mladá fronta.
———. 2018. "Hopes and Tragedy: The Prague Spring from a Polish Perspective." *Sociologický časopis / Czech Sociological Review* 54, no. 3: 423–28.
Wiatr, Jerzy J., and Grzegorz Wysocki. 2021. "Zygmunt Bauman powinien był jasno powiedzieć, że był przekonanym komunistą" [Zygmunt Bauman should have clearly said that he was a convinced communist]. *Gazeta Wyborcza*, 23 April.
Wiatr, Jerzy J., and Paweł Kozłowski. 2022. *O socjologii w Polsce Ludowej rozmów jedenaście* [On sociology in People's Poland: eleven conversations]. Warsaw: Instytut Historii Nauki im. Ludwika i Aleksandra Birkenmajerów Polskiej Akademii Nauk.
Wincławski, Włodzimierz. 2004. "Józef Kądzielski." In *Słownik biograficzny socjologii polskiej* [Biographical dictionary of Polish sociology], vol. 2, edited by Włodzimierz Wincławski, 53–54. Toruń: Wydawnictwo Uniwersytetu Mikołaja Kopernika.
Zysiak, Agata. 2023. *Limiting Privilege: Upward Mobility within Higher Education in Socialist Poland*. West Lafayette, IN: Purdue University Press.

CHAPTER 7

Czechoslovak–Polish Scientific Collaboration on Security in Central and Eastern Europe

The Pugwash Conferences in Czechoslovakia and Poland

Doubravka Olšáková

Throughout the Cold War, scientific collaboration served as an unofficial yet vital channel of Czechoslovak–Polish entanglements, particularly in Central and Eastern Europe, where ideological constraints limited traditional state interactions. The Pugwash Conferences on Science and World Affairs became a key venue for science diplomacy, fostering communication between scientists from opposing blocs despite the extensive state control that characterized scientific work in socialist countries. This chapter examines how Pugwash functioned as an alternative diplomatic space for scientific exchange in Czechoslovakia and Poland, where state oversight heavily influenced international engagement.

While Western historiography has explored Pugwash's role in Cold War diplomacy, its specific impact on scientific communities in the Eastern Bloc—especially Czechoslovakia and Poland—remains underexamined. This chapter contributes to addressing this gap by analyzing the strategies employed by Czechoslovak and Polish scientists within Pugwash as a transnational nonstate actor.

Using archival sources, conference records, and interviews, this study adopts a transnational historical approach to examine how Pugwash provided an avenue for Czechoslovak and Polish researchers to navigate state restrictions. It explores key turning points, such as the occupation of Czechoslovakia in 1968 and the 1982 Warsaw Pugwash Conference, to illustrate how scientific communities adapted to changing political landscapes.

This chapter is organized into several sections, each of which examines a different aspect of the role of the Pugwash Conferences in Czechoslovak–Polish scientific collaboration and Cold War diplomacy.

The first section provides a historical background on Pugwash, tracing its origins from the 1955 Russell–Einstein Manifesto to its establishment as a key platform for informal diplomatic engagement between scientists. It highlights how Pugwash functioned as a nonstate actor, fostering dialogue across ideological divides, and explores the emergence of Pugwash in Central and Eastern Europe, particularly in Czechoslovakia and Poland. It examines the complex relationship between science diplomacy and state control, illustrating how the Czechoslovak and Polish scientific communities navigated the constraints of socialist regimes while engaging with Western counterparts.

The second section focuses on the nuclear-free zone initiative, particularly the Polish-led efforts to implement the Rapacki Plan within the Pugwash framework. It explores how science diplomacy was leveraged to advance disarmament proposals, culminating in significant discussions at the Sopot Conference in 1966.

The third and fourth sections analyze historical turning points, such as the 1964 Karlovy Vary Conference, which marked Czechoslovakia's active participation in international security debates, and the occupation of Czechoslovakia in 1968, which severely disrupted its involvement in Pugwash. The fourth section explores the impact of the Warsaw Pact's occupation of Czechoslovakia on the Pugwash Conferences and the broader dynamics of science diplomacy during the Cold War. It examines the discussions at the 29th Pugwash Continuing Committee meeting in Nice (September 1968), where the occupation was debated alongside other global conflicts, leading to a call for the withdrawal of foreign troops. It then analyzes the 1969 Sochi conference, which revealed stark ideological divisions, as Western participants condemned the invasion while Soviet and Polish representatives sought to suppress the discussion. The section further investigates the decline of Czechoslovakia's role in Pugwash due to political repression, culminating in the 1972 elections that effectively ended its active participation. Finally, it considers Poland's growing prominence in European Pugwash activities during the 1970s and 1980s, particularly in hosting key discussions on security and disarmament, even as earlier efforts like the Rapacki Plan lost relevance. The analysis in this section highlights how geopolitical tensions shaped the role of Central and Eastern European scientists in international disarmament efforts.

The organization of the Pugwash Conference in Warsaw in 1981, during the period of martial law in Poland, and the meeting of the top Pugwash officials with General Jaruzelski is a key point in the fifth section. It addresses the politicization of Pugwash in the 1970s and 1980s, detailing how the organization evolved from a scientific forum into a contested space for debates on human rights and political dissidence. It assesses the impact of key figures such as Andrei Sakharov and František Janouch, whose interventions challenged Pugwash's stance on East–West dialogue.

The conclusion evaluates the long-term significance of Czechoslovak–Polish collaboration within Pugwash and considers its legacy in Cold War diplomacy and its role in shaping international security policies. It reflects on how the movement adapted to shifting geopolitical realities and the challenges it faced after the Cold War.

Through this structure, the chapter highlights both the collaborative and contested nature of science diplomacy in Central and Eastern Europe, emphasizing the strategic role played by Czechoslovak and Polish scientists within Pugwash as a transnational actor.

Although this chapter does not primarily focus on bilateral relations, unlike others in this volume, it highlights how the cooperation between Czechoslovak and Polish scholars within Pugwash played a pivotal role in shaping international debates on Central Europe. While Pugwash is often studied as a transnational movement, the collaboration between the Czechoslovak and Polish Pugwash Committees offers a unique perspective on bilateral science diplomacy within such a specific framework. Their ability to navigate the evolving role of nonstate actors from the 1970s and to foster productive cooperation beyond state-controlled structures underscores how national scientific communities could actively engage in international discourse while simultaneously shaping regional security and policy discussions.

Pugwash: A Brief Introduction

Pugwash was founded in 1957 under the official name the Pugwash Conferences on Science and World Affairs as a platform for discussion on the social responsibility of scientists. It was initiated by Bertrand Russell and the financier Cyrus Eaton (1883–1979). The starting point was the Russell–Einstein Manifesto of 1955, in which Bertrand Russell and Albert Einstein jointly called on scientists to take responsibility for the consequences of their discoveries. Their primary concern was nuclear weapons. In addition to Einstein and Russell, the eleven signatories included Frédéric Joliot-Curie (1900–1958), Linus Pauling (1901–1994), Joseph Rotblat (1908–2005) and others, most of whom were or later became Nobel laureates.[1] The US businessman Cyrus Eaton, following discussions with Bertrand Russell, eventually consented to finance the conferences himself. As a consequence, the funding had been secured and Pugwash meetings regularly took place since 1957, and Pugwash itself had a small secretariat.

Pugwash is one of the most prominent nonstate actors in international relations, having played since its inception in 1957 an absolutely crucial role in informal diplomacy during the Cold War in the negotiation of key security, nuclear weapons–related, and geopolitical issues. Its strong position stems from its role as a platform for informal discussions or prenegotiations of intergovernmental or international arrangements, as well as from the very convenient timing of its emergence. The history of Pugwash is well established in Western historiography,[2] and as an unconventional nonstate actor that spearheaded transnational and cross-bloc exchanges and dialogue, Pugwash has become a significant subject in Cold War historiography. However, research on Pugwash within Central and Eastern European scholarship remains underdeveloped, and the existing literature is both limited and fragmented.[3]

Pugwash has played multiple roles in international affairs. Its analytical function was based on the competence of its members and participants in its deliberations, whose knowledge and experience allowed the formulation of clear positions, free from political pressure or the necessity to compromise in the first phase of free debate. This was a key function of Pugwash, especially in the sensitive area of nuclear weapons and nuclear armament. At the same time, Pugwash and its meetings had a consultative function. This can be seen especially in the case of Czechoslovak engagement in the first half of the 1960s, when some positions and opinions were directly tested at Czechoslovak Pugwash meetings by leaders of the US administration, such as Henry Kissinger, who always stressed that he was in attendance as a private citizen. Another function was that of articulation. With access to a press interested in Pugwash, Pugwash was able to communicate its expertise and opinions around the world, making it an important player on the international scene. It published the opinions and expertise of Pugwashites in various printed sources, including the bimonthly *Pugwash Newsletter*, in which it closely followed the arms situation and most of the issues it addressed in its work.

The most important function, however, remained its ability to provide highly specialized expertise. It was its ability not only to analyze the situation, but, above all, to formulate conclusions and proposals for action on the basis of expertise and social responsibility for development, taking into account the social responsibility motive for the use of scientific knowledge–whether it be nuclear, chemical, or biological weapons.[4]

Early work on Pugwash was produced mainly by those involved in Pugwash. The most notable example is the 1967 volume by one of the founders,

Joseph Rotblat—the publication of which was made possible by the Czech Pugwash group, and it was published in Prague.[5]

Joseph Rotblat (1908–2005) was a Polish-British physicist and a key figure in the international movement against nuclear weapons. Initially involved in the Manhattan Project, he became the only scientist to leave on moral grounds, when it became clear that Nazi Germany was not close to developing an atomic bomb. Deeply committed to nuclear disarmament, Rotblat played a leading role in the Pugwash Conferences on Science and World Affairs, co-founding the movement in 1957 alongside Bertrand Russell and others. For his lifelong efforts in promoting peace and responsible science, he was awarded the 1995 Nobel Peace Prize alongside Pugwash.

Other work by Pugwash members tends to be uncritical and heavily influenced by their own views and experiences of Pugwash as a successful platform for scientists and diplomats to meet across the two blocs[6] or, more moderately, as the main "communication channel" between scientists from East and West.[7] Moreover, the Pugwash story is primarily found as part of literature on wider themes, such as nuclear disarmament, anti-nuclear protest, or the cultural history of the atomic bomb.[8] Such almost uncritical admiration is counterbalanced primarily by contemporary comments on the role and limitations of Pugwash, written by authors publishing in *The Bulletin of the Atomic Scientists*, especially Eugene Rabinowitch, and authors such as Ivan Málek[9] publishing in *The Scientific World*, its parallel journal behind the Iron Curtain, published first in London and then in Prague and Berlin by the World Federation of Scientific Workers. While much has been written about *The Bulletin of the Atomic Scientists* and its role, a critical assessment of the role and importance of *The Scientific World*, the principal organ of the Federation, has been noticeably absent in contemporary Pugwash research.[10]

The only scientist from the Soviet Bloc to sign the manifesto was Leopold Infeld (1898–1968). He was a prominent Polish theoretical physicist, best known for his contributions to the field of relativity and his collaboration with Albert Einstein. Born in Krakow, then part of Austria-Hungary, Infeld continued his study of physics in Lwów (now Lviv, Ukraine), eventually becoming a professor. Driven by an interest in theoretical physics, he traveled to England and Canada before joining Einstein at Princeton University in the 1930s. From 1933 to 1934, he held a prestigious Rockefeller Fellowship at Cambridge University in Great Britain. Subsequently, between 1938 and 1950, he resided in Canada and the United States, during which time he had the opportunity to meet and collaborate with Albert Einstein. Together they developed the Einstein–Infeld–Hoffmann equations, which describe the motion of massive objects in general relativity without relying on gravitational fields, thus advancing the understanding of relativity.

Leopold Infeld returned to Poland in 1950, where he continued his research and became a prominent figure in Polish science, despite the challenges he faced under the communist regime. He published several popular science books, such as *The Evolution of Physics*, co-authored with Einstein, which aimed to explain complex physical theories to a general audience. His work left a significant legacy in the field of theoretical physics, bridging scientific research and public understanding, and his contributions continue to influence studies in relativity and mathematical physics.

Pugwash was founded and engaged with the Soviet Bloc countries during the period of "restricted internationalism," a transitional phase between the Stalinist Sovietization of Central and Eastern European scientific communities in the early 1950s and the multilateral cooperation of Eastern Bloc countries in large-scale programs under the aegis of the UN or UNESCO, especially in the 1960s and 1970s. Initiated after the Geneva Summit in July 1955, Pugwash offered an alternative both to official channels of dialogue between East and West and relations forged in the course of East–West scientific and technological cooperation. The idea of an organization that would provide a cross-bloc platform was welcomed not only by the United States and other Western countries, but also in the East, where the crisis that engulfed the World Peace Council (hereafter WPC) and the World Federation of Scientific Workers (WFSW), which had been unable to advance their international agendas in the 1960s, helped open the doors to Pugwash. For the Soviets, Pugwash was an interesting alternative channel to the WPC and the WFSW, not only as a potential means of discussing with the West issues in which discussion through the usual political or diplomatic channels was difficult or even blocked. To do this, however, the Soviets called on the support of their political allies within the Eastern alliance—who, indeed, had been literally forced by Moscow to join Pugwash.[11]

The absence of distinguished Soviet scientists from the launch of the Russell–Einstein Manifesto is understandable; Stalinism and the Cold War led to the isolation of the scientific community until 1955, when cooperation between the two blocs began at the first major postwar summit of the great powers in Geneva in July 1955. The Soviets, however, remained concerned, mainly because Leopold Infeld happened to attend.

Leopold Infeld was a personality in the Polish postwar community who was comparable in the USSR to another famous physicist, Pyotr Kapitsa, whose biography was very similar. Both Einstein and Russell wanted to include him as one of the signatories, notwithstanding the fact that Infeld had repatriated to Poland from Canada and had assumed a pivotal role in the restoration of the postwar Polish scientific community. Infeld's active participation in Pugwash and his collaboration with the Polish scientific community contradicts prevailing beliefs on the social engagement of Polish scientists and intellectuals, as suggested by Michael D. Kennedy, who claimed that such engagement began only in 1980–81.[12] Though not all East European nations were present at the start of the Pugwash conferences—some joined after destalinization had begun—his participation and involvement remained exceptional.

There were twenty-two participants from ten countries at the first meeting in Pugwash in 1957, which was dedicated to three main fields of discussion: (1) the dangers of radiation, (2) international control of nuclear energy, and (3) the social responsibility of scientists. Leopold Infeld was invited as one of the few delegates from the Soviet Bloc, together with Polish physicist Marian Danysz, who held the prestigious position of vice president of the Joint Institute for Nuclear Research in Dubna, near Moscow, at that time.[13] The USSR delegated Alexander Vasilyevich Topchiev and Andrei Mikhailovich Kuzin.[14] Alexander Vasilyevich Topchiev (1907–1962) was a Soviet chemist and a key figure in Cold War–era science diplomacy.[15] He played a significant role in fostering international scientific cooperation and was actively involved in the Pugwash Conferences on Science and World Affairs as one of the Soviet Union's representatives. As an academician of the Soviet Academy of Sciences, he contributed to discussions on nuclear disarmament and scientific responsibility during the 1950s and early 1960s, representing Soviet scientific interests in international forums. Andrei Mikhailovich Kuzin (1916–2002) was a Soviet physicist specializing in nuclear research and international scientific policy. He was also a representative of the Soviet Union in the Pugwash Conferences, participating in the early meetings in which East–West scientific dialogue on arms control was shaped. As a senior scientist and administrator, he played a role in Soviet nuclear policy discussions and international scientific exchanges, particularly during the period when the USSR sought to balance diplomatic engagement with scientific secrecy in nuclear affairs.

The participation of both Polish scientists was a particular cause of concern in Poland, especially within the ruling Polish United Workers' Party (Polska Zjednoczona Partia Robotnicza, hereafter PZPR).[16] Between 1955 and 1957, Poland and the entire Soviet Bloc were still exploring the possibilities—and therefore the limits—of future scientific cooperation across the Iron Curtain. The uncertainty arising from this new situation was reflected in how the regime treated its scientists when it came to internationalizing. The example of Marian Danysz, a leading Polish expert in nuclear research, illustrates that the participation of Polish scientists was very carefully monitored. In 1962, Joseph Rotblat invited him to attend the Pugwash Conference in Cambridge. Although the PZPR was aware of Leopold Infeld's participation in Pugwash and was able to sell Rotblat's origins propagandistically, the invitation came as a surprise, which alarmed Witold Jarosiński, the secretary of the Polish United Workers' Party's Central Committee (KC PZPR). At least four bodies—the KC PZPR, the PZPR Department of Science, the Polish Academy of Sciences executive board, and the Polish Ministry of Foreign Affairs—evaluated Danysz's invitation and recommended that he ask Rotblat to extend the invitation to theoretical physicist Paweł Szulkin and Witold Stefański.[17]

It was thanks to Leopold Infeld and Marian Danysz that the Polish scientific community was directly involved in the birth of Pugwash. Leopold Infeld was a long-standing member of the Pugwash Continuing Committee. After his death in 1968 he was succeeded by Ignacy Malecki. Ignacy Malecki (1912–2004) was a Polish scientist specializing in electroacoustics and the study of science. He significantly contributed to the development of technical acoustics and was instrumental in rebuilding Poland's energy sector after World War II. In 1973 Maciej Nałecz became a member of the Continuing Committee, which functions as the executive body of the Pugwash Conferences on Science and World Affairs, managing ongoing activities, setting agendas, and guiding policy between the major Pugwash conferences. Maciej Nałecz (1922–2009) was a Polish scientist who specialized in biocybernetics and biomedical engineering, known for his contributions to artificial kidney and pancreas development. He was the founder and longtime director of the Institute of Biocybernetics and Biomedical Engineering at the Polish Academy of Sciences and played a key role in the Pugwash Conferences on Science and World Affairs, serving as chairman of its international council for two decades. Other regular Polish participants were Marian Dobrosielski and Paweł Szulkin. Marian Dobrosielski (1923–1922) was a Polish diplomat, philosopher, and political scientist who actively participated in the Pugwash Conferences on Science and World Affairs, representing Poland in discussions on international security and arms control. Paweł Szulkin (1940–2018) was a Polish physicist specializing in nuclear and theoretical physics. He was also involved in Pugwash initiatives, contributing to science diplomacy and Cold War–era arms control debates.

The first Czechoslovak delegates were among the seventy scientists from twenty countries who met in the Austrian town of Kitzbühel and in Vienna[18] in 1958 and 1959. They attended despite the fact that the Czechoslovak Pugwash committee had not yet been established.[19] Viktor Knapp (1913–1996) and Jaroslav Kožešník (1907–1985), both members of the Czechoslovak Academy of Sciences, were invited through private contacts with Pugwash members and did not attend as official representatives of Czechoslovakia. Viktor Knapp was a Czech legal scholar and expert in civil law who played a key role in drafting the 1968 Czechoslovak federalization reforms. Jaroslav Kožešník later became president of the Czechoslovak Academy of

Sciences (1970–1980) and was instrumental in promoting the development of automation and artificial intelligence in socialist Czechoslovakia.

This rather hesitant participation of Czechoslovak scientists in the Austrian meetings was a result of the destalinization of Czechoslovak and, more generally, Central European science, but it was more directly influenced by the developments of the international situation in Europe, in particular, by the 1955 Geneva Summit.[20] Generally speaking, the first stage of establishing relations between the Pugwash movement and Central and Eastern Europe was rather hesitant and not officially acknowledged by the state and Party. It was only after the Pugwash Conference in Moscow in 1960 that the countries of the Eastern Bloc were encouraged to participate.

The Soviets were very interested in having the full and unanimous support of the Soviet Bloc countries, some of whom were compelled to participate in the Pugwash movement even against the will of the national scientific community. For example, Poland and East Germany had reservations but acted in response to pressure from their national governments, which were in turn pressured by Moscow.[21] Despite these difficult beginnings, the collaboration eventually worked very well.

In July 1960, Bertrand Russell wrote to František Šorm, the president of the Czechoslovak Academy of Sciences, inviting him to the Moscow Pugwash Conference. Despite initial approval from the Presidium of the Czechoslovak Academy of Sciences, Šorm cancelled the trip in early November 1960. It would seem that he had decided to boycott the Moscow Pugwash Conference. But the Soviet Academy of Sciences began to press for Czechoslovak participation as soon as Moscow was informed. After the cancellation, Šorm received several phone calls from Sergei Ivanovich Prasolov, head of the Foreign Committee of the USSR Academy of Sciences.[22] His persistence suggests that the USSR might have considered Czechoslovakia's participation crucial for the success or political goals of the conference. It also highlights that scientific delegations and decisions were subject to political scrutiny and approval, reflecting the controlled environment in which these academies operated. A few days after Šorm's cancellation, a new Czechoslovak delegation was formed at the urging of the USSR Academy of Sciences, which attended the Moscow event.[23]

The Moscow Conference was important for the future of the Pugwash movement in the Eastern Bloc because it inspired the idea of forming structures of national Pugwash committees in the countries of Eastern and Central Europe. Immediately after the conference, this was proposed by three academies of science—the Academy of Sciences of the USSR, the Polish Academy of Sciences, and the Czechoslovak Academy of Sciences—which by law occupied in their respective countries the highest position in science. Discussions about forming national Pugwash committees began at a meeting of representatives of these three academies, which took place after the end of the 6th Pugwash Conference. They were initiated by Evgeny Konstantinovich Fedorov, chief secretary for science of the Academy of Sciences of the USSR. Secretary for Science of the Polish Academy of Sciences and correspondent member Szulkin was sympathetic to the idea and so were the Czechoslovak delegates.[24]

A major difference between the Eastern Bloc and Western Pugwash was structure. The communist countries' national committees were top-down, half-political, half-scientific bodies whose members had to be approved by the relevant Communist Party's Central Committee, which fully controlled them. Czechoslovak and Polish national committees were composed exclusively of those who were approved; an individual seeking conventional members or supporters would be unsuccessful. The Czechoslovak Pugwash Committee was established in 1961 following the Moscow Conference, but the Polish Committee was not formally institutionalized until three years later.

This well illustrates the fact that in the case of the Soviet Bloc the relationship between scientists and the state was much more complicated and, most importantly, tightly controlled by both party and state agencies, including the secret police. The structure of the various Pugwash National Committees was established and administered by the academies of sciences. Pugwash tolerated this situation even though it always claimed that it was a movement of free individuals, because its main goal was to maintain a dialogue between East and West.

Pugwash as a Methodological Challenge

From the point of view of the theory of international relations, liberal theorists see Pugwash as a transnational phenomenon, which began to emerge in the 1970s. They focus on non-governmental actors, the movement and exchange of people, information, knowledge, and ideas across borders, and the circulation of knowledge.[25] The transnational approach to Pugwash is particularly evident in works that address the issue of disarmament from the perspective of externalizing pressure on Soviet leadership.[26] Simultaneously, Pugwash's role is reflected in international agreements such as the Anti-Ballistic Missile Treaty,[27] the 1963 Limited Test Ban Treaty,[28] and the 1968 Non-Proliferation Treaty.[29]

However, international relations theory takes a variety of stances when it comes to Pugwash. For realists, Pugwash is a typical example of informal, or Track II diplomacy, the main aim of which was to maintain stability between the two blocs. In this case, Czechoslovakia and Czechoslovak

Pugwash are absolutely crucial to the understanding of the relationship between Henry Kissinger and the question of security in Central Europe, and especially the German question in the 1960s and 1970s, that is, when challenges to Bonn's Hallstein Doctrine gained traction, fueling growing attempts to secure recognition of the GDR on the international scene.[30] From the perspective of liberal theories of international relations, Pugwash is a key actor whose behavior develops within the discourse defined by the Kantian triad, whereby Pugwash can be seen as a means of "soft power" extending democracy, facilitating economic interdependence and facilitating participation in a network of international institutions.[31] From the perspective of constructivist theories, Pugwash is seen as a crucial process of the Cold War.[32] Throughout its existence, it had been seen as a platform that created a meeting space for analysts, socially engaged scholars and unofficial diplomats to map each other's positions.

From a geopolitical perspective, the role played by the smaller states of the Soviet Bloc in Pugwash, in which the Czechoslovak and Polish Committees were the most active allies of the Soviet Union from the 1960s until 1980s, is taken into account from the decolonizing perspective. The geopolitical dimension is also evident in Czechoslovakia's attempts to institute the principle of balance in the development of international organizations within the Western and Soviet blocs, as opposed to the Soviet Bloc's dispersion of its member states under the pretext of "restricted internationalism."

Another conceptual challenge is the emergence of nonstate actors, which represented a new type of scientific internationalism. Indeed, Pugwash was only the latest institution to emerge between 1947 and 1957 to channel scientific internationalism: UNESCO embodied Joseph Needham's postwar vision of this type of engagement, and the World Federation of Scientific Workers was a product of J. D. Bernal's vision of the role of postwar science in society.

In addition to scientific internationalism, however, Pugwash also focused on the peace agenda, which was claimed by the World Peace Council as part of the peace movement within the Soviet Bloc. The relationship of the WFSW and the WPC to Pugwash was a defining element from which the Soviet international relations strategy was derived in the late 1950s. It aimed to use international engagement as effectively as possible, while ensuring that its own contribution was minimized. In fact, the Soviets' ability and willingness to invest in Pugwash depended on the pursuit of their own objectives—especially in the early 1960s, when their interest was defined by resuming nuclear testing.

From the perspective of the "practice turn," Pugwash represents the preeminent theme of Track II diplomacy.[33] In this regard, attention will focus primarily on the group of Harvard professors who were active in shaping the Pugwash agenda on Central and Eastern Europe in the 1960s. This group included Henry Kissinger (1923–2023), Paul M. Doty (1920–2011), and Marshall Shulman (1916–2001). During the 1960s, Henry Kissinger emerged as a leading strategist in US foreign policy, blending academic theory with real-world diplomacy. As a professor at Harvard University, he gained recognition for his work on nuclear strategy and limited war, particularly through his influential book *Nuclear Weapons and Foreign Policy* (1957). His expertise in deterrence theory and crisis management made him a key advisor to policymakers navigating Cold War tensions, including the Berlin and Cuban Missile crises.

By the late 1960s, Kissinger transitioned from academia to direct political influence, serving as a key foreign policy advisor to Nelson Rockefeller's presidential campaigns before being appointed National Security Advisor by President Richard Nixon in 1969. In this role, he began shaping the policies that would define US engagement with both the Soviet Union and China in the 1970s, laying the groundwork for détente and the eventual opening to Beijing. His mix of realpolitik and backchannel diplomacy positioned him as one of the most influential foreign policy figures of the Cold War era.

Paul M. Doty was a pivotal figure in both nuclear policy and science diplomacy, working at the intersection of academia and arms control. As a biochemist and founder of Harvard's Department of Molecular and Cellular Biology, he was also deeply engaged in Cold War policy debates, advocating for nuclear arms control and scientific cooperation across ideological divides. His influence extended beyond the laboratory, as he played a key role in fostering dialogue between US and Soviet scientists on issues of nuclear non-proliferation.

Doty was a founding member of the Pugwash Conferences on Science and World Affairs. Throughout the decade, he advised policymakers on nuclear arms control, emphasizing the role of scientific engagement in reducing Cold War tensions. His work in the 1960s laid the foundation for later arms control agreements and strengthened the bridge between science and diplomacy, positioning him as a key intellectual force in nuclear policy circles.

The last member of the group, Marshal Shulman, was a key figure in shaping US–Soviet relations, serving as a special assistant to Secretary of State Dean Rusk at the height of the Cold War. As tensions over Berlin, Cuba, and nuclear arms control escalated, Shulman provided critical expertise on Soviet affairs, advocating for strategic diplomacy to manage conflicts and prevent escalation. His influence extended beyond government, as he worked to bridge academia and policy, emphasizing the importance of arms control and sustained dialogue with Moscow.

At Harvard University, Shulman helped shape Cold War scholarship, mentoring future policymakers and contributing to Track II diplomacy—informal but influential discussions between Soviet and US intellectuals. His work in the 1960s laid the groundwork for détente and reinforced the idea that informed diplomacy was essential to Cold War stability. He was the founding director of the W. Averell Harriman Institute for Advanced Study of the Soviet Union at Columbia University. We know that this group was in unofficial contact with Czechoslovak and later Polish scientists. Emmanuel Adler based his analysis of Pugwash[34] on the role of epistemic communities in international relations[35] and emphasized the role of Pugwash as a specific movement of socially engaged scientists and researchers. Haas's concept has been largely adopted by the existing historiography of science diplomacy,[36] which provides a useful analytical tool for understanding the role of Pugwash in the Cold War and international relations. In this context, the question of the "dual identity" of researchers is important.[37] This is particularly true for scientists from the Soviet Bloc countries.[38]

The issue of relations between Pugwash and the question of human rights constitutes a final conceptual challenge. Pugwash was born out of a democratic tradition of debate, but at the same time, as an instrument of science diplomacy (and Track II diplomacy), and therefore it demonstrated a certain degree of realpolitik, which was particularly evident in relation to crises, that is, specifically in the case of the Czechoslovak Committee between 1968 and 1971 and in the case of Poland in the period 1981–82.

Nuclear-Free Zone Proposal

The Moscow Conference was followed by a conference held in London in September 1962, which was preceded by a working meeting held in Cambridge, in which Leopold Infeld participated with a delegation including Marian Danysz and Karol Lapter.[39] The idea of a nuclear weapons–free zone in Central and Eastern Europe gained momentum within the Pugwash Conferences. Leopold Infeld, alongside Marian Danysz and Karol Lapter, used this meeting to advocate for further negotiations on the proposal. The Infeld Plan was a concept advocating for a nuclear-free zone in Central Europe; the roots of this Polish initiative can be traced to an earlier international diplomatic effort—the Rapacki Plan—first presented by Polish Foreign Minister Adam Rapacki at the UN General Assembly on 2 October 1957.[40]

The Rapacki Plan sought to establish a demilitarized zone free of nuclear weapons in Poland, Czechoslovakia, and the two German states (East and

West Germany) as a means of reducing Cold War tensions. Additionally, it proposed that the United States, the Soviet Union, the United Kingdom, and France pledge not to use nuclear weapons against these countries. A control mechanism was to be established to monitor nuclear movements. If full international agreement proved elusive, the plan allowed for bilateral or multilateral agreements to advance its objectives. Despite its rejection by Western powers, the plan remained central to Polish and Czechoslovak efforts within Pugwash.[41]

Adam Rapacki (1909–1970)[42] seems to be a key figure in understanding how Czechoslovak and Polish scientists related to each other within Pugwash. Having served as Poland's Minister of Education (1950–1956) and Minister of Foreign Affairs (1956–1968), he was uniquely positioned to bridge science diplomacy with political strategy. The international status of Czechoslovakia and Poland was particularly precarious in the late 1950s and early 1960s. The existence of two German states on the eastern frontier that were not effectively guaranteed, the invasion of Hungary by Soviet troops in 1956, and the escalation of the Suez crisis in late 1956 suggested that the postwar world order was not yet fully established, and the involvement of the European powers at the global level left the Central European states, although in the Soviet sphere of influence, effectively in a power vacuum. According to archival documents, Adam Rapacki was always very well informed about the state of developments and discussions aimed at reviving his idea of a nuclear-free Central Europe.

The two delegations—Polish and Czechoslovak—began coordinating their steps together with the USSR, nominating Leopold Infeld as a candidate for the Pugwash Continuing Committee. At the 10th Pugwash Conference in London, the Czechoslovak Committee supported Infeld's nomination and he was successfully elected. Behind the scenes, Infeld persuaded fellow scientists to formally introduce a conference specifically dedicated to the nuclear-free zone concept. His initiative led to the scheduling of a follow-up Pugwash meeting in March 1963.[43]

One of the strongest advocates for the total demilitarization of Austria—an idea closely aligned with the nuclear-free zone—was Austrian physicist Walter Thirring (1927–2014), a leading figure in quantum field theory and mathematical physics. Born in Vienna, he studied physics in Innsbruck and later conducted research at prominent institutions, including the Institute for Advanced Study in Princeton and CERN. His support helped broaden the scientific legitimacy of the initiative.

Over time, with evolving geopolitical circumstances and the involvement of Polish leader Władysław Gomułka, the plan was modified and further promoted, sometimes referred to as the Gomułka Plan. While the core objective remained the same—preventing nuclear proliferation in Central

Europe—the plan underwent several adjustments to accommodate the changing political dynamics during the Cold War.

1964: Karlovy Vary Conference, Political Repercussions and the Rise of Poland in Pugwash

Following the Pugwash meeting in 1962, preparations began for the 13th Pugwash Conference, scheduled to take place in Karlovy Vary (Karlsbad), Czechoslovakia, in September 1964. The Czechoslovak Pugwash Committee, under the leadership of Antonín Šnejdárek, played a critical role in organizing the event.

The Czechoslovak Pugwash Committee was headed by Antonín Šnejdárek (1916–1982), a postwar military attaché of Czechoslovakia in Berlin, who worked from 1957 to 1966 at the Institute of History of the Czechoslovak Academy of Sciences, where he focused on international relations, politics, and the German question in Central Europe. In 1966 he became director of the newly established Institute for International Politics and Economics, and in 1969 he emigrated. He went first to Germany, but later settled in Paris, where he became a professor at the Sorbonne.

His extensive political connections—including ties to President Antonín Novotný—enabled him to secure official support for Pugwash activities in Czechoslovakia. Thanks to his initiative, the Czechoslovak Pugwash Committee organized the 13th Pugwash Conference in Karlovy Vary on 13–19 September 1964, which was attended by eighty-six scientists from nineteen states. According to Eugene Rabinowitch,[44] in his article "About Pugwash" in *The Bulletin of the Atomic Scientists* in April 1965, the meeting in Karlovy Vary was a significant event in the history of science and politics, changing the role of science in society.[45]

The Karlovy Vary conference followed a preliminary conference in Prague on 9–12 September 1964, in which thirty-seven international attendees participated.[46] At the Karlovy Vary conference, the first working group focused on reducing tension and preventing conflict in Central Europe dominated the debate and had the largest number of participants. Among the members of this working group were Bentley Glass, advisor to the American Committee for Nuclear Energy, Hans Morgenthau, permanent advisor to the US State Department, head of the Committee for Europe of the Scientific Board of Wall Street, and an important international relations theoretician, and Henry Kissinger, permanent advisor of the Joint Chiefs of Staff of the US Armed Forces. The key topics were Germany's borders, a NATO–Warsaw Pact non-aggression treaty, and nuclear weaponry in this region. It was recommended that the Pugwash Conference should not publicly take a side on difficult issues during the international discussions, such as on the Multilateral Force treaty, a Cold War–era NATO initiative primarily promoted by the United States in the early 1960s.[47] The goal was to create a jointly controlled nuclear deterrent within NATO that would involve European allies, particularly West Germany, in nuclear strategy without giving them direct control over nuclear weapons. However, the initiative never materialized into a formal treaty.

While the conference was considered a success by Czechoslovak scientists, the communist authorities were less enthusiastic. Although the Czechoslovak Pugwash Committee viewed the conference as a major international success, the Department for Education and Science of the Central Committee of the Czechoslovak Communist Party and the Foreign Committee quickly dampened the Pugwash Committee's excitement. The Central Committee of the Czechoslovak Communist Party harshly criticized conference report, stating that the broad findings and resolutions were an unacceptable compromise and that the Central Committee needed to hear the particular perspectives of "Western scientists who are close to the governments of their countries."[48]

In consequence, the Czechoslovak Pugwash Committee had to cancel a meeting with a group of Harvard University professors, including Henry Kissinger, Paul M. Doty, and Marshall Shulman. The International Department of the Central Committee of the Communist Party of Czechoslovakia postponed the Prague summit scheduled for 15–18 May 1965 due to circumstances in Vietnam and the Caribbean.[49]

After the cancellation of the May 1965 Pugwash meeting, Antonín Šnejdárek, a member of the Czechoslovak working group, analyzed the situation pragmatically: (1) The Harvard University Americans would center their German question research on Poland. This group's proposals, notably Kissinger's latest (report, first publication in April 1965), offered a novel, speculative viewpoint on GDR recognition and the German question. Since they could not debate these problems with the Czechoslovaks, they would likely talk to socialist specialists from Poland. (2) The Kissinger-led American group would likely leave Pugwash and return to their roles before the 13th Pugwash meeting in Karlsbad and until spring 1964. Šnejdárek urged that the Polish Pugwash Committee be promptly notified.[50]

Antonín Šnejdárek did not give up. The Czechoslovak Pugwash committee established a Central European security preparation committee in 1965. In 1966, the Pugwash movement declared this group a permanent working group together with the Soviet–American Disarmament Study Group and a study group to prevent bacteriological conflict. In the end, Šnejdárek implemented his original plans only three years later, in Mariánské Lázně in May 1967.

Polish scientists involved in Pugwash supported Czechoslovak plans. Professor Karol Lapter attended the meeting in Mariánské Lázně in May

1967 on behalf of the Polish side.[51] On 17 December 1965 the Polish press reported that a Pugwash working group had been established in Prague to discuss the problems of international security in Europe.[52]

Because of Antonín Šnejdárek's great charisma, he was able to hold the attention of the group of US professors even after the 1965 fiasco, and two years later they attended the meeting of the Pugwash Working Group for the Study of Security Questions in Mariánské Lázně. Once again, political and military security issues in Europe, including the use of biological weapons, were discussed. These meetings were groundbreaking for American policy in this area. As historian J. P. Perry Robinson notes, the gathering in Mariánské Lázně was, for example, Kissinger's first exposure to the idea of rethinking biological weapons control. These meetings also had a profound influence on the attitudes and commitment of the US president, Richard Nixon. In fact, he visited Czechoslovakia at the end of March 1967 as part of his study tour of the socialist bloc, and the only thing we know for certain about this highly classified visit by the future US president is that he had a meeting with Antonín Šnejdárek in Prague.[53]

With Czechoslovakia scaling back its Pugwash activities, the Polish Pugwash group stepped in to assume leadership. The Sopot Conference (11–16 September 1966) became a pivotal moment, solidifying Poland's role in science diplomacy. A few days before the Sopot Conference, Šnejdárek's group, led by the Danish chemist Professor David J. Adler, met in Warsaw. There, in cooperation with Šnejedárek, at whose initiative the Security Group had been established, they selected the most important issues of European security to be discussed. Polish scientists, led by Karol Lapter, ensured that discussions continued regarding the nuclear-free zone in Central Europe, the recognition of German borders and the role of the International Atomic Energy Agency (IAEA) in international relations.[54]

The Sopot Conference played a crucial role in Czechoslovak–Polish cooperation in the field of security in Central Europe, with the Poles taking over some of the Czech Pugwash activities that the Central Committee of the Czechoslovak Communist Party did not want to continue for fear of too democratic a dialogue with Western participants. At Sopot, Joseph Rotblat reaffirmed Poland's central role in security discussions, telling *Trybuna Ludu* that Poland was a natural choice for hosting Pugwash due to its active role in European security affairs.[55]

The conference in Sopot, attended by eighty scientists from thirty-four countries, still echoed the priorities of the Czechoslovak and Polish agenda, which consisted in the creation of a nuclear-free zone in Central Europe. But the plan for such a zone was now directly adopted by the Soviets, and the vice president of the Soviet Academy of Sciences, Mikhail Dmitrievich Millionshchikov (1913–1973), stated that the Rapacki Plan was an excellent starting point for further negotiations on the disarmament of Central Europe. The discussion about German borders and the German question in Central Europe was dominated by Antonín Šnejdárek in Sopot.[56]

Despite setbacks, Šnejdárek maintained his influence in science diplomacy. Šnejdárek's activities culminated at the 17th Pugwash Conference in Ronneby on 3–8 September 1967, where he was elected head of the Second Working Group on the Maintenance of Peace and Security. The agenda of this working group was broad and included the German question, Šnejdárek's area of expertise. In this sense, the Pugwash group fulfilled its function by discussing, at an informal level but with the participation of leading political analysts, possible scenarios for the recognition of the two German states, including the question of the recognition of their borders and admission to the UN, reduction of their arms budgets, and more.

Although these ideas did not immediately translate into policy, they were later reflected in Willy Brandt's Ostpolitik (1970) and the 1972 Basic Treaty between the FRG and GDR. Both states were admitted to the UN in 1973. For the GDR, this was undoubtedly a more than positive outcome of the negotiations. A solution to the problem of the recognition of the GDR's borders was found that was acceptable to the FRG. The treaty recognizing the existing borders was signed on 12 August 1970, and on 8 November 1972 a treaty was signed between the GDR and the FRG, an amendment to which stipulated, among other things, that both states would submit an application for admission to the United Nations. The GDR was admitted as the 133rd UN country on 18 September 1973, the same date as the FRG, but the latter under the serial number 134.[57]

By the late 1960s, Šnejdárek had also become a key informal envoy between East and West. His discussions with Nixon and Kissinger in 1969 helped shape US understanding of Central and Eastern European affairs. Richard Nixon received him twice at the White House in May and June of 1969. But the primary purpose of Šnejdárek's visits was to brief Nixon and Kissinger as thoroughly as possible on the situation in Czechoslovakia and the Soviet Bloc.[58]

1968: Occupation of Czechoslovakia

Two weeks after the Warsaw Pact armies occupied Czechoslovakia, the 29th Pugwash Continuing Committee session took place in Nice on 9–10 and 14–16 September 1968. During this session, the situation in Czechoslovakia was discussed as part of a broader agenda, including issues in Nigeria, the Middle East, and Vietnam. The discussions on Czechoslovakia featured diverse and sometimes conflicting viewpoints, focusing primarily on the presence of foreign troops. The final conference statement called for a swift resolution of the situation, emphasizing the need for foreign troop withdrawal and the restoration of Czechoslovakia's self-governance. Concerns were raised about the impact of these events on disarmament talks, but the statement urged that these negotiations continue unaffected. The Czechoslovak Pugwash Committee, represented by Theodor Němec, Ivan Málek, and František Šorm, had to stay in Prague, as its members were deputies in the National Assembly, where they defended the Czechoslovak policies of the 1960s, implicitly criticizing the USSR by denouncing the use of force and armed interference in sovereign states' affairs.[59]

The Pugwash Conference in Sochi (USSR), which took place on 22–27 September 1969, was a real stress test for relations and cohesion within the Soviet Bloc. Nearly one hundred scholars from twenty-nine countries and ten observers from international organizations, including the UN, WHO, and others, witnessed a diplomatic dance around the issue of the occupation of Czechoslovakia under the title "International Security, Disarmament and Development." In fact, Theodor Němec, the secretary of the Czechoslovak National Committee, did go to Sochi, but he could only read the text prepared for him in Prague: "We appeal to you not to discuss Czechoslovakia as a separate item, because we fear that while doing so you will be hardly able to avoid touching upon our own internal affairs. Such discussion *at this time* would not help Czechoslovakia. . . . We implore you: please, fulfil our humble request."[60]

The difference in how Eastern and Western European journalists covered the discussion of the occupation of Czechoslovakia was striking. The special correspondent of the Polish Communist weekly *Nowe Czasy*, Witold Gawilewski, in his description of the discussions in the working group on "Steps toward the Elimination of Existing Armed Conflicts and the Maintenance of Peace," focused primarily on Vietnam, where he steadfastly called for the withdrawal of US troops from that country. The working group also discussed Nigeria and the Middle East in addition to Vietnam. The Polish Special Rapporteur carefully omitted the "incident" in Czechoslovakia. In fact, Czechoslovakia was not even on his list of key participants, a significant change from the past.[61]

The French special correspondent sent by *Le Monde* described the situation at the Sochi conference in a much more realistic way. The whole group, with the exception of the leaders of two countries, the USSR and Poland, condemned the occupation of Czechoslovakia. He noted that this was the liveliest and largest discussion of the whole conference in this working group. He was surprised by the procedural solution proposed by the representative of Czechoslovakia.[62] The minutes of the 31st meeting, however, stated that "it was finally decided that as a matter of principle the Committee could not agree to the request of the Czechoslovak Pugwash

Committee, though it was sympathetic to it. The text of the statement, however, together with the views of the Committee would be transmitted to the members of Working Group 2 for their consideration. Mr. Němec (who was deputized for by Academician Málek) asked for his dissent from this decision to be recorded in the minutes."[63]

But reports from official news agencies such as Reuters and AFP made no secret of the fact that all the countries in the first group were included: Czechoslovakia, Nigeria, the Middle East, and Vietnam. Reuters pointed out that the Pugwash resolution was received in a very reserved way, given the atmosphere at the conference. Although it was drafted and approved by an absolute majority of the participants, it was not officially adopted, and therefore not included in the official resolutions, because of its rejection by the USSR and Poland. This meant that the Deputy Chairman of the USSR Academy of Sciences, Mikhail Dmitrievich Millionshchikov, was criticized by his colleagues and forced to comment on Soviet policy. The final text, as reported by Reuters, stated that all nations should avoid the threat of military intervention against the territorial integrity and political independence of another country. The armies of foreign states should immediately withdraw from Czechoslovakia. The Czechoslovak people should be given the right to decide their own destiny.[64]

In addition, the special correspondent of *Le Monde* contrasted the behavior of the Polish and Soviet scientists with that of the US scientists when the question of Vietnam was discussed. The latter distanced themselves from Washington's policy, and the journalist asked whether such a gesture could be expected from the Central and Eastern European scientists who were present at the conference.[65]

For many years, the majority of the Czechoslovak Pugwash Committee members were denied the opportunity to travel and attend conferences abroad in response to their participation in the Prague Spring and protests against the occupation. Meetings were sometimes attended by Czechoslovakia's leaders, but after 1972 there was no action of any real importance. The Czechoslovak Pugwash Committee was only involved with the Pugwash movement for a very short time, according to formal records. It covered the period 1962–69, or until 1972, when Ivan Málek formally resigned from the Continuing Committee, albeit that he had not been allowed to participate in it since 1969.

In 1972 elections were held for the Continuing Committee, where two seats were vacant. Academician Angel Balevski (Bulgaria) stood for reelection as a joint candidate of the Bulgarian, Czechoslovak, Hungarian, and Polish Committees and the GDR Committee. However, Ivan Málek, chairman of the Czechoslovak Pugwash Committee, had to resign because Jaroslav Kožešník, the new president of the Czechoslovak Academy of Sciences, had not given him permission to travel or to present his renomination. Therefore, Kožešník first secured for himself the nomination for the Czechoslovak representative on the Continuing Committee and then nominated the relatively unknown and inconspicuous Slovak doctor Karel Šiška, a deputy in the House of Nations of the Federal Assembly of the Czechoslovak Socialist Republic in 1971. This botched election was effectively the end of the most active period of Czechoslovak scientists in Pugwash. In the 1970s and 1980s, the focus for the future shifted to Poland, where three Pugwash Committee meetings were held.

All together, five working meetings were held in Poland: in Radziejowice in 1969, in Kraków in 1974, in Zakopane in 1978, and in Mragowo in 1987, all dedicated to the Working Group on European Security. The fifth was held in Krakow in 1990 and focused on debt and how it affects international stability.[66]

At the Pugwash meeting in Zakopane held under the general title "The Political Situation in Europe: Arms Build-Up and Prospects for Disarmament," the issue of the disarmament in Central Europe was discussed, already in a broader context. However, it turned out that the expanded version of the Rapacki Plan, which took the form of the Infeld–Rapacki–Gomułka Plan, was outdated by then, and the Poles did not try to revive it.

In 1974, a Polish Pugwashite Maciej Nałecz became the Chairman of the Pugwash Continuing Committee and remained in the position until 1990. The question of the relationship between nuclear armament in Europe and nuclear non-proliferation in terms of the dynamics of this mode of armament was also discussed at another symposium in Zakopane in 1978. It was called Security and Cooperation in Europe: Problems and Prospects after Belgrade and the discussion was along the same lines: the issues of complete disarmament and the restoration of trust between nations. The two symposia in the 1970s had a number of common features. They highlighted each bloc's mistrust of the other, their reliance on military force as the main means of ensuring their own security and, not least, the discussions highlighted the enormous pace of armament.[67] Although it may seem that Rapacki's plan—and later Gomułka's—was officially out of the question by the late 1970s, it still appeared in European disarmament thinking, especially in Polish texts and analyses.[68]

1981: Martial Law in Poland and the Warsaw Pugwash Conference

Building on its active role in the 1970s, Poland fully assumed leadership of Pugwash activities in Central and Eastern Europe following the 1968 Warsaw Pact invasion of Czechoslovakia. The invasion not only shattered the political autonomy of Czechoslovakia but also crippled its engagement in international science diplomacy. Most Czechoslovak Pugwash Committee members were denied travel permits, effectively barring them from participating in international conferences.

With Czechoslovak scientists sidelined, Poland became the primary Central and Eastern European voice within Pugwash throughout the 1970s and 1980s. However, while Pugwash as an organization continued its discussions on disarmament and security, it showed little interest in addressing the plight of former Czechoslovak Pugwash members affected by the political crackdown. The post-1968 emigration of Czechoslovak intellectuals, particularly František Janouch, a prominent nuclear physicist who settled in Denmark and later Sweden, shifted the discourse from nuclear diplomacy to broader discussions on dissidence within the Soviet Bloc. Janouch and other exiled scientists played a key role in bringing international attention to political repression in Czechoslovakia, marking a transition from scientific cooperation to activism in exile.

František Janouch (1931–2024) was a Czech nuclear physicist, human rights advocate, and prominent dissident who played an influential role in supporting intellectual and scientific freedom in Central and Eastern Europe.[69] Born in Lysá nad Labem, Czechoslovakia, Janouch studied physics at Charles University in Prague and in Moscow and became a respected nuclear physicist. In the 1970s, he became a vocal critic of the communist regime following the suppression of the Prague Spring, which led to his expulsion from Czechoslovakia in 1974. In exile, Janouch settled in Sweden, where he continued his scientific work while also advocating for political change in his homeland. He founded the Charter 77 Foundation, which supported dissidents and the families of political prisoners in Czechoslovakia. Janouch collaborated closely with other notable figures of the Czech dissident movement, including Václav Havel. His contributions to democracy and human rights in Central and Eastern Europe have been widely recognized, and he was an influential voice in Czech and European intellectual circles. Working with Andrei Sakharov, he challenged Pugwash science diplomacy and its place in international relations and kept the discussion of dissent alive until 1989.

The importance of Pugwash for the intellectual agenda of the dissident movements in Central and Eastern Europe was suggested by Andrei Sakharov in his contribution to the Pugwash Conference in Kyoto in 1975 (mediated and arranged by František Janouch, who read Sakharov's papers in his absence). This was in spite of the fact that the original proposal to have Andrei Sakharov speak at the conference did not meet with the full approval of the Pugwash Committee.[70] In his statement entitled "The Need for an Open World," Sakharov urged scientists and governments to adopt and implement Niels Bohr's idea of an "open world" as a guarantee of peace. In addition to his optimism that an "open society" was possible, he expressed his criticism of Pugwash after 1968, writing "For many years, and with sustained attention, I have been following the work of the Pugwash conferences on the basis of the scant data published in the Soviet Union or obtained through unofficial channels. Unfortunately, one has the impression that the conferences are often no more than a way of unofficially sounding out government positions, and that they have little practical success in enabling the community (including the scientific community) to influence governments on matters so vital to all mankind."

Sakharov's letter, through Janouch's mediation, was at the root of criticism of such controversial decisions as the holding of one of the upcoming Pugwash conferences in Warsaw in August 1982, while martial law was in force in Poland. The meeting of Pugwash representatives with General Wojciech Jaruzelski in particular was criticized by human rights activists.[71]

A large photograph of General Jaruzelski in dark glasses sitting in the Council of Ministers chamber in Warsaw with Pugwash leader Dorothy Hodgkin appeared in *Trybuna Ludu* on 31 August 1981. Jaruzelski said the conference, which brought two hundred scientists and diplomats to Warsaw, including ninety-seven Nobel laureates, was "an expression of respecting the Polish nation and trusting in its wisdom."[72] This sentence was the subject of criticism on the grounds that it implied Pugwash's acquiescence in the imposition of martial law. However, nothing could have been further from Pugwash's thinking.

Pugwash was heavily criticized for organizing the World Conference under martial law, and Dorothy Hodgkin, president of the Pugwash Conferences on Science and World Affairs, and Martin Kaplan, secretary general, responded to this criticism: An invitation to exchange views with Prime Minister General Jaruzelski was accepted by the Pugwash Council, as is customary when requested by a high-level official of a host country. Polish Deputy Prime Minister Rakowski attended a session of one of the working groups in which most of the conference participants participated. "On both occasions there were lengthy discussions during which the two officials were asked tough questions and comments, which were answered openly. Topics included the repression of Solidarity, the alleged brutal treatment of prisoners and other political and economic factors contributing to Poland's current difficulties and their implications for European security."[73]

The meeting with official Polish representatives was counterbalanced by the first unofficial meeting of Pugwashites, that is, individual Pugwash members, with Solidarity representatives, symbolically marking a new epoch in the history of the non-communist peace movement in Central and Eastern Europe. Several Pugwash participants even managed to be doused by water cannons while taking part in a Solidarity demonstration in Warsaw's Old Town, at a time when the rest of the participants were being kept out of harm's way with an excursion to Toruń, the birthplace of Copernicus. It was an important meeting and a result of long-term pressure and lobbying from Central and Eastern European scientific emigration on Pugwash. In fact, Pugwash was very often considered by Western journalists and societies as if it "cowered" to the Soviets, and as Dorothy Hodgkin said, "in some Western circles we have often been considered Soviet dupes."[74]

Pugwash's approach was openly criticized in Andrei Sakharov's letters, which raised the question of whether Pugwash's discussions on peace and security "objectively and impartially exclude criticism of human rights violations."[75] Sakharov wrote three letters to Pugwash: the first in 1975, the second in 1982—on the occasion of the Warsaw Conference—and the third in 1989. However, they were never accepted as part of the official agenda (and printed as such). Sakharov's 1982 letter was brought to the conference by a participant and, in accordance with the rules, was officially distributed as an information document. This raised two fundamental questions about Pugwash, science diplomacy and dissident movements. The first was: did the effort to focus public attention on the threat of nuclear war as an issue of transcendent importance—an issue unlike any other—preclude criticism of human rights abuses? The second question was whether reticence on human rights issues was necessary to maintain the East–West dialogue to which Pugwash was dedicated.

There is a striking difference between Pugwash's neutrality in 1968 and its involvement and activist approach in 1982. There were several reasons for these changes. The first was the Conference on Security and Cooperation in Europe (CSCE) in Helsinki in 1975—the final treaty recognized human rights as part of the agenda of international relations, and the fact that the Soviet Bloc countries signed the final text facilitated and stimulated dissident activities and movements in Central and Eastern Europe. The second was the role and growing impact of scientific emigration on political affairs. The third reason was the activism of individual scientists—in which Janouch and Sakharov played a crucial role. The brain drain and scientific emigration from the Central and Eastern European countries became a significant political force and their voices were heard.

In 1988, the Pugwash Conference was held in Sochi in the Soviet Union. One of the committee's reports was an address sent to Mikhail Gorbachev, General Secretary of the CPSU Central Committee, who had initiated a process of improved relations with the West through *perestroika*, which was welcomed by the participants.[76] The participation of Andrei Sakharov in this conference was a great sign of change not only within the Soviet Bloc but also within Pugwash.

Conclusion

Pugwash maintained a significant presence in Central and Eastern Europe from the late 1950s through the 1990s with two strong national committees, those of Czechoslovakia and Poland. The Czechoslovaks were more active in the 1960s, whereas the Polish Pugwashites took over in the 1970s and 1980s. From the early 1960s to the late 1980s, the nuclear-free zone proposal evolved through Polish and Czechoslovak science diplomacy. Poland ultimately emerged as the leading advocate within Pugwash, particularly after Czechoslovakia's internal restrictions limited its engagement.

What was special about Pugwash was that, in effect, it was a movement of private individuals. Scholars were not supposed to participate as representatives of their states or organizations, but solely as private citizens. Pugwash brought together scientists of all disciplines and orientations and sought the widest possible geographical coverage, and therefore its workshops and conferences were attended by representatives of both blocs as well as, for example, representatives of the Third World. Funding for these conferences came from private sources and was not dependent on institutional or other support from elsewhere. Pugwash's independence and impartiality are also based on these principles.[77]

The cooperation of the Czechoslovak and Polish Pugwash Committees behind the Iron Curtain was an important element in the continuity of the cooperation and security agenda in Central Europe at a time when Czechoslovakia was under occupation by five Warsaw Pact armies in August 1968. Continuity was particularly important after the emigration of leading scholars, political analysts, and science diplomats in Czechoslovakia. The continuity was mainly in terms of personnel, with Maciej Nałecz serving as Chair of the Pugwash Council from 1974 to 1997 and continuing as a full member of the Pugwash Council from 1997 to 2002.[78]

One of the most important Pugwash Conferences took place in Warsaw in 1982, when martial law was in force.[79] It was here that a new stage in the history of pro- and non-communist peace initiatives in Central Europe began, when representatives of the Polish opposition movement Solidarity met with representatives of the Pugwash movement for the first time on unofficial terms.

The discussions at Sopot (1966) and Warsaw (1982) not only reinforced Poland's leadership but also expanded the focus of Pugwash to include human rights and scientific freedom, underscoring the broader role of nonstate actors in Cold War diplomacy. While official contacts stimulated official dialogues on Pugwash platforms, after the CSCE conference in Helsinki in 1975, Pugwash entered into contact with political dissent in Central and Eastern Europe, for example, via Sakharov´s letters read by František Janouch and during the meeting of Pugwash delegates with Solidarity members in Warsaw in 1982. This was a direct consequence of the opening up of new issues in international relations, especially the issue of human rights.

The issue of human rights was very difficult to integrate into the international agenda of the Pugwash Group because of its international ambitions to include all scientific communities, whether they came from the West, the East, the Global North, or the Global South. In this sense the Pugwash legacy is somewhat controversial, as evidenced, for example, by Sakharov's three open letters to the Pugwash Conferences, in which he expressed his doubts about the wisdom of the strategy of negotiating with whomever and under whatever circumstances.[80]

With the fall of the Iron Curtain, the question has naturally arisen as to how much of a need there is for Pugwash after the end of the Cold War. Joseph Rotblat and Maciej Nałecz's answer to this question is that as long as the world is at arms' length, Pugwash's existence will be justified. The most important aspect of Pugwash in the global context has been its ability to open up key issues in a new way. Scientists served as informal diplomats, enabling them to communicate with local governments on issues that would otherwise have remained the exclusive domain of national security agendas.

Author Biography

Doubravka Olšáková is a historian and political scientist, specializing in the history of science and technology and the Cold War. She works currently at the Institute of Political Sciences of the Faculty of Social Sciences at Charles University. She is the founding member of the Standing Commission on Science, Technology and Diplomacy, Division of History of Science and Technology of the International Union of History and Philosophy of Science and Technology, Deputy Editor-in-Chief of *Centaurus*, the journal of the European Society for the History of Science, and a founding member of the Oxford University Press Science, Technology and International Relations series.

List of Abbreviations

AAN	Archiwum akt nowych (Central Archives of Modern Records)
ABS	Archiv bezpečnostních složek (Security Services Archive)
AV ČR	Akademie věd České republiky (Czech Academy of Sciences)
CC CPCz	Central Committee of the Communist Party of Czechoslovakia

CSCE	Conference on Security and Cooperation in Europe
FRG	Federal Republic of Germany (West Germany)
GDR	German Democratic Republic (East Germany)
IAEA	International Atomic Energy Agency
KC PZPR	Komitet Centralny Polskiej Zjednoczonej Partii Robotniczej (Central Committee of the Polish United Workers' Party)
MÚA AV ČR	Masarykův ústav a Archiv AV ČR (Masaryk Institute and Archives of the Czech Academy of Sciences)
NATO	North Atlantic Treaty Organization
PAN	Polska Akademia Nauk (Polish Academy of Sciences)
PZPR	Polska Zjednoczona Partia Robotnicza (Polish United Workers' Party)
WFSW	World Federation of Scientific Workers
WHO	World Health Organization
WPC	World Peace Council

Notes

1. The Russell-Einstein Manifesto was published on 9 July 1955, and signed by Albert Einstein (US), Bertrand Russell (UK), Max Born (DE), Percy W. Bridgman (US), Leopold Infeld (PL), Frédéric Joliot-Curie (FR), Herman J. Muller (US), Linus Pauling (US), Cecil F. Powell (UK), Joseph Rotblat (UK), and Hideki Yukawa (JAP).
2. Rubinson, *Pugwash Literature Review*; Evangelista, *Unarmed Forces*; Wittner, *Rebels against War*; Wittner, *Struggle Against the Bomb,* vol. 1; Wittner, *Resisting the Bomb*; Wittner, *Struggle against the Bomb,* vol. 3; Brown, *Keeper of the Nuclear Conscience*; "Thematic Pugwash Issue"; Kraft, *From Dissent to Diplomacy*; Kraft and Sachse, *Science, (Anti-)Communism, and Diplomacy.*
3. Olšáková, "Czechoslovak Ambitions and Soviet Politics"; Olšáková, "Pugwash in Eastern Europe." Evangelista's and Lüscher's work on the USSR leaving aside here.
4. Kościuk, "Ruch Pugwash a problematyka rozbrojenia."
5. Rotblat, *Pugwash.*
6. Schwartz, "Perspective on Pugwash."
7. Rabinowitch, "Pugwash: History and Outlook"; Scheinman, "Security and the International System."
8. Wittner, *Rebels against War*; Wittner, *Struggle against the Bomb,* vol. 1; Wittner, *Resisting the Bomb*; Wittner, *Struggle Against the Bomb,* vol. 3; Agar, *Science in the Twentieth Century,* 404–6; Boyer, *By the Bomb's Early Light.*
9. Franc, *Ivan Málek a vědní politika.*
10. Olšáková and Barrett, "World Federation of Scientific Workers," 1–14.
11. Olšáková, "Pugwash in Eastern Europe."
12. Kennedy, "Constitution of Critical Intellectuals," 2.
13. List of those attending the first Pugwash Conference at Pugwash, Nova Scotia, 6 to 11 July 1957: M. L. E.Oliphant (physics), Australia; H. Thirring (physics), Austria; G. Brock Chisholm (psychiatry), Canada; John S. Foster (physics), Canada; Cou Pei Yuan (physics), China; A. M. B. Lacassagne (medicine), France; C. F. Powell (phys-

ics), UK; J. Rotblat (physics), UK; I. Ogawa (physics), Japan; H. Yukawa (physics); Japan; S. Tomonaga (physics), Japan; M. Danysz (physics), Poland; D. F. Cavers (law), United States; L. Szilard (physics), United States; H. J. Muller (genetics), United States; P. Doty (chemistry), United States; E. Rabinowitch (biophysics), United States; W. Selove (physics), United States; V. Weisskopf (physics), United States; A. M. Kuzin (biophysics), USSR; D. F. Skobeltzyn (physics), USSR; A. V. Topchiev (chemistry), USSR. See *Bulletin of the Atomic Scientists*, September 1957, 248.

14. Rotblat a nd Nałecz, "Ruch Pugwash," 5.
15. "A. V. Topchiev 1907–1962."
16. AAN, KC PZPR, Wydział Nauki MKF B 59076, syg. 237/XVI–244, f. 69–70, "List Józefa Rotblata do Komisji Pugwash."
17. AAN, KC PZPR, Wydział Nauki, MKF B 59076, syg. 237/XVI–244, f. 68, "Konferencja Pugwash." 14 February 1962.
18. Thirring, *Ein Leben für Physik und Frieden*, 129–30.
19. NA ČR, CC CPCZ, Politbyro 1954–1962, vol. 321, arch. unit 408, point 9, "Vytvoření československého pugwashského výboru." 5 September 1961.
20. Statement of the Third Pugwash Conference, 3: "Although the nations may agree to eliminate nuclear weapons and other weapons of mass destruction from the arsenals of the world, the knowledge of how to produce such weapons can never be destroyed."
21. See Kraft, "Confronting the German Problem"; Olšáková, "Pugwash and the Soviet Peace Agenda."
22. "Dokumenty o rozhodování v oblasti vědy a techniky."
23. MÚA AV ČR, ŘAS ČSAV, I., inv. nr. 83, "Dopis Jaroslava Koželšníka Jiřímu Hendrychovi," 24 November 1960.
24. NA ČR, CC CPCz, Politbyro 1954–1962, vol. 321, arch. unit 408, point 9, "Vytvoření československého pugwashského výboru," 5 September 1961.
25. Keohane and Nye, *Transnational Relations and World Politics*; Nye, "Soft Power"; Nye, *Soft Power*.
26. Evangelista, *Unarmed Forces*.
27. Kubbig, *Communicators in the Cold War*.
28. Rubinson, *Pugwash Literature Review*.
29. Barth, "Catalysts of Change."
30. Kraft, *From Dissent to Diplomacy*; Kraft and Sachse, *Science, (Anti-)Communism, and Diplomacy*; Olšáková, "Pugwash in Eastern Europe"; Sachse, *Wissenschaft und Diplomatie*.
31. Keohane and Nye, *Transnational Relations and World Politics*.
32. Jones, *Track Two Diplomacy*.
33. Ibid.
34. Adler, "Emergence of Cooperation."
35. Haas, "Introduction: Epistemic Communities."
36. Kunkel, "Science Diplomacy in the Twentieth Century"; Sachse, *Wissenschaft und Diplomatie*.
37. Rungius and Flink, "Romancing Science for Global Solutions"; Kraft, *From Dissent to Diplomacy*.
38. Kraft and Sachse, *Science, (Anti-)Communism, and Diplomacy*; Franc, *Ivan Málek a vědní politika*.

39. Karol Lapter (1912–2003) was a Polish historian, diplomat, and communist activist who worked in Poland's Ministry of Foreign Affairs and later as a researcher at the Polish Academy of Sciences before being expelled from the Communist Party in 1968. After immigrating to Sweden in 1969, he continued his academic career at the Swedish Institute of International Affairs in Stockholm.
40. AAN, KC PZPR, Wydział Zagraniczny, sign. 237/XXII–1217, kart. 83, Pilna notatka, 13 October 1962.
41. UN Archives, *Political and Security Council Affairs Reports, Note on the Rapacki Plan,* 21 February 1958, 4.
42. Wandycz, "Adam Rapacki and the Search for European Security."
43. AAN, KC PZPR, Wydział Zagraniczny, sign. 237/XXII–1217, kart. 83, Pilna notatka, 13 October 1962.
44. Eugene Rabinowitch (1901–1973) was a Russian American biophysicist, writer, and advocate for nuclear disarmament. Born in St. Petersburg, he immigrated to Germany to study chemistry and later moved to the United States during the rise of Nazi Germany. Rabinowitch contributed to the Manhattan Project during World War II, focusing on photosynthesis and chemical reactions. However, deeply concerned about nuclear weapons, he co-founded the *Bulletin of the Atomic Scientists* in 1945, famously establishing its Doomsday Clock as a symbol of nuclear threat. He served as the Bulletin's editor, using it to promote global peace, responsible science, and arms control. Rabinowitch's efforts laid the groundwork for science-based activism and informed discussions on ethical responsibility in science.
45. Rabinowitch, "About Pugwash."
46. MÚA AV ČR, Presidium of the Czechoslovak Academy of Sciences, file of the 17th Meeting (30 October 1964), point VII, "Zpráva o průběhu a výsledcích 13. Pugwashské konference a další opatření," 30 October 1964, f. 301–302. A shortened and re-edited version was published as Šorm, "Třináctá pugwashská conference."
47. Rabinowitch, "About Pugwash," 11.
48. NA ČR, CC CPCz, Presidium 1962–1966, vol. 87, arch. unit 91, point 5, f. 8, "Průběh a výsledky 13. Pugwashské konference," 23 November 1964.
49. Karol Lapter, Polish historian and diplomat, probably should have taken part in the meeting as well.
50. MÚA AV ČR, Secretariat of the President Academician František Šorm, file 10, sign. 12, "Zpráva o odsunu data schůzky s profesorem Kissingerem." 4 May 1965.
51. AAN, Pugwash—Prasa 26–975, Aniela Krupińska, "Konferencja Pugwash o bezpieczeństwie europejskim," *Trybuna Ludu,* 5 May 1967.
52. AAN, Pugwash—Prasa 26–975, (W), "Grupa robocza Pugwash powstała w Pradze," *Życie Warszawy,* 17 December 1965.
53. Dawisha, "Soviet Security and the Role of the Military"; Mastny, "Was 1968 a Strategic Watershed."
54. AAN, Pugwash—Prasa 26–975, (PAP), "Prof. David J. Adler o wynikach prac 'grupy europejskiej Pugwash' w Jabłonnie," *Trybuna Ludu,* 10 September 1966.
55. AAN, Pugwash—Prasa 26–975, (PAP), "Konferencja 'Pugwash' w Polsce: Wypowiedź sekretarza generalnego ruchu prof. dr. J. Rotblata," *Trybuna Ludu,* 20 September 1966.
56. AAN, Pugwash—Prasa 26–975, (PAP), "Zakończenie konferencji 'Pugwash'—Wypowiedzi uczonych zagranicznych," *Życie Warszawy,* 17 September 1966, 1, 5.

57. Wentker, *Außenpolitik in engen Grenzen.*
58. ABS, fond I. správa, a.č. 21203, ze svazku r.č.12399/305III Antonín Šnejdárek, "Telegramy z Washingtonu," f. 90, 94–95.
59. Olšáková, "Matter of Courtesy."
60. MÚA AV ČR, File Ivan Málek, Pugwash—T. 2, Inv. No. 3397, "Pugwash Continuing Committee, Minutes of the Thirty-First Meeting Held on the 20th, 21st, 24th and 27th October 1969 at the Intourist Hotel Sochi, U.S.S.R.," 5.
61. AAN, Pugwash—Prasa 26–975, W. Gawilewski, "Konferencja Pugwash w Soczi," *Nowe Czasy*, 10 November 1969.
62. AAN, Pugwash—Prasa 26–975, N.V., "La dix-neuvième conférence du Mouvement Pugwash préconise la limitation des armements stratégiques," *Le Monde*, 18 September 1968.
63. Archiv AV ČR, file Ivan Málek, Pugwash—vol. 2, inv. nr. 3397, "Pugwash Continuing Committee, minutes of the Twenty-Ninth Meeting held on the 9th, 10th, 14th and 16th September 1968 at the Plaza hotel, Nice, France," 4.
64. AAN, Pugwash—Prasa 26–975, "Konferencja Pugwash," *Biuletyn Specjalny*, no. 7082 (17 September 1968), Dział Informacji, 20.
65. AAN, Pugwash—Prasa 26–975, N.V., "La dix-neuvième conférence."
66. Roszak, *Żuraw z origami*, 295 .
67. Kościuk, "Ruch Pugwash a problematyka rozbrojenia," 134.
68. Ibid., 136.
69. Janouch, *Nikdy jsem se nenudil*; Prečan, Slavíčková, *Bez Františka Janoucha.*
70. Janouch, *Případ Andrej Sacharov*, 23–29.
71. See "Wokół Pugwash: Protesty w sprawie warszawskiej konferencji Pugwash"; Ferry, *Dorothy Hodgkin.*
72. AAN, Pugwash—Prasa 26–975, "Oskarżenie przed wojną nuklearną wciąż aktualne—Spotkanie W. Jaruzelskiego z Członkami Rady Pugwash," *Trybuna Ludu*, 31 August 1981, 1–2.
73. Hodgkin and Kaplan, "Warsaw Pugwash Conference."
74. Hodgkin and Kaplan, "Warsaw Pugwash Conference."
75. Brown, *Keeper of the Nuclear Conscience.*
76. AAN, Pugwash—Prasa 26–975, "Posłanie konferencji Pugwash do M. Gorbaczowa—Zapewnić przetrwanie ludzkości," in *Rzeczpospolita*, 7 September 1988.
77. Roszak, *Żuraw z origami*, 289.
78. Weryński, "Biography and Scientific Activity of Prof. Dr. Maciej Nalecz," 16.
79. "Wokół Pugwash: Protesty w sprawie warszawskiej konferencji Pugwash."
80. Olšáková, "Czechoslovak Ambitions and Soviet Politics"; Kraft, *From Dissent to Diplomacy.*

Bibliography

Adler, Emmanuel. 1992. "The Emergence of Cooperation: National Epistemic Communities and the International Evolution of the Idea of Nuclear Arms Control." *International Organization* 46, no. 1 (Winter): 101–45.

Agar, John. 2012. *Science in the Twentieth Century and Beyond.* London: Polity Press.

"A. V. Topchiev 1907–1962." 1963. *Bulletin of the Atomic Scientists* 19, no. 3: 8–9.

Barth, Kai-Henrik. 2006. "Catalysts of Change: Scientists as Transnational Arms Control Advocates in the 1980s." *Osiris* 21, no. 1: 182–206.

Boyer, Paul. 1985. *By the Bomb's Early Light: American Thought and Culture at the Dawn of the Atomic Age.* New York: Pantheon Books.

Brown, Andrew. 2012. *Keeper of the Nuclear Conscience: The Life and Work of Joseph Rotblat.* Oxford: Oxford University Press.

Dawisha, Karen. 1980. "Soviet Security and the Role of the Military: The 1968 Czechoslovak Crisis." *British Journal of Political Science* 10, no. 3: 297–318.

"Dokumenty o rozhodování v oblasti vědy a techniky" [Documents on decision-making in the field of science and technology]. 1991. *Dějiny vědy a techniky* 24, no. 1: 46–60.

Evangelista, Matthew. 1999. *Unarmed Forces: The Transnational Movement to End the Cold War.* New York: Cambridge University Press.

Ferry, Georgina. 2014. *Dorothy Hodgkin: A Life.* London: Bloomsbury Publishing.

Franc, Martin. 2010. *Ivan Málek a vědní politika 1952–1989* [Ivan Málek and the science policy 1952–1989]. Prague: MÚA AV ČR.

Haas, Peter. 1992. "Introduction: Epistemic Communities and International Policy Coordination." *International Organization* 46, no. 1: 1–35.

Hodgkin, Dorothy, and Martin M. Kaplan. 1982. "The Warsaw Pugwash Conference Did Not 'Cower' to the Soviets." *The New York Times* 131, no. 45432 (28 September): 22.

Janouch, František. 2020. *Nikdy jsem se nenudil.* Prague: Torst.

———. 1994. *Případ Andrej Sacharov.* Brno: Atlantis.

Jones, Peter L. 2015. *Track Two Diplomacy in Theory and Practice.* Stanford, CA: Stanford University Press.

Kennedy, Michael D. 1990. "The Constitution of Critical Intellectuals: Polish Physicians, Peace Activists, and Democratic Civil Society." Center for Research on Social Organization, The Working Paper Series. Ann Arbor: The University of Michigan.

Keohane, Robert O., and Joseph S. Nye, ed. 1972. *Transnational Relations and World Politics.* Cambridge, MA: Harvard University Press.

Kissinger, Henry. 1957. *Nuclear Weapons and Foreign Policy.* New York: Council of Foreign Relations.

Kościuk, Lech. 1982. "Ruch Pugwash a problematyka rozbrojenia" [Pugwash movement and the questions of the disarmament]. *Sprawy Międzynarodowe* no. 8–9: 134–37.

Kraft, Alison. 2020. "Confronting the German Problem: Pugwash in West and East Germany, 1957–1964." In *Science, (Anti-)Communism and Diplomacy: The Pugwash Conferences on Science and World Affairs in the Early Cold War,* edited by Alison Kraft and Carola Sachse, 286–323. Leiden: Brill. https://doi.org/10.1163/9789004340176_010.

———. 2022. *From Dissent to Diplomacy: The Pugwash Project During the 1960s Cold War.* Dordrecht: Springer.

Kraft, Alison, and Carola Sachse, eds. 2019. *Science, (Anti-)Communism, and Diplomacy: The Pugwash Conferences on Science and World Affairs in the Early Cold War.* Leiden: Brill.

Kubbig, W. Bernd. 1996. *Communicators in the Cold War: The Pugwash Conferences, the U.S.-Soviet Study Group, and the ABM Treaty.* Natural Scientists as Political Actors: Historical Successes and Lessons for the Future. PRIF Reports No. 44. Frankfurt am Main: PRIF.

Kunkel, Soenke. 2021. "Science Diplomacy in the Twentieth Century: Introduction." *Journal of Contemporary History* 56, no. 3: 473–84.

Mastny, Vojtech. 2005. "Was 1968 a Strategic Watershed of the Cold War?" *Diplomatic History* 29, no. 1: 149–88.

Nye, Joseph S., Jr. 1990. "Soft Power." *Foreign Policy* 80 (Fall): 153–72.

———. 2004. *Soft Power: The Means to Success in World Politics*. New York: Public Affairs.

Olšáková, Doubravka. 2020. "A Matter of Courtesy: The Role of Soviet Diplomacy and Soviet 'System Safeguards' in Maintaining Soviet Influence on Czechoslovak Science before and after 1968." *Berichte zur Wissenschaftsgeschichte* 43, no. 4: 542–59.

———. 2019. "Czechoslovak Ambitions and Soviet Politics in Eastern Europe: Pugwash and the Soviet Peace Agenda in the 1950s and 1960s." In *Science, (Anti)Communism, and Diplomacy: The Pugwash Conferences on Science and World Affairs in the Early Cold War*, edited by Alison Kraft and Carola Sachse, 259–85. Leiden: Brill.

———. 2020. "Pugwash and the Soviet Peace Agenda in the 1950s and 1960s." In *Science, (Anti-)Communism and Diplomacy: The Pugwash Conferences on Science and World Affairs in the Early Cold War*, edited by Alison Kraft and Carola Sachse, 259–85. Leiden: Brill. https://doi.org/10.1163/9789004340176_009.

———. 2018. "Pugwash in Eastern Europe: The Limits of International Cooperation under Soviet Control in the 1950s and 1960s." *Journal of Cold War Studies* 20, no. 1: 210–40.

Olšáková, Doubravka, and Gordon Barrett. 2024. "World Federation of Scientific Workers." In: *The Palgrave Handbook of Non-State Actors in East-West Relations*, edited by Péter Marton, Gry Thomasen, Csaba Békés, András Rácz, 1–14. Palgrave Macmillan: Cham. doi.org/10.1007/978-3-031-40546-4_41.

Prečan, Vilém, and Miloslava Slavíčková, eds. 2016. *Bez Františka Janoucha ani moře nešplouchá: Františku Janouchovi k narozeninám* [František Janouch, influential and essential: In honor of František Janouch on his birthday]. Prague: Československé dokumentační středisko.

Rabinowitch, Eugene. 1965. "About Pugwash." *Bulletin of the Atomic Scientists* 21, no. 4 (April): 9–15.

———. 1957. "Pugwash: History and Outlook." *Bulletin of the Atomic Scientists* 13, no. 7: 243–48.

Rotblat, Joseph. 1967. *Pugwash: A History of the Conferences on Science and World Affairs*. Prague: ČSAV.

Rotblat, Józef, and Maciej Nałecz. 1991. "Ruch Pugwash" [The Pugwash movement]. *Nauka Polska* 38, no. 3 (256) (May–June): 5.

Roszak, Joanna. 2019. *Żuraw z origami: Opowieść o Józefie Rotblacie* [The origami crane: A story about Józef Rotblat]. Sejny: Fundacja Pogranicze.

Rubinson, Paul. 2019. *Pugwash Literature Review*. Washington, DC: Urban Institute.

Rungius, Charlotte, and Tim Flink. 2020. "Romancing Science for Global Solutions: On Narratives and Interpretative Schemas of Science Diplomacy." *Humanities and Social Sciences Communications* 7, 102 (2020).

Sachse, Carola. 2023. *Wissenschaft und Diplomatie: Die Max-Planck-Gesellschaft im Feld der internationalen Politik (1945–2000)* [Science and diplomacy: The Max Planck Society in the field of international politics (1945–2000)]. Göttingen: Vandenhoeck & Ruprecht.

Scheinman, Lawrence. 1971. "Security and the International System: The Case of Nuclear Energy." *International Organization* 25, no. 3: 626–49.
Schwartz, Leonard E. 1967. "Perspective on Pugwash." *International Affairs* 43, no. 3: 498–515.
Šorm, František. 1964. "Třináctá pugwashská konference o vědě a světových problémech" [The thirteenth Pugwash conference on world problems]. *Věstník ČSAV* 73, no. 5: 849–53.
Statement of the Third Pugwash Conference, held at Kitzbühel, Austria, 14–19 September 1958, p. 3.
"Thematic Pugwash Issue." 2018. *Journal of Cold War Studies* 20, no. 1.
Thirring, Hans. 1992. *Ein Leben für Physik und Frieden* [A life for physics and peace]. Cologne: Böhlau.
Wandycz, Piotr. 1994. "Adam Rapacki and the Search for European Security." In *The Diplomats, 1939–1979*, edited by Gordon A. Craig and Francis L. Loewenheim, 289–318. Princeton, NJ: Princeton University Press. https://doi.org/10.2307/j.ctv8pz9nc.15.
Wentker, Hermann. 2007. *Außenpolitik in engen Grenzen: Die DDR im internationalen System 1949–1989*. Munich: Oldenbourg.
Weryński, Andrzej. 2002. "Biography and Scientific Activity of Prof. Dr. Maciej Nalecz." *Biocybernetics and Biomedical Engineering* 22, no. 2–3: 7–68, 16.
Wittner, Lawrence S. 1969. *Rebels against War: The American Peace Movement, 1941–1960*. New York: Cambridge University Press.
———. 1997. *Resisting the Bomb: A History of the World Nuclear Disarmament Movement, 1954–1970*. Stanford, CA: Stanford University Press.
———. 1993. *The Struggle against the Bomb*, vol. 1: *One World or None: A History of the World Nuclear Disarmament Movement through 1953*. Stanford, CA: Stanford University Press.
———. 2003. *The Struggle against the Bomb*, vol. 3: *Toward Nuclear Abolition: A History of the World Nuclear Disarmament Movement, 1971 to the Present*. Stanford, CA: Stanford University Press.
"Wokół Pugwash: Protesty w sprawie warszawskiej konferencji Pugwash" [Around Pugwash: Protests over the Pugwash conference in Warsaw]. 1982. *Kontakt: Miesięcznik redagowany przez członków i współpracowników NSZZ Solidarność*, nos. 5–6 (September–October): 48–50.

AFTERWORD

So, What Was It Like?

The Science Diplomacy of Reciprocity in Eastern and Central Europe during the Cold War

Simone Turchetti

The articles in this volume provide a compelling contribution to our understanding of the collaborations between scholars and scientists working in Eastern and Central Europe during the Cold War period. It is an important addition to the literature because international scientific collaborations in Western Europe have received far more attention before. Moreover, the United States administration's decisive role in sponsoring the revitalization of science and technology in Western Europe after World War II has projected a Western narrative onto the structure of these collaborations.[1] In particular, the weight historians have placed on collaborations in the West has also contributed to forging new narratives and theories modeled on Western examples. For instance, the building of the European Laboratory for Nuclear Research (CERN) in Geneva is perceived as a landmark event, whereas other collaborative initiatives in Eastern and Central Europe have been explored far less. It is true that scholars are still divided on the ambitions behind these schemes: while some claim support for collaborative science was a tool for asserting US political hegemony, others emphasize that it also paved the way for the process of European integration.[2] Even so, they still paint a historical picture that by and large overlooks Eastern and Central Europe.

The chapters in this volume are thus very welcome as they cast light on neglected historical cases worthy of greater attention if one really wants to better understand the structure of international research collaborations during the Cold War era. Beyond recalling the importance of these case studies, this "afterword" seeks to emphasize one central feature of these contributions, namely that they hinged on the notion of reciprocity as uniquely informing the specific form of scientific internationalism (or what we now call "science diplomacy") operationalized in Eastern and Central Europe. After briefly recalling why their countries' collaborations have received less attention in the literature, these conclusive remarks further highlight their unique impact on Cold War affairs in the Eastern Bloc exactly because of the distinctive type of scientific diplomacy they introduced depending on reciprocity. In turn, they emphasize their role also a form of resistance to Soviet-led scientific internationalism perceived instead as an imposition.

There are important historical reasons as to why these research collaborations have not received attention before. From the 1990s onward historians by and large dismissed collaborative projects in the context of the Eastern Bloc as failures mainly because of the collapse of the Soviet Union. The subsequent proliferation of (black) books on communism focused the scholars' attention mostly on crimes perpetrated in the name of Soviet ideology up until the Union's collapse.[3] As a result, the search for crimes and perpetrators overshadowed efforts to gain a deeper understanding of the former Eastern Bloc countries and their political, societal, and scientific infrastructures. Meanwhile, sweeping changes in state and academic structures across the former Eastern Bloc further complicated the completion of historical inquiries. Fundamental tools for historical research, including archival documents, became increasingly unavailable in Russia, making it difficult for historians to reconstruct a coherent, comprehensive, and comprehensible picture of collaborative work in Eastern and Central Europe. Historians often refer to the *glasnost* window under Mikhail Gorbachev's administration as the only historical moment when documents shedding light on the nature of these collaborations finally became accessible in the Soviet Union.[4] Yet since that time, first under the second term of Boris Yeltsin and then under Vladimir Putin, archival documentation became less available, which in turn impeded understanding key aspects of Eastern Bloc countries prior to 1990, especially those related to collaborative frameworks in academic and scientific research. The recent invasion of Ukraine has considerably exacerbated this situation.

While vital archival sources continue to be unobtainable in the former Soviet Union, some became available in some of these countries partly as a result of their integration in the enlarged European Union framework. In the 1990s and beyond, their archival systems have undergone comprehensive and impressive processes of reorganization. Increasing numbers of documents have been declassified. One might even conclude that the chapters of this volume collectively illustrate the breadth and depth of the archival sources that have become available in the last twenty years and how careful scrutiny of these documents can enrich our understanding of collaborative schemes in Eastern and Central Europe.[5]

Even so, the impact of an asymmetric historical coverage of the Cold War Blocs is plain to see especially with respect to "science diplomacy" research. Since 2010, many studies have portrayed scientific collaborations as a key feature of international affairs, arguing that they have forged a parallel diplomatic arena.[6] These studies have shown how the setting up and development of such collaborations between national research groups served as vehicles to administer relations between individual countries. In advancing this narrative they have nonetheless consistently overlooked the experience of the Eastern Bloc, implicitly treating the relevant collaborations as less significant. A clear example of this asymmetry is in the Royal Society and American Association for the Advancement of Science booklet released in 2010 that launched science diplomacy as a new area of practice and investigations. The booklet highlighted the promotion of science and technology in the North Atlantic Treaty Organization (NATO) as a landmark in the history of science diplomacy, but it did not even mention the Eastern Bloc's Council for Mutual Economic Assistance (COMECON). It also emphasized the collaborations on polar research promoted under the US-led Antarctic Treaty, but not those in space research within the Soviet-led Interkosmos. It celebrated CERN and its achievements in collaborative nuclear research (and more recently the Middle Eastern nuclear research center SESAME) but the Joint Institute for Nuclear Research in Dubna received no mention.[7] This uneven coverage contributed to strengthening post-Cold War narratives that implicitly suggest the whole history of science diplomacy to be sited in the West and that science collaborations in the Eastern Bloc were historically insignificant. Moreover, because of the aforementioned historiographical bias, new narratives tend to frame the relevant questions regarding collaborative research endeavors from the perspective of a Western paradigm, thus taking for granted Soviet hegemony in the Eastern Bloc on the basis of the parallel US hegemony in the Western Bloc. Moreover, research developments in the Eastern Bloc are often interpreted in terms of transnational transfer from the West (or even technological espionage) translated into a socialist framework. This is not to deny that transfers and translations of this kind occurred,[8] but rather to argue that presenting the Eastern Bloc scientific production only through this lens inevitably portrays it as merely passive acceptance and reproduction of Western practices and models. This framing has thus negatively affected our understanding of international collaborations within the Eastern Bloc and has contributed to undervalue its specificity as a heterogeneous and diverse group of transnational experiences with a range of different cultural, social, and historical legacies.

This edited volume should therefore be praised as an addition to literature not only as casting new light on episodes of scientific collaboration in Eastern and Central Europe, but also providing a decisive contribution to reshaping the debate on the historical origins and development of science diplomacy. It implicitly challenges the prevailing notion of science diplomacy as an inherently Western phenomenon by displaying its Eastern and Central European lineage. The volume allows therefore decentering science diplomacy studies and exploring competing science diplomacy patterns as they emerged in the East.[9] It also offers a detailed examination of cases that do not fit the traditional geographical, historical, or ideological frameworks of science diplomacy, thereby deepening its exploration from a fresh perspective. The collection has therefore the merit of shining a light on case studies elaborating a potentially new and different perspective on science diplomacy, which, at the very least, emphasizes a pluralistic narrative and stresses the distinctiveness of the Eastern and Central European experience. This volume marks the beginning rather than the completion of this investigation, and a great deal more research is needed to fully appreciate the significance of these Eastern Bloc collaborations in appraising conventional science diplomacy accounts. This is especially true given that, for instance, this volume's primary emphasis is on the history of the social sciences and therefore more research is needed to understand whether this picture can also apply to the natural sciences. Even so, the volume paves the way to a non-Western understanding of science diplomacy and should be welcomed for that.

What, then, was it like to develop international research collaborations in the Eastern Bloc? How did these collaborations shape distinctive science diplomacy trajectories? The chapters in this volume are particularly illuminating on these alternative pathways because their narratives pivot around two notions, "scientific internationalism" and "reciprocity," construing Eastern Bloc science diplomacy practices as not only distinctive but unique. These new narratives thus document science diplomacy initiatives that have largely remained unexamined in the existing literature. To appreciate the significance of the chapters of this volume, we need first to deepen the analysis of the two notions, the meanings they convey, and how they challenge traditional perspectives on science diplomacy.

Scientific internationalism has been the subject of several historical reconstructions. Some trace its origins back to the tensions and aspirations of the early twentieth century. This is the time when, for instance, the Dutch physician Pieter Eijkman envisioned scientific (and medical) internationalism as means of uniting the world. According to Eijkman, future global societies would have transcended nationalism through the international exchange of scientific and medical knowledge. Eijkman futuristically argued that this exchange was integral to the establishment of a world administration comprising a World Capital, an International Senate, and a World

Government.[10] His internationalist dream ultimately came to nothing, expectedly, also shattered by the nationalistic sentiment that it sought to overcome, but the appeal of international collaborations founded on science (and technology) survived Eijkman. The beginning of the Cold War also marked a decisive shift, as competing internationalist models emerged in the Western and the Soviet Blocs, each embracing antithetical ideological frameworks aligning to the tense debate on the freedom and planning of science.[11]

In the Eastern Bloc, internationalism expectedly adapted to the socialist ideology, especially in terms of the application of socialist planning to international collaborative operations. This socialist version of scientific internationalism combined with the tightening of Soviet control in Eastern and Central Europe, especially at the end of World War II, which meant that planning aligned with the decisions taken centrally by Moscow's apparatchik.[12] At the height of Stalin's era, internationalism was in fact redefined as "restricted" as specific constraints on autonomy and sovereignty in shaping new collaborative initiatives played a central role in their development.[13] This strong influence continued for several decades and was still prevalent under Leonid Brezhnev, whose doctrine postulated that the Soviet Union had the right to intervene in the affairs of other Bloc countries and define what they should contribute.[14] Overall, a Soviet centralization model of collaboration was thus imposed on research groups in Eastern and Central Europe. This explains why in recent years a growing body of literature has emphasized that Soviet-led internationalism gave rise to a distinctive but equally problematic form of imperialism. During the Cold War era, Soviet leaders such as Nikita Khrushchev championed the rights of newly decolonized countries at the United Nations, while simultaneously imposing on Eastern Bloc countries social and scientific principles and practices alien to their culture and customs. Scholars such as Dorota Kołodziejczyk and Cristina Şandru contend, for instance, that "there are, then, numerous grounds for treating the communist imposition in East-Central Europe as a particular historical embodiment of a persistent and widespread imperial drive."[15]

These impositions had implications for scientific collaborations within Eastern Bloc countries. Major events in the history of these collaborations, such as the International Geophysical Year (1957/8), exhibited similar impositions in that Soviet administrators often dictated what research groups in other countries should do.[16] Likewise, decisions regarding the development of the multilateral space program Interkosmos offered little room for individual input from Soviet allies. Indeed, Soviet authorities instructed the selection of astronauts from Eastern Bloc countries to pit national space science groups against one another (for example, East Germans against

Czechoslovaks) through propaganda operations aimed at strengthening imperialist control over Central and Eastern European countries.[17] In essence, the core philosophy of this Soviet approach to scientific internationalism was more alike the old Machiavellian principle of *divide et impera* than the ambition congenial to socialist thinking of uniting scientific workers through planning.

The volume's chapters make a decisive contribution to scholarship, especially on science diplomacy, precisely because they do not simply recount the details of these Soviet impositions with respect to Bloc-wide collaborations. They explore instead alternative modes of collaboration elaborated in the Eastern Bloc countries to counter this imperial drive. In particular, they show that Eastern and Central European scholars and scientists did not passively accept that all forms of international collaboration elaborated within the Eastern Bloc had to conform to the Soviet Union's dictates or take the Soviet Union as an inescapable reference. They instead elaborated competing forms of international collaboration characterized by different patterns of membership, which in turn can even be construed, implicitly, as way to use research collaborations to resist these impositions.

One concept—reciprocity—directly evokes the underlying philosophy of collaborations that challenged Soviet hegemony. The *Oxford English Dictionary* defines it as "reciprocation of cooperative or altruistic behaviour" and the "mutual recognition by two parties of certain rights and privileges" as result of this behavior.[18] In this respect, scholars and scientists in Eastern and Central European countries subtly (astutely, one might say) redefined socialist scientific internationalism as something that hinged on reciprocity. In essence, any collaborative endeavor required a fundamental understanding among all parties on their ability to reciprocate in all aspects, whether it be project design or the necessary exchanges of materials, personnel, and ideas.

One could even contend that reciprocity offers a fresh perspective on the circumstances of the Eastern Bloc precisely because it identifies an underlying criticism of Soviet-led collaborative initiatives. The historical moments when this criticism erupted into open contestation, such as the 1956 Budapest uprising, the 1968 Prague Spring, or the 1980 Solidarność strikes, are, of course, well documented. However, the notion of reciprocity alludes to a continuing resistance that did not end in major revolts and was never formally and explicitly articulated as such. While it cannot be unequivocally asserted that reciprocity always shaped acts of resistance to restricted internationalism, it conveyed an understanding of international collaborations as premised upon mutual contribution and collective decision-making. Reciprocity demarcated a space in which collaboration yields mutual benefit rather than being forced upon scientists and scholars through top-down mechanisms of supposedly socialist coordination directed entirely from Moscow. As an act of resistance, it could even be likened to the role that collectives in *dacha* kitchens played in the Soviet Union's dissent movement. Slava Gerovitch, for instance, convincingly shows how the *dachas* helped to build "a network of connections through which one could be introduced to an exclusive informal circle."[19]

Reciprocity collaborations challenged Soviet hegemony in equally interesting ways, especially when explored from a science diplomacy perspective. In Western Europe, the response to US hegemony involved the gradual consolidation of a new supranational sovereign entity, the would-be European Union. In contrast, the response in Eastern and Central European countries was to redirect seemingly subordinate collaborative projects into a sphere of past and long-standing regional collaboration often justified through belonging to the group of Visegrád countries. In this respect, the strategy diverged from the integrative approach taken in Western European countries. Rather than reinforcing and emphasizing cross-border ties, this form of collaboration was based on a projection of less formalized but durable historical ties, which were able to withstand Soviet pressures during the Cold War era.

The introduction to this volume already explores the historical background of the Visegrád countries, so this chapter will not revisit this ancestry. However, it is worth pausing to reflect on the somewhat anachronistic nature of this appeal to Visegrád unity. It bears a resemblance (at least to the author of this chapter) to how scholars of international relations often cite the 1648 Treaty of Westphalia as the constitutive moment of the modern state, despite the historians' objections to this characterization as a myth.[20] Similarly, the introduction to this volume highlights the fundamental ambivalence surrounding the real meaning of Visegrád and the historical oscillation "between friendship and conflict, shaped—but not defined—by shifting political prerogatives."[21] In reality, what the Visegrád countries really shared was a fourteenth-century resolution of disputes between the royal houses of Poland, Hungary, and Bohemia. However, the ad hoc projection of this settlement into the Cold War never meant to make it historically valid. It was instead an instrumental use of Visegrád countries' history to project unity and reciprocity for collaboration purposes. This may well be the reason the appeal to Visegrád unity remained popular after the Cold War. For instance, it was recalled again in the context of joint positions regarding the accession of Poland, Hungary, Czech Republic, and Slovakia into NATO and the EU. In other words, it does not matter whether such unity really existed in the fourteenth century.[22] It matters instead that it reaffirmed the unity of Visegrád nations during the Cold War to further strengthen a reciprocity stance.

During the Cold War, the notion of Visegrád "reciprocity" emerged as a decisive narrative element in the pursuit of research collaborations in Eastern and Central European countries. It allowed for the construction of an alternative collaborative model in the second half of the twentieth century that could be used to indirectly bypass the Bloc-wide collaborative model the Soviet Union sought to impose. Its application was far from uniform though. It included many national groups in Poland and Czechoslovakia, while effectively keeping the Hungarians by and large at its margins, thus revealing once again the selective use of Visegrád "reciprocity." Moreover, it was relatively easier to pursue in the social sciences, but far less practiced in the natural sciences, which have an inherent level of division of labor that lends itself to closer control and centralization. This is particularly the case with Big Science projects in atomic energy, space research, and other branches of the hard sciences, which are conducive to imposing principles of centralization precisely because their processes of labor division cater to separating research tasks and sequencing them in finite steps.[23] In contrast, the social sciences are characterized by a more horizontal modus operandi. The degree of engagement of individual researchers with the subject matter obstructs attempts to regiment research around finite procedures or processes conducive to centralization. This volume demonstrates that Visegrád coordination found fertile ground in the social sciences because of the prominent role individual researchers played in shaping collaborative endeavors and the greater independence with which they could be organized in comparison to the natural sciences.

Overall, by focusing especially on the social sciences, the cases presented in this volume shine a light on an alternative model of research collaboration that existed across the Eastern Bloc. Importantly, reciprocity highlighted that international collaboration entailed mutual assistance and reciprocal exchange. The term was thus an implicit reminder that such reciprocity was absent in collaborative initiatives led by the Soviets. This is not to say that the Soviets ignored the concept altogether. Indeed, the principal organization tasked with coordinating collaboration in the Eastern Bloc, COMECON (Rada vzájemné hospodářské pomoci in Czech), explicitly emphasized reciprocity (*vzájemnost* in Czech; *wzajemność* in Polish). However, although reciprocity was stated in Soviet-led collaborative science, it was not truly practiced. The appeal of reciprocity to the scholars discussed in the volume's contributions therefore challenges the propagandized version of reciprocity of Soviet-led international collaborations.

Vzájemnost was an actor category. In her contribution to the volume, Michaela Šmidrkalová reveals its use in 1965 by historians seeking to establish a Polish–Czech collaborative project to extend the study of the history of science in both countries. Even more significantly, these historians organized a conference explicitly celebrating reciprocity, framing their reciprocity-based collaboration as more durable and effective than Soviet-led projects. In Šmidrkalová's account these historians deliberately made instrumental use of *vzájemnost* to strengthen the perception that their academic collaboration differed from those the Soviets had in mind. By referring to a shared historical legacy perceived as dating back to Visegrád times, they sought to bolster the socialist collaborative initiatives they were developing. This legacy projected the idea that regional collaboration produced unity that had withstood the test of time, rooted in a thousand-year-old tradition of friendship between the people of the Visegrád countries. By implication, it could also be successfully applied to socialist internationalism as practiced in Cold War Poland and Czechoslovakia. It is worth noting once more that it did not matter that this reference was not historically sound, since it aimed to convey a political message about the manner of execution of collaborative research.

This interpretation helps us appraise other notable cases discussed in the volume, for instance, the career trajectory and propositions of the Czech philosopher Radovan Richta. His work on the Scientific and Technological Revolution (STR) has received considerable attention in the past, and at one point was even (naively) considered the socialist response to Thomas Kuhn's *The Structure of Scientific Revolution.* Richta's STR resonated in the Eastern Bloc (but also in Western countries) because it offered a similar understanding of waves of change across history connected to scientific and technological transformation. However, in contrast with Kuhn, Richta's theory focused on transitions in practices rather than ideas and methods. It emphasized how technological progress increasingly replaced obsolete manual activities with intellectual ones, thereby opening a path of liberation from manual labor. It also distinctively envisioned on that basis a fundamental role for the scientific worker in future societies, who would be increasingly occupying a central role in the production process.[24] The theory reached its height of influence just before the 1968 Prague Spring, and this is important since competing portrayals of Richta's work have emerged since then. Some perceived it as emblematic of the Prague Spring and a socialist effort to move beyond the Soviet centralization model, whereas others claimed Richta sought to mitigate such tensions and appease the Soviets.

Jan Surman's chapter invites us to reexamine Richta's research through the lens of the evolution of competing socialist models of collaboration, one premised on Soviet control and the other on Visegrád reciprocity. It suggests that STR should be reinterpreted because of the shift between these two collaborative models during Richta's career. In particular, the suppression of the Prague Spring prevented the establishment of a broader Czech–Polish STR group, while the post-1968 normalization forced Richta to align his writings more closely with Soviet ideological ambitions. Surman's article thus suggests that propagandized reciprocity outside the bounds of orthodoxy threatened the Soviet socialist model. This is an insight that has not been widely discussed in international scholarship. The chapter also helps to make familiar lesser-known scholars and intellectuals that Soviet censorship rendered internationally invisible. For instance, Ladislav Tondl was one of the scholars that the Soviets succeeded in marginalizing, also to break up a potentially independent experiment materializing a Czech–Polish collaborative centers developing STR studies, taking their development away from Soviet ideology, and prompting exchanges across the Iron Curtain.

The significance of marginalization is further discussed in Luboš Studený and Lukas Becht's analysis of future studies research, which is another area associated with and derived from STR research in Eastern and Central Europe. Future studies fueled optimism about the value of forecasting in the social studies, which further encouraged the development of research schools that brought together scholars in Poland and Czechoslovakia with a *vzájemnost* ethos. However, Studený and Becht go in a direction different from Surman. Instead of focusing on prominent scholars marginalized on account of their enthusiasm for Visegrád collaborations, they highlight the strategies the Soviet authorities used to reduce the impacts of these collaborations and integrate them in Eastern Bloc–wide projects. They also claim that the novel research discipline of future studies was subsumed into COMECON, thereby watering down its original *vzájemnost* agenda. Although the resulting research had far greater reach, extending across the whole of the Eastern Bloc, it had lesser value from an intellectual viewpoint. Indeed, forecasting exercises became too formalized to achieve meaningful results for either the Visegrád countries or the Eastern Bloc as a whole.

This edited collection also reveals that the concept of reciprocity as resistance may not always be applicable, particularly as the failure of projects such as those in future studies often led to negative assessments of some of the historical experiences described. Even so, the notion lends itself to the exploration of historical "paths not taken," which are also deserving attention. A notable example is the search for synergies between Polish and Czech economists, which Maciej Górny describes in his contribution. In this case, the pursuit of collaboration appears as a genuine form of reciprocity, which went beyond the merely declarative Soviet version. According to Górny, academic exchanges between economists from Visegrád countries sought viable alternatives and recognized the shortcomings of

Soviet economic thinking, especially to certain dogmatic approaches tying work operations to specific salary levels. Górny argues that the economic "path not taken," that of economic revisionism, emphasized the merits of decentralizing economic decision-making and linking wages to actual production values. There is a similar path not taken in the field of sociology, explored in Kornelia Kończal's chapter. While the scholarly tradition and character differed in each of the countries in the region, and Polish sociology had far a greater influence and tradition than its Czechoslovak counterpart, there was significant cross-fertilization before the Prague Spring, which was abruptly halted in the normalization period. The Spring's quashing also prevented one of the most influential sociologists of the 1970s, the Polish scholar Zygmunt Bauman, from relocating to the Czechoslovak capital rather than moving to the West.

This example illustrates that an important but unstated consequence of the defeat of reciprocity collaborative schemes, also understood as implicit resistance, is that these setbacks played a role in shaping new migration flows from Eastern Bloc countries. Another notable aspect is that while most chapters in this volume explore cooperation between the Visegrád countries as a way to resist Soviet imposition, a few simply emphasize the existence of a tradition of collaboration in the sciences, both natural and social, which provided fertile ground for this strategy. This is evident in Tomáš Pavlíček's account on the flows of mathematical knowledge across the region, which over the years facilitated meetings, seminars, and joint activities.

Doubravka Olšáková extends these considerations beyond academic and state domains, further complicating the collective picture that the volume presents concerning the significance of reciprocity as pioneered within some of the Visegrád countries. Her chapter, placed at the end of the volume, serves not to reiterate the merits of reciprocity but to address its limitations too. The case in point is the Pugwash Conference, the transnational network of scientists eager to promote nuclear disarmament, which also played a key role in Poland and Czechoslovakia. Olšáková's chief claim is that this advocacy persisted even in the darkest periods of these countries' recent history. In particular, the urgency of connecting East and West for the sake of disarmament required Pugwash representatives in Poland and Czechoslovakia to engage with Soviet-controlled state officials, some of whom were complicit in human rights violations. This has implications that clearly go beyond the scope of the volume, but compel readers to rethink science diplomacy, not only through the Western-centric frameworks promoted by current literature, but also within the distinctive Eastern Bloc context, where the linkages between science and international relations were very different and, in many ways, unique.

To sum up, the chapters of this volume help us recognize that incorporating this Eastern Bloc legacy into future accounts of how international collaborations shape international affairs would greatly improve our understanding of what we now call science diplomacy. At the very least, the relevant case studies help us to understand that our current understanding remains problematic, given the prevailing picture of science diplomacy as a globally unifying phenomenon rooted in the Western experience (and partly poorly emulated in the Soviet one during the Cold War). This interpretation is unsatisfactory precisely because countries in the former Eastern Bloc, especially Eastern and Central European ones, experienced and developed competing approaches to international collaboration to those imposed by the Soviet. The dominant approach was based on a rigid definition of scientific internationalism that glorified socialist science but effectively imposed a Soviet model of centralized control on other nations. In contrast, the subordinate and less visible approach emphasized the merits of regional coordination grounded in the principle of reciprocity. The emerging picture is thus one of a complex entanglement of projects and schemes that either aligned with or circumvented Soviet hegemony. So, thanks to the contributions to this volume, we can better understand what it was like and why it matters to operate within a collaborative framework emphasizing reciprocity in Eastern and Central Europe. Indeed, this new perspective has the potential to open many fruitful avenues for further inquiry on aspects of the Eastern Bloc that so far have been looked at almost exclusively through the lens of the imposition of a Soviet collaborative model.

Author Biography

Simone Turchetti researches twentieth-century science diplomacy, Cold War geoscience, surveillance, and nuclear history. He is Professor of History of Science and Technology at the Centre for the History of Science, Technology and Medicine (CHSTM) of the University of Manchester, leads the ERC project NEWORLD@A on global research data networks and presides over the European Society for the History of Science. He previously led the ERC-funded TEUS project and co-founded the IUHPST Commission on Science, Technology and Diplomacy.

Notes

1. On this, see Krige, *American Hegemony*; De Maria, Grilli, and Sebastiani, *The Restructuring of Physical Sciences*. On biology, see de Chadarevian, *Designs for Life*.

2. For competing perspectives, see, for instance, Lalli, "Crafting Europe from CERN to Dubna."
3. Courtois and Kramer, *The Black Book of Communism*. For useful reflections on the implications, see Altrichter, "Russia: On-Going Construction Site"; Tismaneanu, "Communism and the Human Condition."
4. See, for instance, Raleigh, "Doing Soviet History."
5. Another relatively recent contribution along similar lines is Betts and Smith, *Science, Religion and Communism*.
6. Fedoroff, "Science Diplomacy;" Ruffini, *Science and Diplomacy*. Historically informed critical reflections on this definition (and on science diplomacy more generally) can be found in Adamson and Lalli, "Global Perspectives on Science Diplomacy"; Turchetti et al., "Introduction."
7. Royal Society/American Association for the Advancement of Science (AAAS), *New Frontiers in Science Diplomacy*. The booklet was prepared following a two-day meeting of the same title hosted by the Royal Society on 1–2 June 2009.
8. For instance, see Aronova, "Scientometrics with and without Computers."
9. On the importance of decentering science diplomacy narratives based on Asian cases, see Barrett and Homei, "Decentering Histories of Science Diplomacy."
10. Van Acker and Somsen, "A Tale of Two World Capitals." For an overview, see also Crawford, *Nationalism and Internationalism in Science*.
11. A good example of this confrontation in atomic energy research is in Krige, "Atoms for Peace, Scientific Internationalism, and Scientific Intelligence." For Antarctic research, see Turchetti et al., "On Thick Ice." On freedom and the planning of science, see Nye, *Michael Polanyi and His Generation*.
12. Some relevant examples are discussed in the collection of essays in Guth et al., "Soviet Nuclear Modernity." See especially Khandozhko, "Quantum Tunnelling through the Iron Curtain." For the social sciences, see the relevant chapters in Aronova and Turchetti, *Science Studies during the Cold War and Beyond*. See also those by Babintseva and Aronova in Solovey and Dayé, *Cold War Social Science*.
13. For an overview, see Imlay, *The Practice of Socialist Internationalism*. For an analysis of its implications in science and technology, see Olšáková, "Between Stalinism and Infrastructural Globalism."
14. On the doctrine, see Ouimet, *Rise and Fall of the Brezhnev Doctrine*.
15. Kołodziejczyk and Şandru, "On Colonialism, Communism and East-Central Europe," 115.
16. Olšáková, "Between Stalinism and Infrastructural Globalism," 111. See also Turchetti, "The World United in Studying Itself?"
17. Dubrovina, "Intercosmos and the Practice of Soviet Space Diplomacy."
18. "Reciprocity" as defined by the *Oxford English Dictionary*.
19. Gerovitch, "The Kitchen and the Dacha."
20. See, for instance, Beaulac, "The Westphalian Model."
21. Introduction, in this volume.
22. Koziej, "The Visegrad Group"; Latawski, "NATO's 'Near Abroad.'"
23. For an overview on Big Science, see the essays in Charitos et al., *Big Science*.
24. Richta, *Civilization at the Crossroads*. See also Sommer, "Scientists of the World, Unite!"

Bibliography

Adamson, Matthew, and Roberto Lalli. 2021. "Global Perspectives on Science Diplomacy: Exploring the Diplomacy-Knowledge Nexus in Contemporary Histories of Science." *Centaurus* 62: 1–16.

Altrichter, Helmut. 1999. "Russia: On-Going Construction Site. Thoughts on the Black Book of Communism." *Vierteljahrshefte für Zeitgeschichte* 47, no. 3: 321–62.

Aronova, Elena. 2021. "Scientometrics with and without Computers: The Cold War Transnational Journeys of the Science Citation Index." In *Cold War Social Science: Transnational Entanglements*, edited by Marc Solovey and Christian Dayé. Cham: Palgrave Macmillan, 73–98. https://doi-org.manchester.idm.oclc.org/10.1007/978-3-030-70246-5_3.

Aronova, Elena, and Simone Turchetti. 2016. *Science Studies during the Cold War and Beyond: Paradigms Defected*. New York: Palgrave Macmillan.

Barrett, Gordon, and Aya Homei. 2024. "Decentering Histories of Science Diplomacy: Cases from Asia." *The British Journal for the History of Science* 57, no. 2: 165–73. https://doi.org/10.1017/S0007087424000578.

Beaulac, Stephane. 2004. "The Westphalian Model in Defining International Law: Challenging the Myth." *Australian Journal of Legal History* 8: 181–213. Retrieved 6 April 2026 from https://ssrn.com/abstract=672241.

Betts, Paul, and S. A. Smith, eds. 2016. *Science, Religion and Communism in Cold War Europe*. London: Palgrave Macmillan. https://doi.org/10.1057/978-1-137-54639-5.

Charitos, Panagiotis, Theodore Arabatzis, Harry Cliff, Günther Dissertori, Juliette Forneris, and Jason Li-Ying. 2023. *Big Science in the 21st Century*. Bristol: IOP Publishing.

Courtois, Stéphane, and Mark Kramer, eds. 1999. *The Black Book of Communism: Crimes, Terror, Repression*, transl. Jonathan Murphy. Cambridge, MA: Harvard University Press.

Crawford, Elisabeth T. 1992. *Nationalism and Internationalism in Science, 1880–1939: Four Studies of the Nobel Population*. Cambridge: Cambridge University Press.

De Chadarevian, Soraya. 2002. *Designs for Life: Molecular Biology after World War II*. New York: Cambridge University Press.

De Maria, Michelangelo, Mario Grilli, and Fabio Sebastiani, eds. 1989. *The Restructuring of Physical Sciences in Europe and the United States, 1945–1960: Proceedings of the International Conference, Università "La Sapienza," Rome, Italy, 19–23 September 1988*. Singapore: World Scientific.

Dubrovina, Olga. 2025. "Intercosmos and the Practice of Soviet Space Diplomacy." Presentation at the workshop Rethinking Space Diplomacy in the Cold War, Centre Lopez Pinero, Valencia, 5 June.

Fedoroff, Nina. 2009. "Science Diplomacy in the 21st Century." *Cell* 136: 9–11.

Gerovitch, Slava. 2024. "The Kitchen and the Dacha: Productive Spaces of Soviet Mathematics." *Studia Historiae Scientiarum* 23: 657–83.

Guth, Stefan, Klaus Gestwa, Tanja Penter, Julia Richers, and Dominic Bonfiglio. 2019. "Soviet Nuclear Modernity Transnational Dimensions, Decentering Dynamics and Enduring Legacies." *Cahiers Du Monde Russe* 60, no. 2–3: 229–56.

Imlay, Talbot C. 2017. *The Practice of Socialist Internationalism: European Socialists and International Politics, 1914–1960*. Oxford: Oxford University Press.

Khandozhko, Roman. 2019. "Quantum Tunnelling through the Iron Curtain: The Soviet Nuclear City of Dubna as a Cold War Crossing Point." In "Soviet Nuclear Modernity Transnational Dimensions, Decentering Dynamics and Enduring Legacies," Special issue, *Cahiers Du Monde Russe* 60, no. 2–3: 369–96.

Kołodziejczyk, Dorota, and Cristina Şandru. 2012. "On Colonialism, Communism and East-Central Europe: Some Reflections." *Journal of Postcolonial Writing* 48, no. 2: 113–16.

Koziej, Stanislaw. 2018. "The Visegrad Group and Europe's Security System: a story to watch." GISonline, July 26. https://www.gisreportsonline.com/r/visegrad-group/.

Krige, John. 2006. *American Hegemony and the Postwar Reconstruction of Science in Europe*. Cambridge, MA: MIT Press.

———. 2006. "Atoms for Peace, Scientific Internationalism, and Scientific Intelligence." *Osiris* 21: 161–81. https://doi.org/10.1086/507140.

Kuhn, Thomas. 1962. *The Structure of Scientific Revolutions*. Chicago: University of Chicago Press.

Lalli, Roberto. 2021. "Crafting Europe from CERN to Dubna: Physics as Diplomacy in the Foundation of the European Physical Society." *Centaurus* 63: 103–31. https://doi.org/10.1111/1600-0498.12304.

Latawski, Paul. 1993. "NATO's 'Near Abroad': The Visegrad Group and the Atlantic Alliance." *The RUSI Journal* 138, no. 6: 39–43. https://doi.org/10.1080/03071849308445764.

Nye, Mary Jo. 2011. *Michael Polanyi and His Generation: Origins of the Social Construction of Science*. Chicago: University of Chicago Press.

Olšáková, Doubravka. 2017. "Between Stalinism and Infrastructural Globalism: The IGY in Czechoslovakia, Poland and the German Democratic Republic." *Acta Poloniae Historica* 115: 97–122.

Ouimet, Matthew J. 2003. *Rise and Fall of the Brezhnev Doctrine in Soviet Foreign Policy*. Chapel Hill, NC: The University of North Carolina Press.

Oxford English Dictionary. 2000. Oxford, UK: Oxford University Press.

Raleigh, Donald J. 2002. "Doing Soviet History: The Impact of the Archival Revolution." *The Russian Review* 61, no. 1: 16–24.

Richta, Radovan. (1969) 2018. *Civilization at the Crossroads: Social and Human Implications of the Scientific and Technological Revolution*. Abingdon: Routledge. https://doi.org/10.4324/9781315177663.

Royal Society/American Association for the Advancement of Science (AAAS). 2010. *New Frontiers in Science Diplomacy*. London: Royal Society.

Ruffini, Pierre-Bruno. 2017. *Science and Diplomacy: A New Dimension of International Relations*. Cham: Springer. https://doi.org/10.1007/978-3-319-55104-3.

Solovey, Mark, and Christian Dayé, eds. 2021. *Cold War Social Science: Transnational Entanglements*. New York: Palgrave Macmillan.

Sommer, Vítězslav. 2016. "Scientists of the World, Unite! Radovan Richta's Theory of Scientific and Technological Revolution." In *Science Studies during the Cold War and Beyond: Paradigms Defected*, edited by Elena Aronova and Simone Turchetti, 177–204. New York: Palgrave Macmillan.

Tismaneanu, Vladimir. 2001. "Communism and the Human Condition: Reflections on the Black Book of Communism." *Human Rights Review* 2, no. 2: 125–34. https://doi.org/10.1007/s12142-001-1027-1.

Turchetti, Simone. 2026. “The World United in Studying Itself? Data and the Coloniality of the International Geophysical Year, 1957–58.” *Isis* 117, no. 1: 98–119.

Turchetti, Simone, Matthew Adamson, Giulia Rispoli, Doubravka Olšáková, and Sam Robinson. 2020. “Introduction: Just Needham to Nixon? On Writing the History of ‘Science Diplomacy.’” *Historical Studies in the Natural Sciences* 50, no. 4: 323–39.

Turchetti, Simone, Simon Naylor, Katrina Dean, and Martin Siegert. 2008. “On Thick Ice: Scientific Internationalism and Antarctic Affairs, 1957–1980.” *History and Technology* 24, no. 4: 351–76. https://doi.org/10.1080/07341510802357419.

Van Acker, Wouter, and Geert Somsen. 2012. “A Tale of Two World Capitals: The Internationalisms of Pieter Eijkman and Paul Otlet.” *Revue belge de philologie et d’histoire* 90, no. 4: 1389–409.

Index